BARRON'S

The Leader in Test Preparation

AP®

ENVIRONMENTAL SCIENCE

2ND EDITION

Gary S. Thorpe, M.S.
AP Environmental Science Teacher
Beverly Hills High School
Beverly Hills, CA

Acknowledgments

I would like to thank my wife, Patti, and my two daughters, Kris and Erin, for their patience and understanding while I was writing this book. A special thanks goes to Dr. Jerry Bobrow of Bobrow Test Preparation Services and Sarah Utley (AP College Board Reader) for coordinating and editing this project. And most of all, special thanks goes to the many professional and dedicated high school APES teachers across the United States who contributed their finest work in making this book possible.

All inquiries should be addressed to:
Barron's Educational Series, Inc.
250 Wireless Boulevard
Hauppauge, New York 11788
www.barronseduc.com

ISBN-13: 978-0-7641-3643-6
ISBN-10: 0-7641-3643-7

Library of Congress Control Number: 2006034340

Library of Congress Cataloging-in-Publication Data
Thorpe, Gary S.
AP environmental science / Gary S. Thorpe.—2nd ed.
p. cm.
Includes index.
ISBN-13: 978-0-7641-3643-6
ISBN-10: 0-7641-3643-7
1. Environmental sciences--Examinations--Study guides. 2. Advanced placement programs (Education)—Examinations—Study guides. I. Title.

GE76.T479 2007
363.70076—dc22 2006034340

Printed in the United States of America

9 8 7 6 5 4 3 2 1

Contents

UNIT IV: LAND AND WATER USE (10–15%)

UNIT V: ENERGY RESOURCES AND CONSUMPTION (10–15%)

UNIT VI: POLLUTION (25–30%)

UNIT VII: GLOBAL CHANGE (10–15%)

PRACTICE EXAMS

What's New in the 2nd Edition?

Barron's 2nd Edition of this *APES* review has undergone massive changes. The book has been condensed, but not to the point where it is superficial. Only information relevant to the APES exam as reflected by the most current curriculum developed by the AP Environmental Science Curriculum Committee has been included. Second, an attempt has been made to compartmentalize information into grids and charts to make the information and its relationships more understandable. Finally, the number of multiple-choice questions has been doubled. Each chapter now contains 20 or more multiple-choice questions with full explanations—there are now over 400 multiple-choice questions with answers in the *NEW Barron's 2nd Edition APES Review*! No other prep book on the market even comes close!

Introductory Material

The introductory material contains information on the AP exam itself. It includes commonly asked questions, strategies for taking the exam, and techniques for writing outstanding essays.

Specific Topics

The majority of the book is divided into seven units. Each contains specific chapters on material covered by the APES exam. Each chapter provides a condensed review of key information, case studies, relevant environmental laws, multiple-choice questions and explanations, and free-response essays. These have been contributed by APES instructors and AP readers from across the United States.

Practice Exams

The last part of the book contains two complete practice APES exams. All questions are thoroughly explained and include pointers for the free-response questions.

Format of the APES Exam

The APES exam is three hours long and is divided equally in time between a multiple-choice section and a free-response section. The multiple-choice section, which makes up 60% of the final score and lasts 90 minutes, consists of 100 multiple-choice questions and is designed to cover the breadth of your knowledge and understanding of environmental science. Thought-provoking problems and questions based on fundamental ideas are included along with questions based on the recall of basic facts and major concepts. The number of multiple-choice questions taken from each major topic area is reflected in the percentages shown in "Topics Covered" later in this section. For example, in Earth Systems and Resources, you will see "(10–15%)." Expect to find 10 to 15 multiple-choice questions from this area on the actual APES exam. Therefore, spend about 10–15% of your time reviewing this area.

The free-response section, which is also 90 minutes long, makes up 40% of your final score. It contains one data set question, one document-based question, and two synthesis and evaluation questions. In this section, you must organize your answers to demonstrate reasoning and analytical skills as well as the ability to write clearly and concisely.

Test-Taking Strategies

The Educational Testing Service will send you your AP Environmental Science score in July. Depending upon your choice, the scores are also sent to colleges and universities. The scores are reported on the following scale:

- 5—Extremely well qualified
- 4—Well qualified
- 3—Qualified
- 2—Possibly qualified
- 1—No recommendation

Most colleges and universities accept a score of 4 or 5 for credit and placement. Many colleges and universities may accept a score of 3 for credit and/or placement. Scores of 1 or 2 are not accepted by colleges and universities for either credit or placement.

The rule of thumb in determining how well you will probably do on the AP Environmental Science Exam is to look at the number of multiple-choice questions you answer correctly on the practice exams in this book. If you consistently get between 50 and 60% correct, you should be able to score a minimum of a 3. If you score consistently between 65 and 75% of the multiple-choice questions correct, you should be able to achieve a 4, and if you score 80 to 100 percent correctly, you should be looking at a 5. This assumes, of course, that you adequately answer the questions in the free-response section.

> **TIP**
>
> Always read the entire question. Underline key words in the question such as:
>
> - *all of the following EXCEPT*
> - *which of the following*
> - *increases*
> - *decreases*
> - *are commonly used*
> - *is responsible*
> - *principles*
> - *most accurately compares*
> - *is recognized as*
> - *best describes*
> - *is correct*
> - *results*
> - *reflects*
> - *is most likely*
> - *all the following are true EXCEPT*
> - *least likely*
> - *which best DEFINES*

The Multiple-Choice Questions

- Go through the entire set of multiple-choice questions answering *only* the ones you know for sure. Remember, there is a penalty for wild guessing.
- When you reach a question that you are not quite sure of and need more time, place a "+" next to the number.
- When you reach a question that you cannot answer, place a "–" next to the number.
- After you have answered all of the ones you know for sure, go back to only your "+" questions. If you can narrow the answer choices down to two, go ahead and choose (or guess) between the best answers. If you cannot narrow it down to two choices, skip the question and go to the next "+" question.
- If time remains, scan through the "–" questions one more time and see if you can answer any of them.

TYPES OF MULTIPLE-CHOICE QUESTIONS

The AP Environmental Science Exam relies on a variety of multiple-choice questions including identification and analysis. Those kinds of questions can be further identified as generalizations, comparing and contrasting concepts and events, sequencing a series of related ideas or events, cause-and-effect relationships, definitions, solutions to a problem, hypothetical situations, chronological problems, multiple correct answers, and negative questions. More than 75% of the multiple-choice questions fit into these categories.

In addition to identification and analysis questions, there are stimulus-based questions that rely on your interpretation and understanding of maps, graphs, charts, tables, pictures, flowcharts, photographs or sketches, cartoons, short narrative passages, surveys and poll data, quotations that come from primary source documents, and passages from environmental legislation and statutes, as well as questions that relate directly to laboratory and field investigations that you were *supposed* to have done.

Identification and Analysis Questions

Question Type	Example
Definitional	Any factor that influences a natural process under study is a(n) (A) independent variable (B) dependent variable (C) control (D) placebo (E) experimental value
Cause-and-effect relationships	______________ contributes to the formation of ______________ and thereby compounds the problem of ______________. (A) ozone, carbon dioxide, acid rain (B) carbon dioxide, carbon monoxide, ozone depletion (C) sulfur dioxide, acid deposition, global warming (D) nitrous oxide, ozone, industrial smog (E) nitric oxide, ozone, photochemical smog
Sequencing a series of related ideas or events	Which of the following statements describe the process of how environmental legislation would pass through Congress? I. Reports the bill out of the appropriate committee II. Debates the bill on the floor of the respective houses III. Rejects or accepts amendments to the bill IV. Resolves any differences in a conference committee (A) I only (B) I and II only (C) I, II, and III only (D) I, III, and IV only (E) I, II, III, and IV

Question Type	Example
Generalization	What is generally considered to be the most significant factor in terms of being a causative agent for cancer? (A) Smoking (B) Diet (C) Stress (D) Heredity (E) Pollution
Solution to a problem	A field biologist had been keeping density counts of the number of coastal side-blotched lizards (*Uta stansburiana hesperis* Richardson) that had recently been introduced into a large section of California desert. The yearly counts per acre were: 1998: 4; 1999: 8; 2000: 16. Assuming that the carrying capacity had not been reached, which of the following would be true? (A) In 2001, there should be 20 lizards due to arithmetic growth. (B) In 2001, there should be 20 lizards due to exponential growth. (C) In 2001, there should be 32 lizards due to exponential growth. (D) In 2001, there should be around 16 lizards due to population stability. (E) It is impossible to predict how many lizards there would be in 2001.
Hypothetical situation	Converting to a solar-hydrogen energy source could theoretically be achieved by (A) attracting private investors. (B) passing legislation that would fund "seed money" for entrepreneurs. (C) passing legislation that would discontinue government subsidies of fossil fuels. (D) educating the public as to the environmental benefits of solar-hydrogen fuel sources. (E) all of the above.

Question Type	Example
Chronological problem	The following events are related to major case studies of environmental pollution. Place the events in chronological order. I. Bhopal, India II. *Exxon Valdez* III. Donora, Pennsylvania IV. Three Mile Island V. Chernobyl, Ukraine (A) I, II, III, IV, V (B) V, IV, III, II, I (C) I, III, V, II, IV (D) I, V, II, IV, III (E) III, IV, V, I, II
Comparing and contrasting concepts and events	Which of the following acts or treaties accurately compares the political, environmental, and economic goals of the participating nations to the goal of reducing global greenhouse gas emissions? (A) Clean Air Act of 1955 (B) National Environmental Policy Act (NEPA) 1969 (C) Comprehensive Environmental Response, Compensation, and Liability (Superfund) Act (CERCLA) 1980 (D) Pollution Prevention Act of 1990 (E) Kyoto Protocol of 1997
Multiple correct answers	Nitrogen is assimilated in plants in what form? (A) NO_2^- (B) NH_3 (C) NH_4^+ (D) NO_3^- (E) Choices B, C, and D
Negative questions	An effective method to decrease the amount of pesticide use would include all of the following EXCEPT (A) using monoculture techniques (B) rotating crops (C) using pheromones (D) using polyculture techniques (E) using insect-resistant crops

A WORD OF CAUTION: If you can reasonably attack the question through a process of elimination or if you have factual knowledge regarding the content of the question, then you should attempt to answer it. You may be able to identify the type of question, but if you don't understand the issue or vocabulary that is used, you are not going to be able to find the correct choice. In that case, skip the question and avoid the possibility of a "guessing penalty."

Stimulus-Based Multiple-Choice Questions

Question Type	Example
Short narrative passage	"Human beings are adaptable. They have survived and conquered when other species have become extinct. Granted, that we as humans have a long way to go in order to live in harmony with nature, nevertheless, the future looks bright for man to conquer all environmental problems. There is no limit to what problems humans can solve. If we all work together, we can solve any problem. The future is bright." The quote above would be most closely associated with (A) Malthusian principles. (B) neo-Luddites. (C) cornucopian fallacy. (D) nihilists. (E) utilitarianism.
Short quotation	As the troposphere warms, the stratosphere begins to cool and causes the formation of ice-clouds above the South Pole. The ice-clouds tend to accelerate the decrease of ozone above the South Pole by providing a surface for chemical reactions to occur. This is an example of (A) serendipity. (B) a positive feedback loop. (C) a negative feedback loop. (D) synergy. (E) chaos.
Environmental case law	Which act established the first clear water purity standards? (A) Water Resources Planning Act (B) Water Resources Development Act (C) Safe Drinking Water Act (D) Surface Water Treatment Rule (E) Water Quality Act
Sketch or diagram	The following diagram indicates 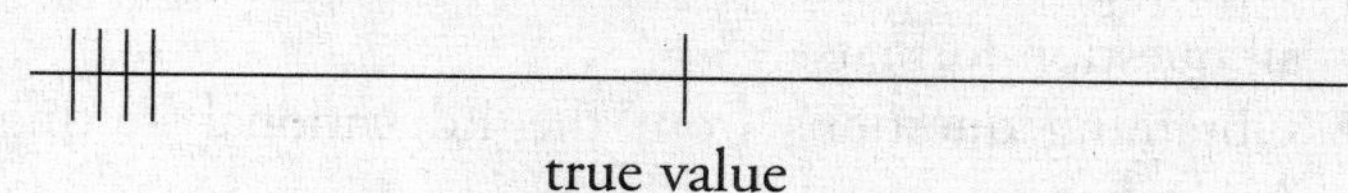 (A) a measurement that is accurate and precise. (B) a measurement that is accurate but not precise. (C) a measurement that is precise but not accurate. (D) a measurement that is neither accurate nor precise. (E) a conclusion cannot be made regarding this distribution.

THE ELIMINATION STRATEGY

Take advantage of being able to mark in your test booklet. As you go through the "+" questions, eliminate choices from consideration by marking them out in your question booklet. Mark with question marks any choices you wish to consider as possible answers. See the following example:

~~A.~~
B. ?
~~C.~~
~~D.~~
E. ?

This technique will help you avoid reconsidering those choices that you have already eliminated and will thus save you time. It will also help you narrow down your possible answers.

If you are able to eliminate all but two possible answers, such as B and E in the previous example, you may want to guess. Under these conditions, you stand a better chance of raising your score by guessing than by leaving the answer sheet blank.

Question Type	Example
Graph or table interpretation	Two varieties of the same species of voles (meadow mice), albino and red-backed, were used in an experiment. Both varieties were subjected to the predation of a hawk, under controlled laboratory conditions. During the experiment, the floor of the test room was covered on alternate days with white ground cover that matched the albino voles and red-brown cover that matched the red-backed voles. The results of fifty trials are shown in the following tabulation.

NUMBER OF VOLES CAPTURED

Variety	White Cover	Red-Brown Cover	Total
Albino	35	57	92
Red-backed	60	40	100
Total	95	97	192

Which result would have been most likely if only red-brown floor covering had been used?

(A) Ninety-seven red-backed voles would have survived.
(B) Ninety-five albino voles would have survived.
(C) The survival rate of the albino voles would have decreased markedly.
(D) A greater number of red-backed voles would not have survived.
(E) There would have been no change in the results.

A FINAL WORD ABOUT THE MULTIPLE-CHOICE QUESTIONS

Throughout the APES course, your teacher will be giving you a variety of sample multiple-choice questions with five possible answers. Other methods include:

- Building up your own database of questions by using questions from this book.
- Sharing and collecting questions that other students come up with (the Internet is wonderful for this).
- Creating a class database where everyone contributes ten questions per chapter in all question formats.
- Obtaining questions from the Educational Testing Service as found in their Acorn book.
- Searching the Internet for home pages from other APES classes (I counted over 60 when I searched). Also searching the Internet for home pages from students who have already taken the AP Environmental Science Exam and can share with you their thoughts, tips, and experiences.
- Making your own class web page that includes multiple-choice questions and links for more information.
- Creating reciprocal agreements with other schools for material ("we'll give you 300 of our questions if you give us 300 of yours").

- Visiting the many college-text publishers that have websites where you can get more multiple-choice questions and working through these questions.
- Debriefing. Right after you finish the AP exam come immediately back and let your teacher know the type of questions you found easy and what type you found more difficult; let them know how well-prepared you were and where more attention could have been placed. This will help your teacher and the class next year to know where to place more emphasis. After a few years of student feedback, your teacher will have a pretty good idea of what to expect.

> **TIP**
>
> The best ways to earn the highest score possible on the AP Exam is to look at other students' essays from past APES exams and see how their essays were graded by the College Board. Yes, you can do this! Go to *www.apcentral.collegeboard.com* Now, click on EXAMS in the left column. Select "Environmental Science." You're there—WOW!! This is the best kept secret there is for getting a 5. There are many sample essays to look at—from students who got 5's to students who got 1's. And, you've got access to every essay question that was ever asked on an APES exam.

The Free-Response Questions

There are three types of free-response questions: (1) data analysis, (2) document based, and (3) synthesis and evaluation.

The data analysis (or data set) question presents data in tabular or graphical form. It measures your ability to interpret and analyze data. The document-based question (DBQ) presents material in the form of real-life documents (newspaper clippings, product advertisements, etc.). The synthesis and evaluation questions are in-depth essay questions, often with multiple parts.

In scoring the free-response essays, points are awarded only for arguments that are supported by scientific facts and principles. Each question is scored on a scale from 1 to 10.

All four essay questions must be answered within 90 minutes. If you spend an equal amount of time on each essay, this averages to about 23 minutes per essay. Lists, outlines, and unlabeled drawings are not acceptable and will not be awarded credit.

The following table outlines essay skills and how they relate to the AP score.

Free-Response Scoring

Students Earning	Description
5	Excellent thesis or introductory statement. Good explanation of data. Responds to every section of the question using specific and accurate examples to support his/her answer.
4	A thesis statement is evident in form and relates to the question. An introductory statement presents some examples in a logical manner to support the data but may present weaker or factually incorrect arguments or fewer examples to support the answer.
3	Presents a week thesis or introductory statement. Shows some relationship of examples to thesis statement. May leave out pertinent information or may present incorrect examples.
2	Has no organized thesis or introductory statement. Few examples are provided. May contain incomplete or factually incorrect information and does not answer the entire question.
1	Thesis or introductory statement that cannot be proven or is not related to the question. Examples do not support thesis or data and are factually incorrect. Information is irrelevant to the question.

The free-response questions often include phrases that should guide how you write your essays. The following table lists many of these key terms and what they actually mean.

Key Phrases

Key Term	Meaning	Example
Analyze the effect . . .	Evaluate the impact	Analyze the effects of major global wind patterns in determining the distribution of biomes.
Assess the accuracy . . .	Determine the truth	Assess the accuracy of the following statement, "A high birth rate corresponds to a high standard of living."
Compare the strengths and weaknesses . . .	Show differences	Compare the strengths and weaknesses of three government intervention programs that have been adopted into law to combat pollution.
Critically evaluate evidence that both supports and refutes . . .	Give examples that agree and/or disagree	Critically evaluate evidence that both supports and refutes the claim that greenhouse gases are responsible for the hole in the ozone layer.
Define and evaluate the contention . . .	Give a definition and analyze the point of view	Define ecofeminism and evaluate the contention that male-dominated societies are responsible for environmental imbalance and social oppression.
Discuss . . .	Give examples that illustrate	Discuss the rapid increase of African "killer bees" in the United States as an example of the process of natural selection.
Evaluate the claim . . .	Determine the validity	Evaluate the claim that increasing the standard of living and improving the role of women will result in a lower birthrate.
Explain . . .	Offer the meaning, cause, effect, and influence	Explain three different theories of moral responsibility to the environment. Include in your discussion the major proponents of each theory and their specific ideas(s) or "school."
To what extent…	Explain the relationship and role	To what extent does moisture determine a soil profile? Provide specific soil types, overview of soil horizons, and geographic distribution.

Constructing Your Essay

Construct your essay like the letter "T." The top of the "T" represents your thesis statement, and it is broader than the evidence below it. Notice that each box below the thesis statement decreases in size. This means you should have the bulk of your essay in Evidence I and the least amount of your essay devoted to the conclusion.

THESIS STATEMENT

- Tells the reader *how you will interpret* the significance of the subject matter under discussion.
- Is a road map for the essay; in other words, it tells the reader *what to expect* from the rest of your essay.
- Is usually a single sentence somewhere in your first paragraph that *presents your argument* to the reader. The rest of the essay gathers and organizes evidence that will persuade the AP grader of the logic of your interpretation.

EVIDENCE I

Include historical context, facts, dates, people, environmental concepts, etc. Draw labeled diagrams if possible. Analysis of data. Pros and cons of issue.

EVIDENCE II

Case studies. Relate case study to question.

EVIDENCE III

Relevant laws. Cite parties affected, approximate time frame, consequences, arguments pro and con for legislation.

CONCLUSION

Summarize in a few sentences what you discussed.

Topics Covered on the Exam

I. Earth Systems and Resources (10–15%)

A. **Earth Science Concepts**—Geologic time scale, plate tectonics, earthquakes, volcanism, seasons, solar intensity and latitude.
B. **The Atmosphere**—Composition, structure, weather, climate, atmospheric circulation and the Coriolis effect, atmosphere-ocean interactions, and ENSO.
C. **Global Water Resources and Use**—Freshwater, saltwater, ocean circulation, agriculture, industrial and domestic use, surface and groundwater issues, global problems, and conservation.
D. **Soil and Soil Dynamics**—Rock cycle, formation, composition, physical and chemical properties, main soil types, erosion and other soil problems, and soil conservation.

II. The Living World (10–15%)

A. **Ecosystem Structure**—Biological populations and communities, ecological niches, interactions among species, keystone species, species diversity, and edge effects, and major terrestrial and aquatic biomes.
B. **Energy Flow**—Photosynthesis and cellular respiration, food webs, and trophic levels, and ecological pyramids.
C. **Ecosystem Diversity**—Biodiversity, natural selection, evolution, and ecosystem services.
D. **Natural Ecosystem Change**—Climate shifts, species movement, and ecological succession.
E. **Natural Biogeochemical Cycles**—Carbon, nitrogen, phosphorus, sulfur, water, and conservation of matter.

III. Population (10–15%)

A. **Population Biology Concepts**—Population ecology, carrying capacity, reproductive strategies, and survivorship.
B. **Human Population**

1. Human population dynamics—Historical population sizes, distribution, fertility rates, growth rates and doubling times, demographic transition, and age-structure diagrams.
2. Population size—Strategies for sustainability, case studies, and national policies.
3. Impacts of population growth—Hunger, disease, economic effects, resource use, and habitat destruction.

IV. Land and Water Use (10–15%)

A. **Agriculture**

1. Feeding a growing population—Human nutritional requirements, types of agriculture, green revolution, genetic engineering and crop production, deforestation, irrigation, and sustainable agriculture.
2. Controlling pests—Types of pesticides, costs and benefits of pesticide use, integrated pest management, and relevant laws.

B. **Forestry**—Tree plantations, old-growth forests, forest fires, forest management, and national forests.

C. **Rangelands**—Overgrazing, deforestation, desertification, rangeland management, and federal rangelands.

D. **Other Land Use**

1. Urban land development—Planned development, suburban sprawl, and urbanization.
2. Transportation infrastructure—Federal highway system, canals and channels, roadless areas, and ecosystem impacts.
3. Public and federal lands—Management, wilderness areas, national parks, wildlife refuges, forests, and wetlands.
4. Land conservation options—Preservation, remediation, mitigation, and restoration.
5. Sustainable land use strategies.

E. **Mining**—Mineral formation, extraction, global reserves, and relevant laws and treaties.

F. **Fishing**—Fishing techniques, overfishing, aquaculture, and relevant laws and treaties.

G. **Global Economics**—Globalization, World Bank, *Tragedy of the Commons*, and relevant laws and treaties.

V. Energy Resources and Consumption (10–15%)

A. **Energy Concepts**—Energy forms, power, units, conversions, and laws of thermodynamics.

B. **Energy Consumption**

1. History—Industrial Revolution, exponential growth, and energy crisis.
2. Present global energy use.
4. Future energy needs.

C. **Fossil Fuel Resources and Use**—Formation of coal, oil, and natural gas; extraction/purification methods; world reserves and global demand; synfuels; and environmental advantages/disadvantages of sources.

D. **Nuclear Energy**—Nuclear fission process, nuclear fuel, electricity production, nuclear reactor types, environmental advantages/disadvantages, safety issues, radiation and human health, radioactive wastes, and nuclear fusion.

E. **Hydroelectric Power**—Dams, flood control, salmon, silting, and other impacts.

F. **Energy Conservation**—Energy efficiency, CAFE standards, hybrid electric vehicles and mass transit.

G. **Renewable Energy**—Solar energy, solar electricity, hydrogen fuel cells, biomass, wind energy, small-scale hydroelectric, ocean waves and tidal energy, geothermal, and environmental advantages/disadvantages.

VI. Pollution (25–30%)

A. **Pollution Types**

1. Air pollution—Sources (primary and secondary) major air pollutants, measurement units, smog, acid deposition—causes and effects, heat islands and temperature inversions, indoor air pollution, remediation and reduction strategies, Clean Air Act, and other relevant laws.
2. Noise pollution—Sources, effects, and control measures.
3. Water pollution—Types, sources, causes and effects, cultural eutrophication, groundwater pollution, maintaining water quality, water purification, sewage treatment/septic systems, Clean Water Act, and other relevant laws.
4. Solid waste—Types, disposal, and reduction.

B. **Impacts on the Environment and Human Health**

1. Hazards to human health—Environmental risk analysis, acute and chronic effects, dose-response relationships, air pollutants, smoking, and other risks.
2. Hazardous chemicals in the environment—Types of hazardous waste, treatment/disposal of hazardous waste, cleanup of contaminated sites, biomagnification, and relevant laws.

C. **Economic Impacts**—Cost-benefit analysis, externalities, marginal costs, and sustainability.

VII. Global Change (10–15%)

A. **Stratospheric ozone**—Formation of stratospheric ozone, ultraviolet radiation, causes of ozone depletion, effects of ozone depletion, strategies for reducing ozone depletion, and relevant laws and treaties.

B. **Global Warming**—Greenhouse gases and the greenhouse effect, impacts and consequences of global warming, reducing climate change, and relevant laws and treaties.

C. **Loss of Biodiversity**

1. Habitat loss—Overuse, pollution, introduced species, and endangered and extinct species.
2. Maintenance through conservation.
3. Relevant laws and treaties.

Questions Commonly Asked About the APES Exam

Q: How are the questions graded?

A: Section I, worth 60% of the total score, is 90 minutes long and consists of 100 multiple-choice questions. The total score for Section I is the number of correct answers minus ¼ for each wrong answer. If you leave a question unanswered, it does not count at all. You will need to answer around 50% of the multiple-choice questions correctly to obtain a 3 on the exam. The multiple-choice questions are based on recall of basic facts and major concepts of environmental science. Section II is worth 40% of your score, is also 90 minutes long, and consists of four essay questions.

Q: What do the scores mean?

A: The APES exam is graded on a 5-point scale:

5: Passing. ~9% earn this score.
4: Passing. ~23% earn this score.
3: Passing. ~19% earn this score.
2: Not passing. ~18% earn this score.
1: Not passing. ~31% earn this score.

In summary, about half pass and half do not.

Q: What materials do I take with me to the exam?

A: Bring your admission ticket, an official photo including signature I.D., your social security number, several sharpened #2 pencils with erasers, and a watch. Calculators are NOT allowed.

Q: Should I guess?

A: There is a penalty for guessing. I recommend NOT guessing if you do not know the answer. However, if you can narrow down the multiple-choice answers to two possibilities, then your odds are better if you do guess.

Q: Can I cancel my scores?

A: Yes. Your request must be received in writing by June 15. You may also request that one or more of your AP grades NOT be sent to colleges.

Q: Can I write on the test booklet?

A: Yes. The next section will show you how to take advantage of this.

Q: How do I get more information?

A: Log on to *http://apcentral.collegeboard.com*

UNIT I: EARTH SYSTEMS AND RESOURCES (10–15%)

Areas on Which You Will Be Tested

A. **Earth Science Concepts**—geologic time scale, plate tectonics, earthquakes, volcanism, seasons, solar intensity, and latitude.
B. **The Atmosphere**—composition, structure, weather, climate, atmospheric circulation and the Coriolis effect, atmosphere-ocean interactions, and ENSO.
C. **Global Water Resources and Use**—freshwater, saltwater, ocean circulation, agriculture, industrial and domestic use, surface and groundwater issues, global problems, and conservation.
D. **Soil and Soil Dynamics**—rock cycle, formation, composition, physical and chemical properties, main soil types, erosion and other soil problems, and soil conservation.

The Earth

CHAPTER 1

The poetry of the earth is never dead.

—John Keats

GEOLOGIC TIME SCALE

Two time scales are used to measure the age of Earth. One is a relative time scale based on the sequence of layering of the rocks and the evolution of life. The other is the radiometric time scale, based on the natural radioactivity of chemical elements in rocks. Earth's past has been organized into various units according to events that took place in each period. Different spans of time on the time scale are usually separated by major geologic or paleontological events, such as mass extinctions. For example, the boundary between the Cretaceous period and the Paleogene period is defined by the extinction of the dinosaurs and many marine species. The largest defined unit of time is the eon. Eons are dived into eras, which are in turn divided into periods, epochs, and stages (Eon → Eras → Periods → Epochs → Stages). Key principles of the geological time scale:

1. Rock layers (strata) are laid down in succession with each strata representing a "slice" of time.
2. The principle of superposition—any given stratum is probably older than those above it and younger than those below it.

Several factors complicate the geologic time scale:

1. Sequences of strata are often eroded, distorted, tilted, or even inverted after deposition.
2. Strata laid down at the same time in different areas can have entirely different appearances.
3. Strata of any given area represent only part of Earth's history.

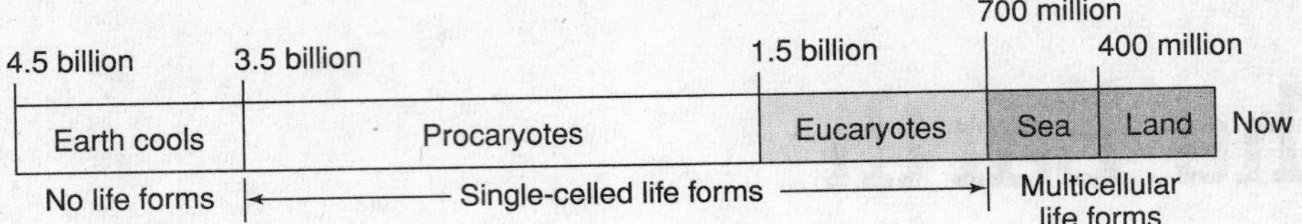

Figure 1.1 Timeline of life development

THE GEOLOGIC TIME SCALE

Era	When period began (millions of years ago)	Period	Animal life	Plant life	Major geologic events
Cenozoic	2	Neogene	Rise of civilizations	Increase in number of herbs and grasses	Ice Age
	65	Paleogene	Appearance of first men; dominance on land of mammals, birds, and insects	Dominance of land by flowering plants	
Mesozoic	135	Cretaceous		Dominance of land by conifers; first flowering plants appear	Building of the Rocky Mountains
	180	Jurassic	Age of dinosaurs		
	225	Triassic	First birds		
Paleozoic	275	Permian	Expansion of reptiles		Building of the Appalachian Mountains
	350	Carboniferous	Age of amphibians	Formation of great coal swamps	
	413	Devonian	Age of fishes		
	430	Silurian	Invasion of land by invertebrates	Invasion of land by primitive plants	
	500	Ordovician	Appearance of first vertebrates (fish)	Abundant marine algae	
	570	Cambrian	Abundant marine invertebrates	Appearance of primitive marine algae	
Precambrian		—		Primitive marine life	

Figure 1.2 Geologic time scale

EARTH STRUCTURE

> **TIP**
>
> When writing your FRQ essays, a picture is worth a thousand words! If you can sketch out a labeled diagram of what you are describing it will go a long way in improving your score.

Earth, which formed about 4.6 billion years ago, is the third planet away from the sun in the solar system and is the only planet known to support life. Earth can be divided into three sections: the biosphere, the hydrosphere, and the internal structure. The biosphere includes all forms of life (plants and animals) both on land and in the sea. The hydrosphere includes all forms of water (fresh and saltwater, snow, and ice). The internal structure of Earth is divided into the crust, mantle, and core.

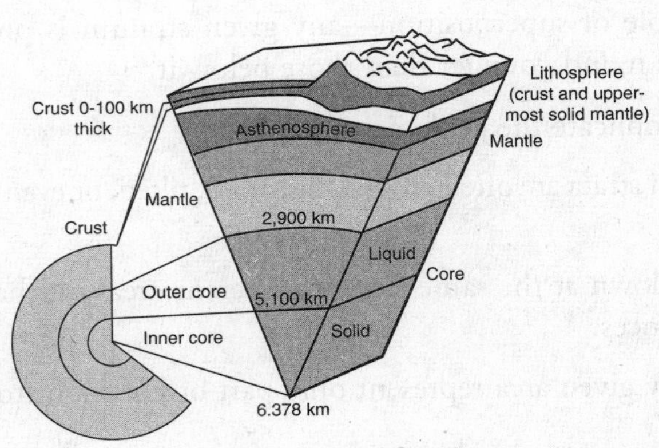

Figure 1.3 Layers of the Earth

Crust

The crust makes up only 0.5 % of Earth's total mass and can be subdivided into two main parts: basalt-rich continental crust and granite-rich oceanic crust. The crust floats on top of the mantle. Oceanic crust covers about two-thirds of Earth's surface but comprises about only one-third of the crustal mass, as the continental crust is much thicker. Being relatively cold, the crust is rocky and brittle, so it can fracture in earthquakes.

CONTINENTAL CRUST

The continental crust extends from the surface of Earth down to 20–30 miles (30–50 km). The exposed parts of the continental crust are less dense than oceanic crust. This is because oceanic crust contains minerals rich in heavier elements such as iron and magnesium. However, continental crust appears to be stratified (layered) and becomes denser with depth. It is largely composed of volcanic, sedimentary, and granite-type rocks, although the older areas are dominated by metamorphic rocks.

OCEANIC CRUST

From the surface of Earth down to about 7 miles (11 km) is the oceanic crust. The oceanic crust can be divided into ocean basins where water depth exceeds 2 miles (3 km), and the crust is layered and very uniform.

MANTLE

Most of Earth's mass is in the mantle, which is composed of iron, magnesium, aluminum, and silicon-oxygen compounds. At over 1800°F (1000°C), most of the mantle is solid. However, the upper third (known as the asthenosphere) is more plastic-like in nature.

CORE

The core is composed mostly of iron and is so hot that the outer core is molten. The inner core is under such extreme pressure that it remains solid.

PLATE TECTONICS

Plate tectonic theory arose out of two separate geological observations: continental drift and seafloor spreading.

The Continental Drift Theory

In 1915, Alfred Wegener proposed that all present-day continents originally formed one landmass (Pangaea). Wegener believed that this supercontinent began to break up into smaller continents around 200 million years ago. He based his theory on five factors:

1. Fossilized tropical plants were discovered beneath Greenland's icecaps.
2. Glaciated landscapes occurred in the tropics of Africa and South America.

3. Tropical regions on some continents had polar climates in the past, based on paleoclimatic data.
4. The continents fit together like pieces of a puzzle.
5. Similarities existed in rocks between the east coasts of North and South America and the west coasts of Africa and Europe.

Continental drift gained acceptance in the 1960s when the theory of plate tectonics provided a mechanism that would account for the movement of the continents.

The Seafloor Spreading Theory

During the 1960s, alternating patterns of magnetic properties were discovered in rocks found on the seafloor. Similar patterns were discovered on either side of mid-oceanic ridges found near the center of the oceanic basins. Dating of the rocks indicated that as one moved away from the ridge, the rocks became older. This suggested that new crust was being created at volcanic rift zones.

The lithosphere (crust and upper mantle, approximately 62 miles (100 km) thick) is divided into massive sections known as plates. These float and move on the viscous asthenosphere.

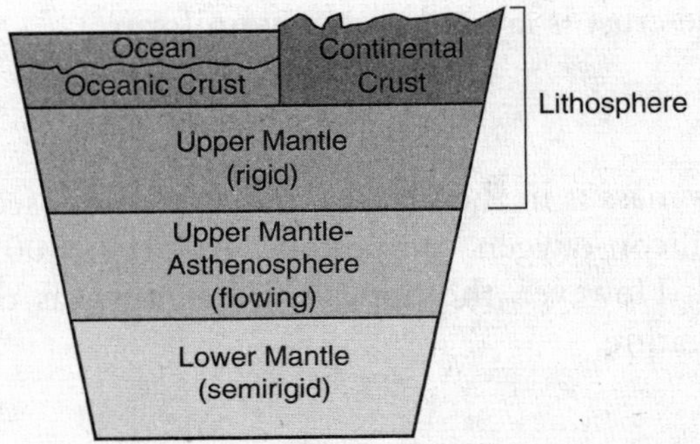

Figure 1.4 The outer layers of Earth

The plates move slowly over time. They sink in areas of volcanic island chains, folded mountain belts, and trenches. They rise up from ridges and rift valleys. These plates move in relation to one another at one of three types of plate boundaries: transform, divergent, and convergent.

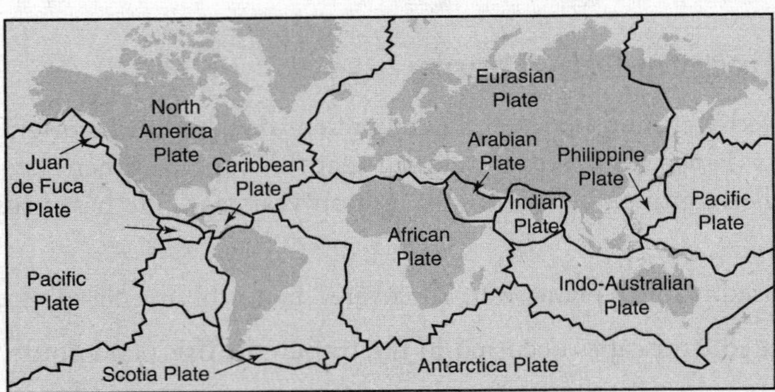

Figure 1.5 Earth's major plates

TRANSFORM BOUNDARIES

Transform boundaries occur where plates slide *past* each other. The friction and the stress buildup from the sliding plates frequently cause earthquakes—a common feature along transform boundaries. The San Andreas Fault, which is found near the western coast of North America, is where the Pacific and North American plates move relative to each other such that the Pacific plate is moving northwest with respect to North America. In about 50 million years, the part of California that is west of the San Andreas Fault will be a separate island near Alaska.

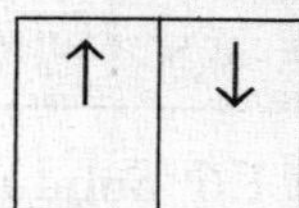

DIVERGENT BOUNDARIES

Divergent boundaries occur where two plates slide *apart* from each other with the space that was created being filled with molten magma from below. Examples of areas of oceanic divergent boundaries include the Mid-Atlantic Ridge and the East Pacific Rise. Examples of areas of continental divergent boundaries include the East African Great Rift Valley. Divergent boundaries can create massive fault zones in the oceanic ridge system and are areas of frequent oceanic earthquakes.

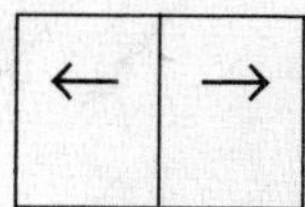

CONVERGENT BOUNDARIES

Convergent boundaries occur where two plates slide *toward* each other, commonly forming either a subduction zone (if one plate moves underneath the other) or an orogonic belt (if the two plates collide and compress). When a denser oceanic plate moves underneath (subducts) a less-dense continental plate, an oceanic trench is produced on the ocean side and a mountain range on the continental side. One example is the Cascade Mountain range. It extends north from California's Sierra Nevada Mountains and includes Mount Saint Helens.

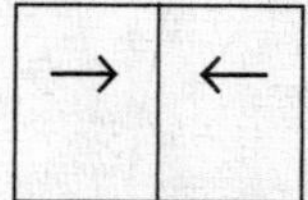

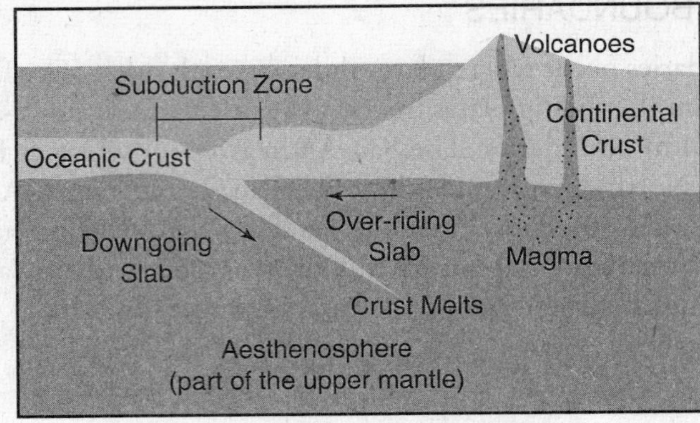

Figure 1.6 Subduction

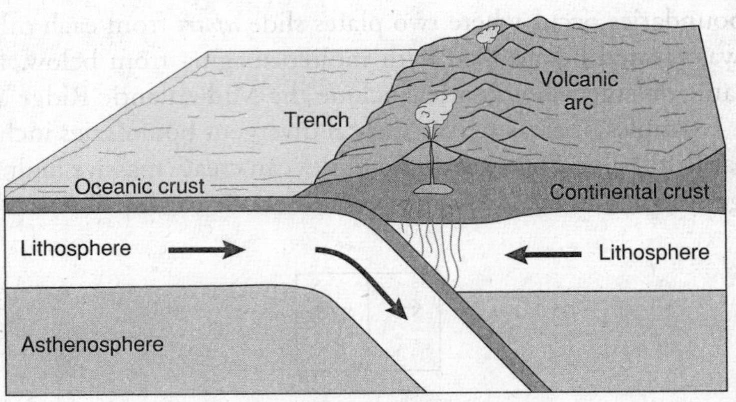

Figure 1.7 Oceanic-continental convergence

When *two oceanic plates converge*, they create an island arc—a curved chain of volcanic islands rising from the deep seafloor and near a continent. They are created by subduction processes and occur on the continent side of the subduction zone. Their curve is generally convex toward the open ocean. A deep undersea trench is located in front of such arcs where the descending plate dips downward. Examples include Japan and the Aleutian Islands in Alaska.

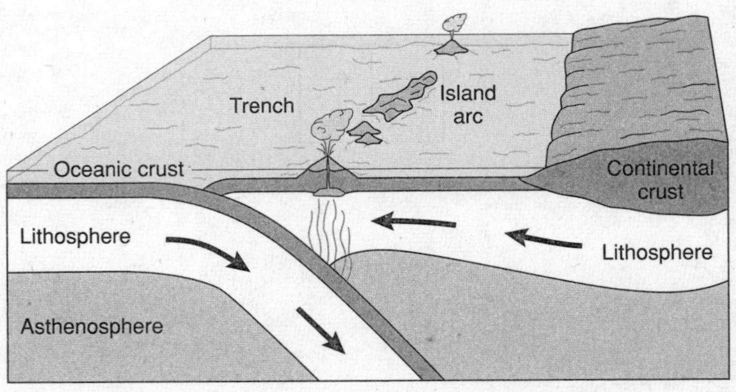

Figure 1.8 Oceanic-oceanic convergence

When *two continental plates collide*, mountain ranges are created as the colliding crust is compressed and pushed upward. An example is the northern margins of the Indian subcontinental plate being thrust under a portion of the Eurasian plate, lifting it and creating the Himalayas.

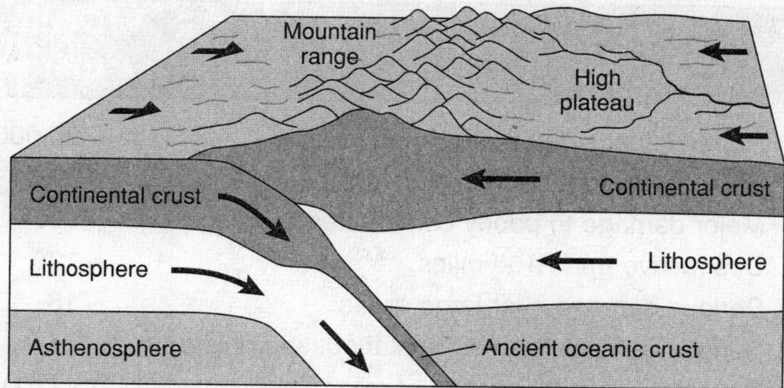

Figure 1.9 Continental-continental convergence

EARTHQUAKES

Earthquakes occur during abrupt movement on an existing fault, along tectonic plate boundary zones or along mid-oceanic ridges. A massive amount of stored energy, held in place by friction, is released in a very short period of time. The area where the energy is released is called the focus. From the focus, seismic waves travel outward in all directions. Directly above the focus, on Earth's surface, is the epicenter.

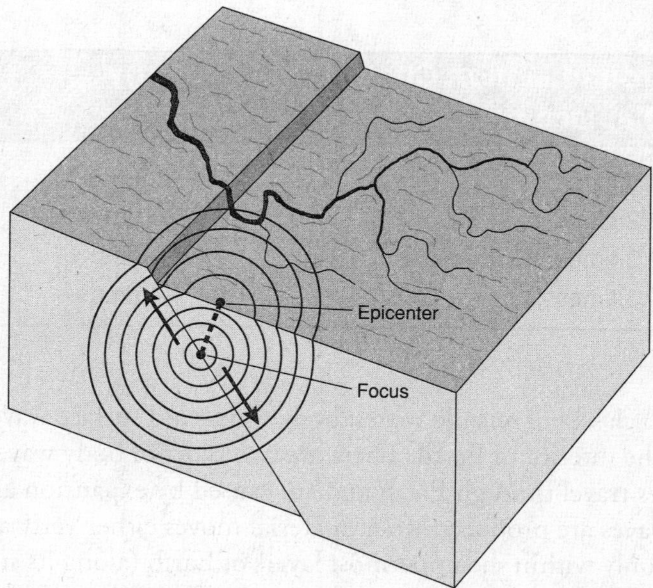

Figure 1.10 Relationship of an epicenter to a focus

The strength or magnitude of an earthquake is commonly measured by the logarithmic Richter scale and is recorded by a seismograph.

The Richter Scale

Severity	Richter scale	Description	Occurrence (per year)
Minor	3.0–3.9	Rarely causes damage.	~ 50,000
Light	4.0–4.9	Shaking of indoor items. No significant damage.	~ 6,000
Moderate	5.0–5.9	Major damage to poorly constructed buildings.	~ 800
Strong	6.0–6.9	Destructive up to 100 miles.	~ 120
Major	7.0–7.9	Serious damage over large areas.	~ 18
Great	8.0–8.9	Serious damage over several thousand miles.	~ 1
Extreme	9.0+	Devastating for thousands of miles.	1 in 20 years

However, the Richter magnitude scale is really comparing amplitudes of waves on a seismogram, not the strength (energy) of the quakes. It is the energy or strength of the earthquake that knocks down buildings and causes damage. The following chart relates changes in the Richter magnitude scale with the strength or energy of earthquakes:

Richter Magnitude

Richter Magnitude Change	Ground Motion Change (Displacement)	Energy Change
1.0	10.0 times	~ 32 times
0.5	3.2 times	~ 5.5 times
0.3	2.0 times	~ 3 times
0.1	1.3 times	~ 1.4 times

There are two classes of seismic waves: body waves and surface waves. Body waves travel through the interior of Earth. There are two types of body waves: P waves and S waves. P waves travel through Earth and are caused by expansion and contraction of bedrock. S waves are produced when material moves either vertically or horizontally and travel only within the uppermost layers of Earth (along its surface). Surface waves produce rolling and/or swaying motion and are slower than P or S waves. Surface waves cause ground motion and damage.

The severity of an earthquake depends upon:

1. The amount of potential energy that had been stored
2. The distance the rock mass moved when the energy was released

3. How far below the surface the movement occurred
4. The makeup of the rock material

Primary effects are due to the shaking and resulting damage to buildings and infrastructure and due to the loss of life or injury. Secondary effects include rock slides, flooding due to subsidence (sinking) of land, liquefaction of recent sediments, fires, and tsunamis. Damage due to earthquakes can be reduced through mapping of faults, preparing computer models and simulations, strengthening building codes, preparing emergency teams with adequate training, upgrading communication technology and availability, storing emergency supplies, and educating the public.

TIP

Information in the "Factoids" throughout this book can often be helpful to you when writing your essays. However, be sure the factoid fits into the answer and applies to the question. Throwing in facts that don't apply will not raise your score.

FACTOIDS

- The largest earthquake recorded in the United States occurred in Prince William Sound (the site of the *Exxon Valdez* oil spill) in 1964. It measured 9.2 on the Richter scale.
- The number one state for the quantity of earthquakes is Alaska—followed by California and Hawaii.
- The earthquake that caused the most deaths (about 830,000) occurred in Shansi, China, in 1556. In the United States, the earthquake that caused the most deaths occurred in San Francisco, California, in 1906 (about 3,000, which included deaths due to the fires that resulted).

VOLCANISM

Active volcanoes produce magma (melted rock) at the surface. Other types of volcanoes are classified as intermittent, dormant, or extinct. The majority of volcanoes—95%—occur at subduction zones and mid-oceanic ridges. The remaining 5% occur at hot spots, areas where plumes of magma come close to the surface. Volcanoes may produce ejecta (lava rock and/or ash), molten lava, and/or toxic gases. The most common gases released by volcanoes are steam, carbon dioxide, sulfur dioxide, and hydrogen chloride. Volcanoes affect the climate by introducing large quantities of sulfur dioxide (SO_2) into the atmosphere that is later converted to sulfate ions (SO_4^{2-}) in the stratosphere. The sulfate particles reflect shorter wavelengths of solar radiation and serve as condensation nuclei for high clouds. In 1992, the year after the Mt. Pinatubo eruption, the effect of stratospheric sulfate particles decreased the average global temperature by as much as 1°F (0.5°C) by decreasing the amount of sunlight reaching Earth. The particles settle out of the atmosphere usually within two years and contribute to acid rain.

Eruptions occur when pressure within a magma chamber forces molten magma up through a conduit (pipe) and out a vent at the top of the volcano. The type of eruption depends on the gases, the amount of silica in the magma (which determines viscosity), and how free the conduit is (whether the volcano flows or explodes). Correlation exists between seismic and volcanic activity. Benefits of volcanic eruptions include producing new landforms as seen in the Hawaiian Islands and increased soil nutrient levels produced from erosion of lava rock. Methods of dealing with volcanoes include:

1. Modeling and data analysis for better volcanic activity prediction
2. Better evacuation plans
3. Study of precursors such as changes in the cone
4. Measuring changes in temperature and gas composition
5. Magnetic changes
6. Changes in seismic activity.

Major Types of Volcanoes

There are four major types of volcanoes. They vary in size, shape, and how they are formed.

CINDER CONES

A cinder cone is a steep, conical hill formed above a vent. Cinder cones are among the most common volcanic landforms found in the world and usually do not cause significant loss of life. The cones usually grow up in groups, and they often occur on the flanks of composite and shield volcanoes. They are built from lava fragments called cinders. The lava fragments are ejected from a single vent and accumulate around the vent when they fall back to Earth. Cinder cones grow rapidly and soon approach their maximum size. They rarely exceed 800 feet (250 m) in height and 1,600 feet (500 m) in diameter. One example is Paricutín in Mexico, which grew to 300 feet (90 m) in 5 days.

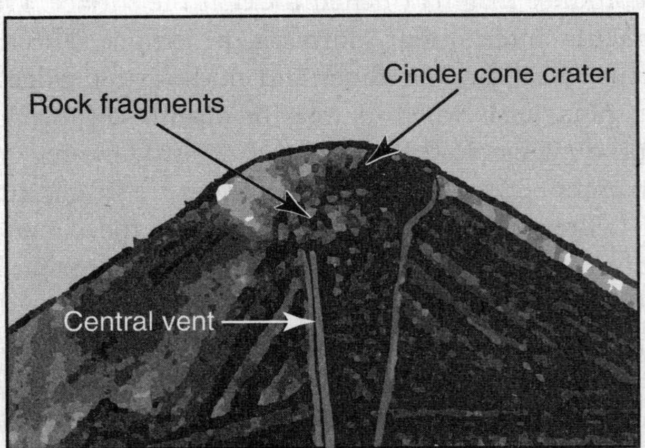

COMPOSITE (STRATO)

Composite volcanoes, also known as strato volcanoes, are formed by alternating layers of lava and rock fragments, are constructed along subduction zones, and often form impressive, snow-capped peaks that often exceed 1.5 miles (2,500 m) in height. Between eruptions they are often so quiet they seem extinct. When very viscous magma rises to the surface, it usually clogs the crater pipe. Gases build up pressure and result in an explosive eruption. Examples of composite volcanoes include Mount Hood, Mount Rainier, Mount Shasta, Mount Pinatubo, Mount Fuji, and Mount Vesuvius.

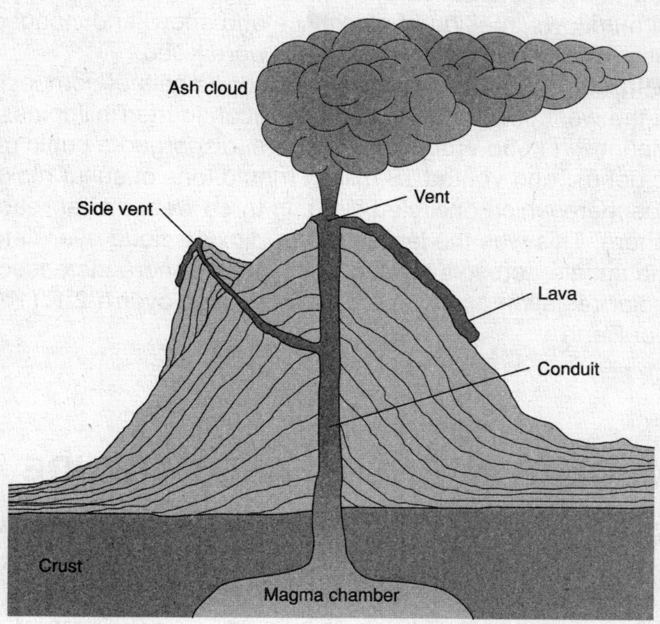

Figure 1.11 Parts of a volcano

Lava domes are a principal structural feature of many composite volcanoes. The large amount of silica in the lava does not allow the lava to flow far from the vent. Lava domes grow slowly and steadily for months or years and are prone to collapsing. When part of a lava dome collapses while it still contains molten rock and gases, it produces a pyroclastic flow—one of the most lethal volcanic events possible (Examples include Mount Saint Helens in Washington State and Mount Lassen in California).

SPATTER CONES

When hot, erupting lava contains just enough explosive gas to prevent the formation of a lava flow but not enough to shatter it into small fragments, the lava is torn by expanding gases into fluid hot clots. These range in size from half an inch (1 cm) to 20 inches (50 cm) across and are called spatter. Because spatter is not fully solid when it lands, the spatter welds itself together. Spatter cones are typical of volcanoes with highly fluid magma, such as those found in the Hawaiian Islands.

SHIELD

Shield volcanoes are very large and are built by many layers of runny lava flows. As the very fluid lava spills out of a central vent or group of vents, a broad-shaped, gently sloping cone is formed. Shield volcanoes may be produced by hot spots that lie far away from the edges of tectonic plates. Shield volcanoes also occur along mid-oceanic ridges, where seafloor spreading is in progress, and along subduction-related volcanic arcs. Examples of shield volcanoes are Mauna Loa and Kilauea in Hawaii.

CASE STUDIES

- **Mount Saint Helens:** Located in Washington State, Mount Saint Helens erupted in 1980. The earthquake removed trees, increased soil erosion, destroyed wildlife, and polluted the air with gases and ash. Other effects included mudflows, melting of glacial ice and snow, and clogged rivers that caused flooding. Fifty-seven people were killed.
- **Mount Pinatubo:** Mount Pinatubo is part of a chain of composite volcanoes on the west coast of the island of Luzon in the Philippines. In June 1991, Mount Pinatubo erupted for 9 hours, disgorged a cubic mile of volcanic debris, and vented 18 million metric tons of sulfur dioxide into the atmosphere which encircled Earth in three weeks after reaching the stratosphere. This was the largest sulfur dioxide cloud ever detected to date. The sulfate aerosols formed in the stratosphere increased reflection of solar radiation and within 3 years caused over a 2°F (1°C) overall cooling of Earth.

SEASONS, SOLAR INTENSITY, AND LATITUDE

Factors that affect the amount of solar energy at the surface of Earth (which is directly correlated with plant productivity) include Earth's rotation (once every 24 hours), Earth's revolution around the sun (once per year), tilt of Earth's axis (23.5°), and atmospheric conditions. Summer (the period of greatest solar radiation) occurs in the Northern Hemisphere when Earth is oriented more toward the sun. The sun rises higher in the sky and stays above the horizon longer, with the rays of the sun striking the ground more directly (at less of an angle). Likewise, in the Northern Hemisphere winter, the hemisphere is tilted away from the sun. The Sun rises lower in the sky and stays above the horizon for a shorter period, with the rays of the sun striking the ground more obliquely (at a greater angle).

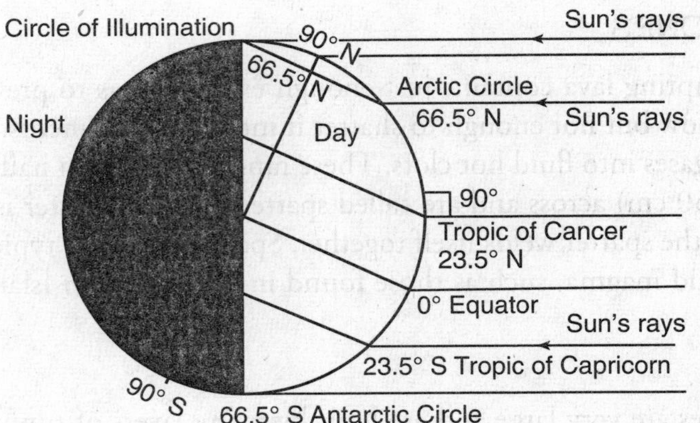

Figure 1.12 Summer solstice

Earth is actually closer to the sun in the Northern Hemisphere during the winter than in the summer. Earth is at its closest approach to the sun (perihelion) in January of each year (the middle of the Northern Hemisphere winter) and the farthest away in July (aphelion) during the middle of the Northern Hemisphere summer. Seasons are NOT caused by Earth's distance from the sun.

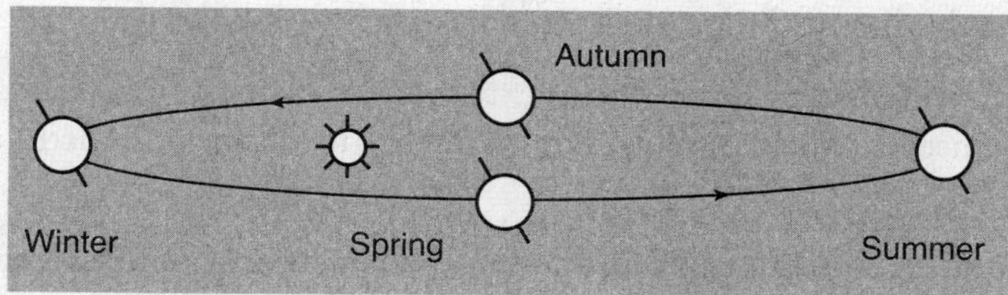

Figure 1.13 Seasons in the Northern Hemisphere

SOIL

Soils are a thin layer on top of most of Earth's land surface. This thin layer is a basic natural resource and deeply affects every other part of the ecosystem. For example, soils hold nutrients and water for plants and animals; water is filtered and cleansed as it flows through soils; and soils affect the chemistry of water and the amount of water that returns to the atmosphere to form rain. Soils are composed of three main ingredients: minerals of different sizes, organic materials from the remains of dead plants and animals, and open space that can be filled with water or air. A good soil for growing most plants should have about 45% minerals (with a mixture of sand, silt, and clay), 5% organic matter, 25% air, and 25% water.

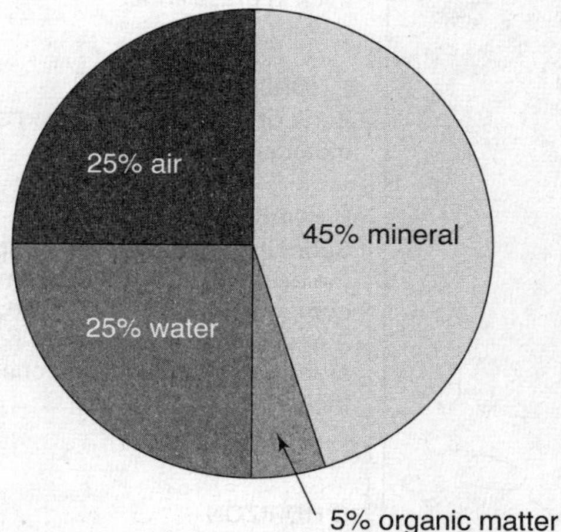

Figure 1.14 Soil components

Soils develop in response to several factors:

1. **Parent material.** This refers to the rock and minerals from which the soil derives. The nature of the parent rock, which can be either native to the area or transported to the area by wind, water, or glacier, has a direct effect on the ultimate soil profile.
2. **Climate.** This is measured by precipitation and temperature. It results in partial weathering of the parent material, which forms the substrate for soil.
3. **Living organisms.** These include the nitrogen-fixing bacteria *Rhizobium*, fungi, insects, worms, snails, etc. that help to decompose litter and recycle nutrients.
4. **Topography.** This refers to the physical characteristics of the location where the soil is formed. Topographic factors that affect a soil's profile include drainage, slope direction, elevation, and wind exposure.

With sufficient time, a mature soil profile reaches a state of equilibrium. Feedback mechanisms involving both abiotic and biotic factors work to preserve the mature soil profile. The relative abundance of sand, silt, and clay is called the soil texture.

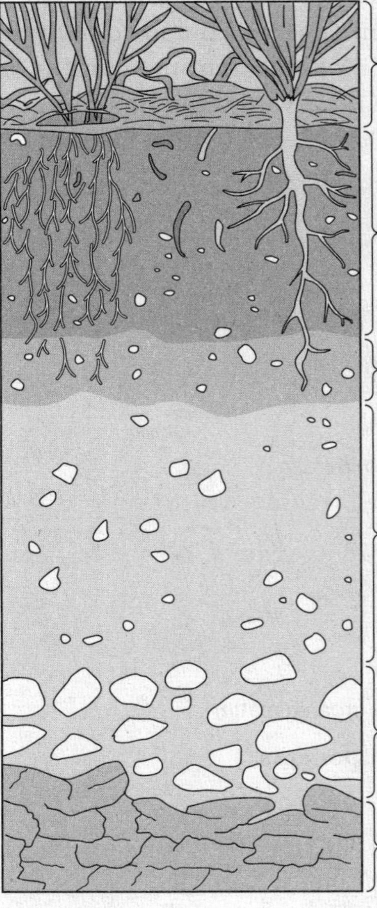

Figure 1.15 Soil profile

Soil Components

Component	Description
Clay	Very fine particles. Compacts easily. Forms large, dense clumps when wet. Low permeability to water; therefore, upper layers become waterlogged.
Gravel	Coarse particles. Consists of rock fragments.
Loam	About equal mixtures of clay, sand, silt, and humus. Rich in nutrients. Holds water but does not become waterlogged.
Sand	Sedimentary material coarser than silt. Water flows through too quickly for most crops. Good for crops and plants requiring low amounts of water.
Silt	Sedimentary material consisting of very fine particles between the size of sand and clay. Easily transported by water.

EROSION

Soil erosion is the movement of weathered rock or soil components from one place to another. It is caused by flowing water, wind, and human activity such as cultivating inappropriate land, burning of native vegetation, deforestation, and construction. Soil erosion destroys the soil profile, decreases the water-holding capacity of the soil, and increases soil compaction. Because water cannot percolate through the soil, it runs off the land, taking more soil with it (positive feedback loop). Because the soil cannot hold water, crops grown in areas of soil erosion frequently suffer from water shortage. In areas of low precipitation, erosion leads to significant droughts. Poor agricultural techniques that lead to soil erosion include monoculture, row cropping, overgrazing, improper plowing of the soil, and removing crop wastes instead of plowing the organic material back into the soil. There are three common types of soil erosion:

1. **Sheet erosion**—soil moves off as a horizontal layer
2. **Rill erosion**—fast-flowing water cuts small channels in the soil
3. **Gully erosion**—extreme case of rill erosion, where over time, channels increase in size and depth.

Soil erosion causes damage to agriculture, waterways (canals), and infrastructures (dams). It interferes with wetland ecosystems, reproductive cycles (as in salmon), oxygen capacity, and the pH of the water.

FACTOIDS

- One inch (2.5 cm) of topsoil can take up to 1,000 years to form.
- Due to extensive soil erosion in the United States, more soil is lost each year than during the 1930s dust bowl.
- In 1850, soil in the prairie states averaged 14 inches (35 cm) of topsoil. Today it is 7 inches (18 cm).
- More than half of the fertilizer used today is required to replace organic nutrients lost through soil erosion.

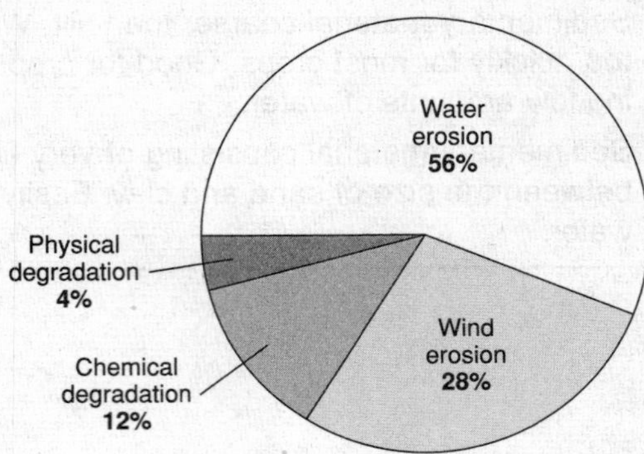

Figure 1.16 Soil degredation by type

THE ROCK CYCLE

There are three main categories of rocks: metamorphic, igneous, and sedimentary. Metamorphic rock has been subjected to tremendous heat and pressure, such as slate and marble. Igneous rock has been formed by cooling, such as granite, basalt, and quartz. Sedimentary rock (sandstone, shale, limestone) has been formed by piling of materials (diatoms, weathered chemical precipitates, etc.) over time.

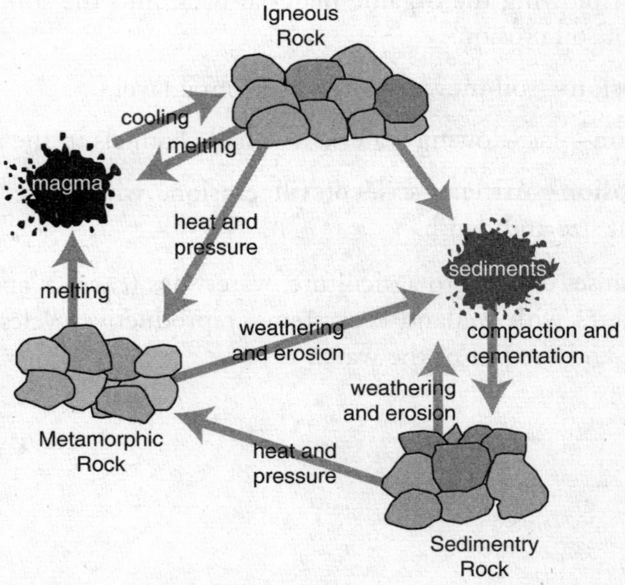

Figure 1.17 The rock cycle

CASE STUDY

Dust Bowl: The Dust Bowl occurred during the 1930s in Oklahoma, Texas, and Kansas. It was caused by plowing the prairies and resulted in the loss of natural grasses that rooted the soil. Drought and winds that occurred blew most of the topsoil away, causing people to leave the area.

RELEVANT LAW

1935 Soil Erosion Act: Established the Soil Conservation Service. Mandates the protection of the nation's soil reserves. Deals with soil erosion problems, carries out soil surveys, and does research on soil salinity. Provides computer databases for scientific research.

MULTIPLE-CHOICE QUESTIONS

1. The majority of the rocks in the Earth's crust are:

   (A) igneous
   (B) metamorphic
   (C) sedimentary
   (D) basalt
   (E) volcanic

2. Which of the following is an example of an igneous rock?

   (A) Marble
   (B) Slate
   (C) Limestone
   (D) Granite
   (E) Sandstone

3. The smallest particle of soil is known as

   (A) clay
   (B) sand
   (C) silt
   (D) gravel
   (E) humus

TIP

When answering multiple choice questions, there is a penalty for guessing. If you are not sure on any of the questions—DO NOT GUESS. But, if you can rule out one or more wrong answer choices, then go ahead and take an educated guess.

4. Acid rain affects soil by

(A) decreasing soil porosity
(B) decreasing the pH
(C) decreasing soil aeration
(D) lowering nutrient capacity
(E) all of the above

5. The higher the amount of _____ in the soil, the better its nutrient-holding capacity.

(A) clay
(B) silt
(C) sand
(D) gravel
(E) loam

6. An example of a volcano with broad, gentle slopes and built by the eruption of runny, fluid-type basalt lava would be

(A) Mount Saint Helens
(B) Krakatau
(C) Kilauea
(D) Vesuvius
(E) Mount Rainier

7. Which of the following is at a convergent boundary where two continental plates are presently colliding?

(A) The Appalachian Mountains
(B) The Himalayas
(C) The Andes Mountains
(D) The Rocky Mountains
(E) None of the above

8. The Dust Bowl of the 1930s resulted in the passage of what legislation?

(A) Endangered American Wilderness Act
(B) Soil and Water Conservation Act
(C) Federal Land Management Act
(D) Public Rangelands and Improvement Act
(E) Soil Erosion and Conservation Act

9. Poor nutrient-holding capacity, good water infiltration capacity, and good aeration properties are examples of what type of particle found in soil?

(A) Clay
(B) Silt
(C) Sand
(D) Loam
(E) Humus

10. An alkaline, dark soil, rich in humus, found in a semiarid climate would be most characteristic of

    (A) deserts
    (B) grasslands
    (C) tropical rain forests
    (D) deciduous forests
    (E) coniferous forests

11. Which period in geological time describes the following: "Development of flowering plants. Large diversity in dinosaurs but ending with their sudden extinction approximately 65 million years ago. Formation of the Andes Mountains. African and South American plates begin to separate. Climate is cooling. Shallow seas are prominent."?

    (A) Paleogene
    (B) Neogene
    (C) Jurassic
    (D) Cretaceous
    (E) Permian

12. The process of weathering produces what type of rock?

    (A) Igneous
    (B) Metamorphic
    (C) Sedimentary
    (D) Volcanic
    (E) None of the above

13. A rock that would most likely contain a fossil would be

    (A) igneous
    (B) metamorphic
    (C) sedimentary
    (D) volcanic
    (E) all of the above

14. The most common element found in Earth's crust is

    (A) oxygen
    (B) hydrogen
    (C) iron
    (D) silicon
    (E) aluminum

15. The horizon of soil also known as the topsoil layer, that contains humus, minerals, and roots, and that is rich in living organisms is known as the

    (A) A layer
    (B) B layer
    (C) C layer
    (D) D layer
    (E) O layer

16. Earth is closest to the sun in the Northern Hemisphere during

    (A) winter
    (B) summer
    (C) spring
    (D) fall
    (E) all seasons

17. An earthquake of Richter magnitude 5 is how many times larger on the Richter scale than an earthquake of Richter magnitude 3?

    (A) One-fourth
    (B) One-half
    (C) Twice
    (D) Four times
    (E) One hundred times

18. An example of a primary earthquake effect would be

    (A) soil liquefaction
    (B) tsunamis
    (C) fires
    (D) collapse of buildings
    (E) liquefaction

19. The San Andreas Fault in California occurs at

    (A) a convergent boundary
    (B) a divergent boundary
    (C) a transform boundary
    (D) a subduction zone
    (E) an oceanic ridge

20. Earth's surface is part of the

    (A) asthenosphere
    (B) lithosphere
    (C) benthosphere
    (D) troposphere
    (E) stratosphere

FREE-RESPONSE QUESTION

By: Dr. Ian Kelleher, Brooks School, North Andover, MA
B.S. University of Manchester, England;
Ph.D. Cambridge University

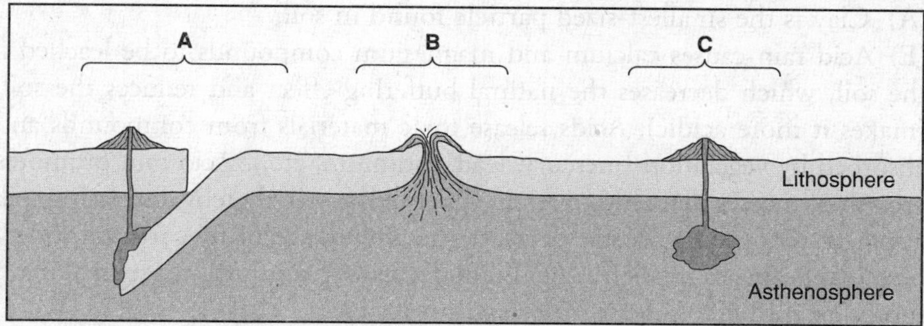

(a) Geological features A–C above are formed as a consequence of plate tectonics. Chose two of these features and
   (i) describe what is occurring there
   (ii) give an actual geographic example

(b) Charles Darwin was the geologist, botanist, and zoologist on the research vessel *Beagle* when he made observations that lead to his book, *The Origin of Species*, in which the theory of evolution and natural selection was first introduced. A century later, scientists developed the theory of plate tectonics, describing how the solid Earth formed. Describe two ways in which evolution may occur as a consequence of plate tectonics.

(c) Mount Pinatubo in the Philippines erupted in 1991. Examine the temperature graph below and answer the following questions.

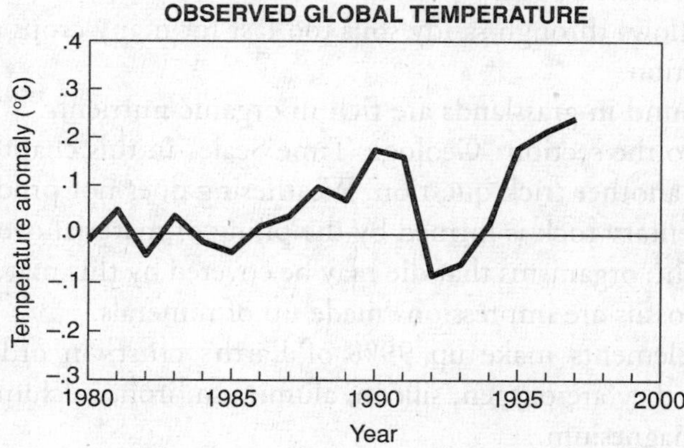

   (i) Compare Earth's climate before after the eruption of Mount Pinatubo.
   (ii) Explain how the eruption of Mount Pinatubo might affect short-term and long-term climate change.

MULTIPLE-CHOICE ANSWERS AND EXPLANATIONS

1. **(A)** Igneous rocks are solidified from magma. If the magma cools slowly, the rocks are coarser in nature. Had the question been worded "the majority of the rocks on the surface of Earth," the answer would have been sedimentary.
2. **(D)** Other examples of igneous rocks include basalt and quartz.
3. **(A)** Clay is the smallest-sized particle found in soil.
4. **(E)** Acid rain causes calcium and magnesium compounds to be leached from the soil, which decreases the natural buffering effect and reduces the soil pH (makes it more acidic). Acids release toxic materials from compounds and are absorbed by vegetation (mercury, lead, cadmium, etc.). Acid rain promotes the growth of mosses that tend to retain water in the soil, thereby decreasing soil aeration (waterlogged). Mosses decrease the abundance of mycorrhizal fungi that help plants absorb nutrients. Acid rain decreases plants' resistance, making them more susceptible to disease, insects, drought, etc.
5. **(C)** This is a trick question. The question was NOT asking for what is the best type of soil to grow crops in—that would be loam. It was asking which type of soil particle would retain nutrients best. The small size of clay particles causes clay to clump. As a result, the soil becomes waterlogged with poor aeration. Since clay particles do not allow water and nutrients to percolate through the soil, the water and nutrients are retained in the soil and often do not reach the roots.
6. **(C)** Kilauea is a shield volcano that is characterized by basalt lava building enormous, low-angle, gently sloping cones. The fluid nature of the lava prevents it from piling in steep mounds. Shield volcanoes occur along the mid-oceanic ridge, where seafloor spreading is in progress and along subduction zones related to volcanic arcs. The largest volcanoes on Earth are shield volcanoes.
7. **(B)** Notice the word "presently." The Appalachian and Rocky mountains were formed at *ancient* convergent plate boundaries. The Andes lie at a convergent boundary where oceanic lithosphere is being subducted under the South American continent.
8. **(B)** Refer to the case study presented in this chapter.
9. **(C)** Water flows through sandy soils too fast for many crops and requires frequent irrigation.
10. **(B)** Soils found in grasslands are rich in organic nutrients.
11. **(D)** Refer to the section "Geologic Time Scale" in this chapter.
12. **(E)** This is another trick question. Weathering does not produce rock.
13. **(C)** Sedimentary rock is formed by the piling of material over time. If conditions are right, organisms that die may be covered by this material and become fossilized. Fossils are impressions made up of minerals.
14. **(A)** Eight elements make up 99% of Earth's crust. In order of decreasing abundance, they are oxygen, silicon, aluminum, iron, calcium, sodium, potassium, and magnesium.
15. **(A)** If the topsoil is brown or black, it is rich in nitrogen and is good for crops. If the topsoil is gray, yellow, or red, it is low in organic matter and poor for crops.
16. **(A)** The angle of sunlight determines the season, not how close Earth is to the sun.

17. **(E)** The Richter scale is a log-based scale. $5 - 3 = \underline{2}$ and $10^{\underline{2}} = 100$. However, the Richter scale does NOT measure the energy of an earthquake. The energy of a Richter magnitude 5 has 32 times more energy than a Richter magnitude of 4.
18. **(D)** The collapse of buildings is a direct effect caused by earthquakes. All other choices are consequences of earthquakes.
19. **(C)** Places where plates slide past each other are called transform boundaries. The most famous transform boundary in the world is the San Andreas fault.
20. **(B)** The asthenosphere is below the lithosphere.

FREE-RESPONSE ANSWER

(a) Feature *A* is a subduction zone. One lithospheric plate is subducting (sinking) below another, largely due to differences in density (the denser plate sinks). This is an example of a convergent plate boundary. As the subducted plate sinks to greater depths, the temperature increases to the point where it begins to melt. This molten magma is less dense than the solid rock around it, so it rises up and forms a chain of volcanic mountains parallel to the plate boundary. The Cascade Mountains in Washington State are examples of a volcanic arc. When two oceanic plates converge, they create an island arc—a curved chain of volcanic islands rising from the deep seafloor and near a continent. They are created by subduction processes and occur on the continent side of the subduction zone. Japan is an example of an island arc.

Feature *B* is a divergent plate boundary. Lithospheric plates are moving apart. The space created between them is filled by hot molten magma coming up from the asthenosphere that then cools and adds to the crust. In oceanic crust, they are known as a mid-oceanic ridge. The Mid-Atlantic Ridge is an example. When they form on continental crust, they are known as rift valleys, such as the African Rift Valley.

Feature *C* is a hot spot. This is a place in the asthenosphere where the temperature is higher than average such that localized melting occurs. This molten rock, being less dense due to its temperature and state of matter, rises up. It forms a volcano on Earth's surface. Over geologic time, the location of the hot spot remains constant, whereas the lithospheric plate moves over it. This causes a chain of volcanoes to form over time from a single hot spot. The Hawaiian Islands are an example of the consequences of hot spots.

(b) Evolution may occur as a consequence of geographic separation of one population of a species into two or more populations. Plate tectonics may cause this separation by either of two methods.

First, a divergent (constructive) plate boundary could cause one land mass to be divided into two or more distinct parts, perhaps even separated by an ocean. For example, identical fossils can be found on the east cost of South America and the west coast of Africa, indicating that these were once the same connected landmass. After these two continents diverged, different species would evolve from this common ancestor as a reaction to the different environments and consequent environmental pressures on the different landmasses.

Second, faulting occurring as a consequence of plate tectonics may cause a river to be diverted. The new path of the river could divide a population into two and serve as a geographic barrier preventing gene flow. Different conditions in geographically separated regions would eventually lead to the evolution of different species as each population adapted to its environment in different ways.

Plate tectonics may result in climatic change in one of the following ways: First, Earth's atmosphere has changed considerably throughout geological history, largely as a consequence of gases emitted through volcanic activity caused by plate tectonics. These changes in the atmosphere have caused climate change. For example, there is evidence that the Earth was much warmer hundreds of millions of years ago.

Second, lithospheric plates move over the surface of Earth at speeds of a few centimeters a year. Individual plates have moved thousands of miles over geologic history. As the latitude of the plate changes, so will its climate. For example, some rocks in Alaska indicate that they were originally deposited at a time when the plate had a tropical climate and so must have been closer to the equator.

Species evolve in reaction to these climate changes. For example, animals will adapt to shifting food sources as different plants grow in different climates. A species of animal may develop fur over time (through natural selection) as temperatures decrease over time.

As lithospheric plates move to latitudes farther from the equator, climates will have greater seasonal variations. This could lead to evolutionary adaptations such as plants shedding their leaves to conserve resources or animals hibernating during winter months to conserve energy.

Evidence suggests that some mass extinctions in Earth's history may have been caused by large-scale volcanic activity, such as flood basalts, which occur as a consequence of plate tectonics. These mass extinctions tend to be followed by periods with high rates of evolution (punctuated) and increase in species diversity (adaptive radiation) as ecological niches are filled.

(c) After the eruption of Mount Pinatubo, Earth's temperature was approximately 0.3°C lower for the next two years. Earth's temperature rose by approximately 0.1°C during the third year. By the fourth year after the eruption, Earth's temperature had returned to the pre-eruption level.

In the short term, dust and other particulates released into the atmosphere from the eruption would block the sun's rays. This decrease in energy reaching Earth would result in lower global temperatures.

In the long term, gases such as carbon dioxide released during the eruption would accumulate in the stratosphere. There they would contribute to the greenhouse effect. They would absorb energy radiated back from the Earth, leading to an increase in global temperature. The degree of temperature change from the Mount Pinatubo eruption via this mechanism would be much less than that by dust blocking the sun, but the effect would last much longer.

The Atmosphere

CHAPTER 2

A frozen Continent . . . beat with perpetual storms . . . the parching Air Burns frore, and cold performs th' effect of Fire.

—Milton, *Book II—Paradise Lost*

COMPOSITION

Earth's atmosphere is composed of seven different compounds:

Component

NITROGEN (N_2) 78%

Fundamental nutrient for living organisms. Deposits on Earth through nitrogen fixation and reactions involving lightning and subsequent precipitation. Returns to the atmosphere through combustion of biomass and denitrification.

OXYGEN (O_2) 21%

Oxygen molecules are produced through photosynthesis and are utilized in cellular respiration.

WATER VAPOR (H_2O) 0–4%

Largest amounts occur near equator, over oceans, and in tropical regions. Areas where atmospheric water vapor can be low are polar areas and deserts.

CARBON DIOXIDE (CO_2) <<1%

Volume of CO_2 has increased about 25% in the last 300 years due to the burning of fossil fuels and deforestation. CO_2 is produced during cellular respiration and the decay of organic matter. It is a reactant in photosynthesis. CO_2 is also a major greenhouse gas. Humans are responsible for about 5,500 million tons of CO_2 per year. The average time of a CO_2 molecule in the atmosphere is approximately 100 years.

METHANE (CH_4) <<<1%

Methane contributes to the greenhouse effect. Since 1750, methane has increased about 150% due to the use of fossil fuels, coal mining, landfills, grazers, and flooding of rice fields. Human activity is responsible for about 400 million tons per year as compared with approximately 200 million tons per year produced naturally. Average cycle of a methane molecule in the atmosphere is approximately 10 years.

> **TIP**
>
> Don't confuse weather and climate. Weather describes whatever is happening outdoors in a given place at a given time. Weather is what happens from minute to minute. Climate describes the total of all weather occurring over a period of years in a given place. This includes average weather conditions, regular weather sequences (like winter, spring, summer, and fall), and special weather events (like tornadoes and floods.)

NITROUS OXIDE (N_2O) <<<1%

Concentration increasing about 0.3% per year. Sources include burning of fossil fuels, fuels, use of fertilizers, burning biomass, deforestation, and conversion to agricultural land. Humans are responsible for about 6 million tons per year. N_2O is a contributor to the greenhouse effect. Average time of a N_2O molecule in the atmosphere is approximately 170 years.

OZONE (O_3) <<<1%

97% of ozone is found in the stratosphere (ozone layer) 9–35 miles (15–55 km) above Earth's surface. Ozone absorbs UV radiation. Ozone is produced in the production of photochemical smog. A "hole" in the ozone layer occurs over Antarctica. Chlorofluorocarbons (CFCs) are the primary cause of the breakdown of ozone.

STRUCTURE

The atmosphree consists of several different layers:

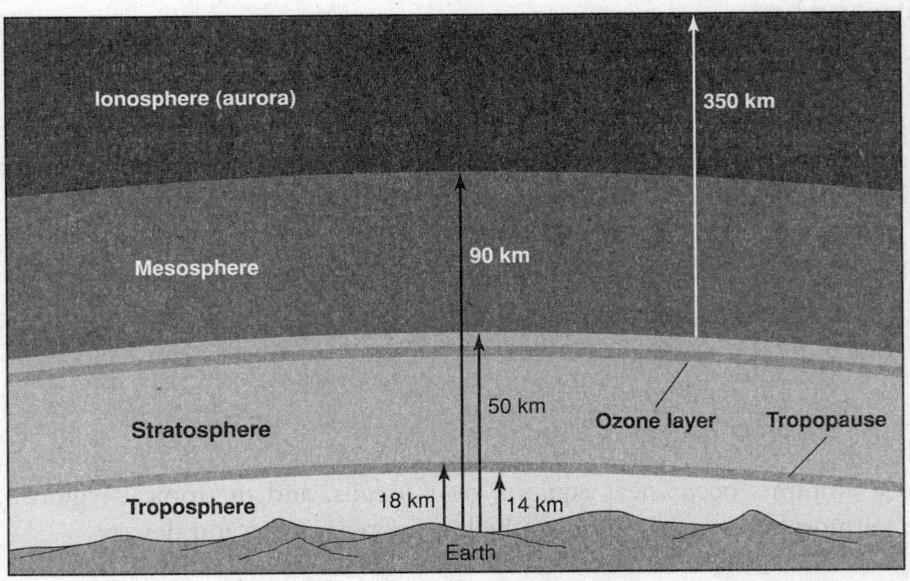

Figure 2.1 Layers of the atmosphere

Layers

TROPOSPHERE

0–7 miles (0–11 km) above surface. 75% of atmosphere's mass is in the troposphere. Temperature decreases with altitude, reaching –76°F (–60°C) near the top. Weather occurs in this zone.

STRATOSPHERE

Temperature increases with altitude due to absorption of heat by ozone. Ozone is produced by UV radiation and lightning. Contains the ozone layer.

MESOSPHERE

Temperature decreases with altitude. Coldest layer. Ice clouds occur here. Meteors (shooting stars) burn up in this layer.

THERMOSPHERE (IONOSPHERE)

Temperature increases with height due to gamma rays, X-rays, and UV radiation. Molecules are converted into ions which results in the aurora borealis (northern lights) in the northern hemisphere and the aurora australis (southern lights) in the southern hemisphere. The aurora borealis most often occurs from September to October and from March to April.

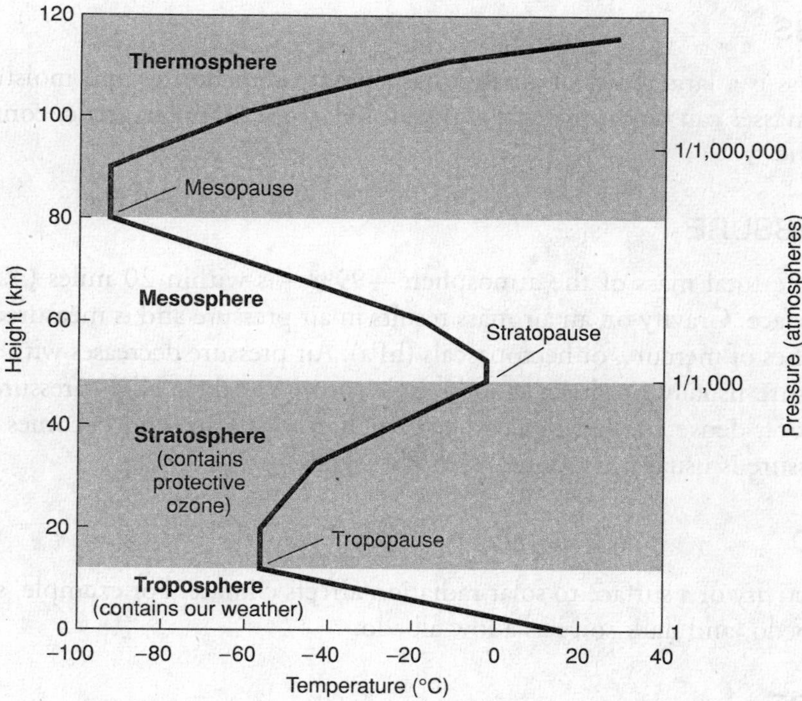

Figure 2.2 Changes in temperature in the atmosphere

> **TIP**
>
> Several different factors influence climate:
>
> - Air mass
> - Air pressure
> - Albedo
> - Altitude
> - Angle of sunlight
> - Clouds
> - Distance to oceans
> - Fronts
> - Heat
> - Land changes
> - Latitude
> - Location
> - Humidity or moisture content of air
> - Mountain ranges
> - Pollution
> - Rotation
> - Wind patterns
> - Human activity

WEATHER AND CLIMATE

Weather is caused by the movement or transfer of heat energy. Energy can be transferred wherever there is a temperature difference between two objects. Energy can be transferred through radiation, conduction, and convection.

Radiation is the flow of electromagnetic radiation. It is the method by which Earth receives solar energy.

Conduction is energy. It is transferred by the collisions that take place between heat-carrying molecules.

Convection is the primary way energy is transferred from hotter to colder regions in Earth's atmosphere and is the primary determinant of weather patterns. Convection involves the movement of the warmer and therefore more energetic molecules in air. Convection takes place both vertically and horizontally. When air near the ground becomes warmer and therefore less dense than the air above it, the air rises. Pressure differences that develop because of temperature differences, result in wind or horizontal convection.

Regions nearer the equator receive much more solar energy than regions nearer the poles and are consequently much warmer. These latitudinal differences in surface temperature create global-scale flows of energy within the atmosphere, giving rise to the major weather patterns of the world. Without convection and the transfer of energy, the equator would be about 27°F (15°C) warmer and the Arctic would be about 45°F (25°C) colder than they actually are.

FACTORS THAT INFLUENCE CLIMATE

AIR MASS

An air mass is a large body of air that has similar temperatures and moisture content. Air masses can be categorized as equatorial, tropical, polar, arctic, continental, or maritime.

AIR PRESSURE

Most of the total mass of the atmosphere—99%—is within 20 miles (32 km) of Earth's surface. Gravity on an air mass results in air pressure and is measured in millibars, inches of mercury, or hectopascals (hPa). Air pressure decreases with altitude. Low pressure usually produces cloudy and stormy weather. High pressure masses contain cool, dense air that descend toward Earth's surface and becomes warmer. High pressure is usually associated with fair weather.

ALBEDO

The reflectivity of a surface to solar radiation affects climate. For example, snow has a high albedo, and dark soil has a low albedo.

ALTITUDE

For every 1,000 feet (300 m) rise in elevation, there is a 3°F (1.5°C) drop in temperature. Every 300 feet (90 m) rise in elevation is equivalent to a shift of 62 miles (100 km) north in latitude and biome similarity.

ANGLE OF SUNLIGHT

In the Northern Hemisphere winter, Earth is closest to the sun. The angle of sunlight reaching Earth affects the climate. Areas closest to the equator receive the most sunlight and therefore higher temperatures.

CLOUDS

Clouds are collections of water droplets or ice crystals suspended in the atmosphere. As warmer air rises, it drops in temperature and therefore cannot hold as much water vapor. The vapor begins to condense forming tiny water particles or ice crystals. High-level clouds (prefix *cirr*) are primarily ice crystals. Midlevel clouds (prefix *alto*) and low-level clouds (prefix *strat*) are composed primarily of water droplets but may also contain ice particles or snow.

DISTANCE TO OCEANS

Oceans are thermally more stable than landmasses; the specific heat (heat-holding capacity) of water is five times greater than air. Because of this, changes in temperature are more extreme in the middle of the continents than on the coasts.

FRONTS

When two different air masses meet, the boundary between them forms a front. The air masses can vary in temperature, dew point, or wind direction. A warm front is the boundary between an advancing warm air mass and the cooler one it is replacing. Since warm air is less dense, it rises and cools, and the moisture it contains is released as rain. A cold front is the leading edge of an advancing mass of cold air. Cold fronts are associated with thunderhead clouds, high surface winds, and thunderstorms. After a cold front passes, the weather is usually cool with clear skies.

HEAT

Climate is influenced by how heat energy is exchanged between air over the oceans and the air over land.

LAND CHANGES

Climate is influenced by urbanization and deforestation.

LATITUDE

The higher the latitudes, the less solar radiation. This affects the climate.

LOCATION

Climate is influenced by the location of high and low air pressure zones and where landmasses are distributed.

MOISTURE CONTENT OF AIR (HUMIDITY)

The moisture content of air is a primary determinant of plant growth and distribution and is a major determinant of biome type (desert vs. tropical forest). Atmospheric water vapor supplies moisture for clouds and rainfall, and it plays a role in energy exchanges within the atmosphere. Water vapor is also a greenhouse gas as it traps heat energy leaving Earth's surface. The dew point is the temperature at which condensation takes place.

MOUNTAIN RANGES

The presence or absence of mountain ranges affects the climate. Mountains influence whether one side of the mountain will receive rain or not (rain shadow effect). The side facing the ocean is the windward side and receives rain; the side of the mountain opposite the ocean is the leeward side and receives little rain. Temperatures decrease as the altitude increases.

POLLUTION

Greenhouse gases from pollution and volcanoes affects the climate.

ROTATION

Daily temperature cycles are primarily influenced by Earth's rotation on its axis (once every 24 hours). At night, heat escapes from the surface. Daily minimum temperatures occur just before sunrise.

WIND PATTERNS

Wind patterns are influenced by temperature and pressure differences (gradients).

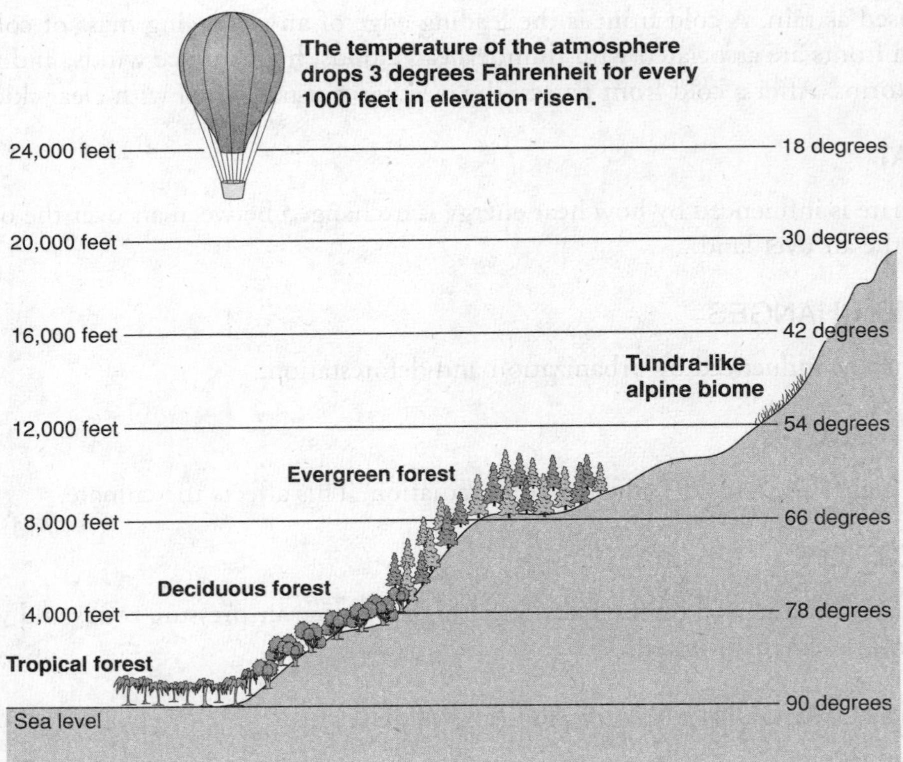

Figure 2.3 Change in temperature in response to change in altitude.

HUMAN ACTIVITY

Climate can also be influenced by human activity. Deforestation, urbanization, heat island effects, release of pollutants including greenhouse gases and the burning of fossil fuels, and the production of acid rain are examples of how humans have altered climatic patterns. Increased pollution alone, combined with an increase in convectional uplift in urban areas, tends to increase the amount of rainfall in urban areas as much as 10% when compared with undeveloped areas.

ATMOSPHERIC CIRCULATION AND THE CORIOLIS EFFECT

Due to the rotation of Earth on its axis, rotation around the sun, and the tilt of Earth's axis, the sun heats the atmosphere unevenly. Air closer to Earth's surface is the warmest and rises. Air at higher elevations is cooler and, as such, more dense and sinks. This sets up convection processes and is the primary cause for winds. Global air circulation is also affected by uneven heating of Earth's surface, seasons, the Coriolis effect, the amount of solar radiation reaching the Earth over long periods of time, convection cells created by warm ocean waters that commonly leads to hurricanes, and ocean currents, which are caused by differences in water density, winds, and Earth's rotation.

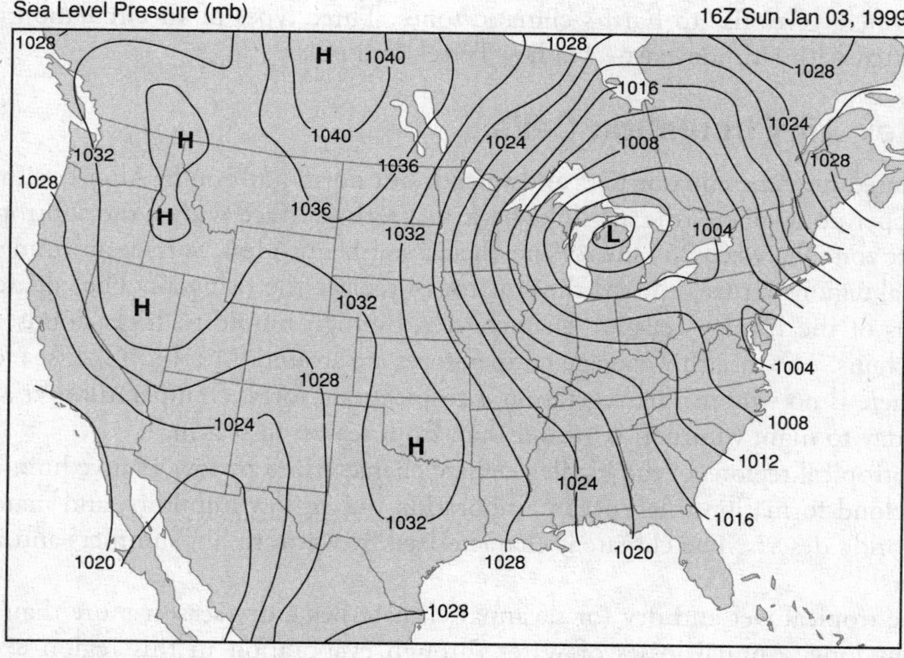

Figure 2.4 The solid contours represent pressure contours (isobars) in millibars. The isobars have an interval of 4 mb. Wind speed is directly related to the distance between the isobars. The closer together they are, the stronger the wind.

Horizontal winds move from areas of high pressures to areas of low pressures. Wind speed is determined by pressure differences between air masses. The greater the pressure difference is, the greater the wind speed. Wind direction is based upon from where the wind is coming. A wind coming from the east is called an easterly.

Wind speed is measured with an anemometer, and wind direction is measured with a wind vane.

Earth's rotation on its axis causes winds not to travel straight. This is called the Coriolis effect. It causes prevailing winds in the Northern Hemisphere to spiral clockwise out from high-pressure areas and spiral counterclockwise in toward low-pressure areas.

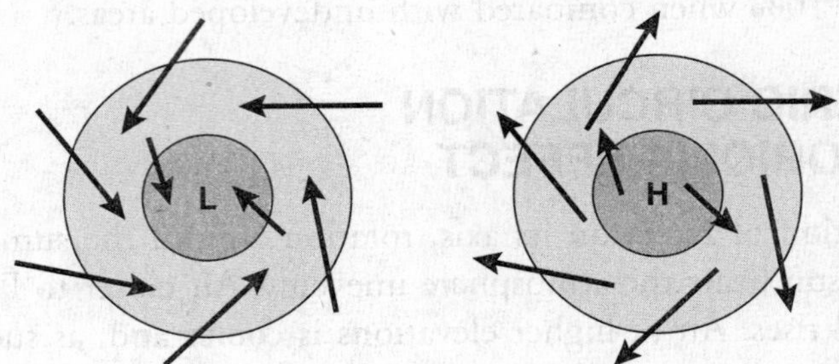

Figure 2.5 Circulation wind patterns of high- and low-pressure systems in the Northern Hemisphere. The pattern reverses in the Southern Hemisphere.

The worldwide system of winds, which transports warm air from the equator where solar heating is greatest toward the higher latitudes where solar heating is diminished, gives rise to Earth's climatic zones. Three types of air circulation cells associated with latitude exist—Hadley, Ferrel, and polar.

Hadley Air Circulation Cells

Air heated near the equator rises and spreads out north and south. After cooling in the upper atmosphere, the air sinks back to Earth's surface within the subtropical climate zone (between 25° and 49° north and south latitudes). Surface air from subtropical regions returns toward the equator to replace the rising air. The equatorial regions of the Hadley cells are characterized by high humidity, high clouds, and heavy rains. The monthly average temperatures are around 90°F (32°C) at sea level, and there is no winter. The vegetation is tropical rain forest. Temperature variation from day to night (diurnal) is greater than from season to season.

Subtropical regions of the Hadley cell are characterized by low relative humidity, little cloud formation, high ocean evaporation due to low humidity, and many of the world's deserts. The climate is characterized by warm to hot summers and mild winters.

The tropical wet and dry (or savanna) climate has a dry season more than two months long. Annual losses of water through evaporation in this region exceed annual water gains from precipitation.

Ferrel Air Circulation Cells

Ferrel cells develop between 30° and 60° north and south latitudes. The descending winds of the Hadley cells diverge as moist tropical air moves toward the poles in winds known as the westerlies. Midlatitude climates can have severe winters and cool summers due to midlatitude cyclone patterns. The western United States is drier in summer than the eastern United States due to oceanic high pressures that brings cool, dry air down from the north. The climate of this area is governed by both tropical and polar air masses. Defined seasons are the rule, with strong annual cycles of temperature and precipitation. The seasonal fluctuation of temperature is greater than the change in temperature occurring in a 24-hour cycle. Climates of the middle latitudes have a distinct winter season. The area of Earth controlled by Ferrel cells contains broadleaf deciduous and coniferous evergreen forests.

Polar Air Circulation Cells

The polar cells originate as icy-cold, dry, dense air that descends from the troposphere to the ground. This air meets with the warm tropical air from the midlatitudes. The air then returns to the poles, cooling and then sinking. Sinking air suppresses precipitation; thus, the polar regions are deserts (deserts are defined by moisture, not temperature). Very little water exists in this area because it is tied up in the frozen state as ice. Furthermore, the amount of snowfall per year is relatively small.

In general, climates in the polar domain are characterized by low temperatures, severe winters, and small amounts of precipitation, most of which falls in summer. The annual fluctuation of temperature is greater than the change in temperature occurring in a 24-hour cycle. In this area where summers are short and temperatures are generally low throughout the year, temperature rather than precipitation is the critical factor in plant distribution and soil development. Two major biomes exist—the tundra and the taiga.

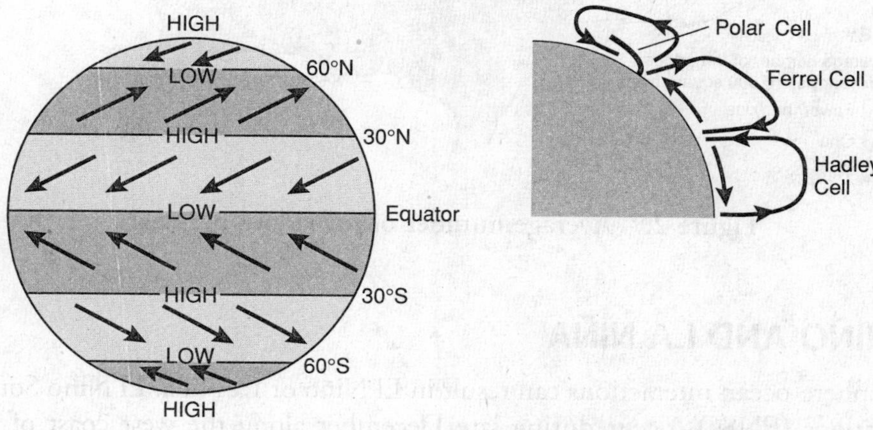

Figure 2.6 The Hadley, Ferrel, and polar cells.

Hurricanes and Tornadoes

Hurricanes are the most severe weather phenomenon on the planet. Hurricane Katrina that hit New Orleans, Louisiana, in 2005 was responsible for $75 billion in damage and approximately 1,830 deaths. Hurricanes begin over warm oceans in areas where the trade winds converge. A subtropical high-pressure zone creates hot daytime temperatures with low humidity that allows for large amounts of evaporation. The Coriolis effect initiates the cyclonic flow.

The stages of hurricane development include the presence of separate thunderstorms that have developed over tropical oceans, and cyclonic circulation that begins to cause these thunderstorms to move in a circular motion. This cyclonic circulation allows them to pick up moisture and latent heat energy from the ocean. In the center of the hurricane is the eye, an area of descending air and low pressure. The energy of a hurricane dissipates as it travels over land or moves over cooler bodies of water. Rainfall can be as much as 24 inches (0.6 m) in 24 hours. A storm surge, which results from the increase in the height of the ocean near the eye of a hurricane, can cause extensive flooding.

Tornadoes are swirling masses of air with wind speeds close to 300 miles per hour (485 kph). Like hurricanes, the center of the tornado is an area of low pressure. In the United States, tornadoes are frequent from April through July and occur in the center of the United States in an area known as "Tornado Alley." Due to advances in weather forecasting, modeling, and warning systems, the death rate due to tornadoes has decreased significantly.

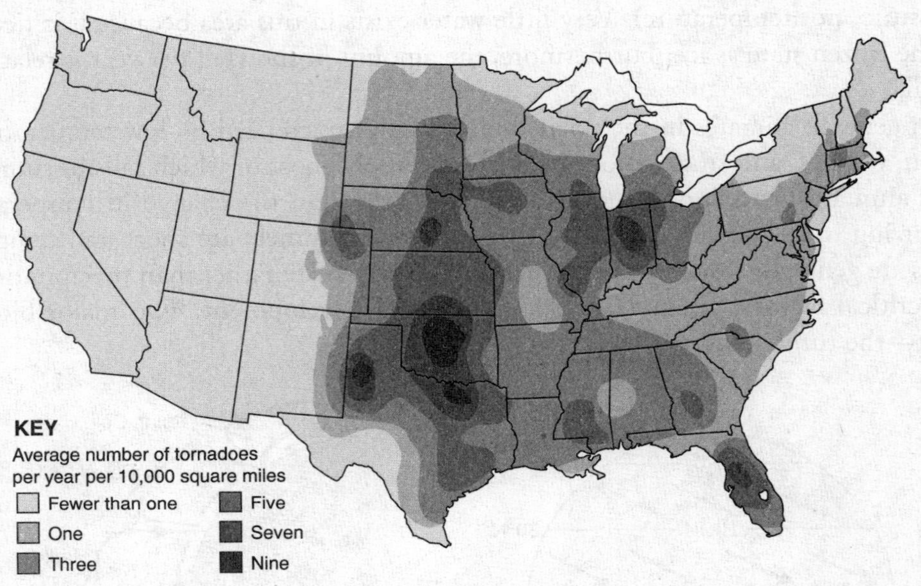

Figure 2.7 Average number of tornadoes per year.

EL NIÑO AND LA NIÑA

Atmosphere-ocean interactions can result in El Niño or La Niña. El Niño Southern Oscillations (ENSO) occur during late December along the west coast of South America. During an ENSO year, air pressure drops along the west coast of South America, while in the western Pacific, high-pressure areas develop. The result is that normal trade winds are reduced causing warmer-than-normal ocean water to develop off the west coast of South America. This prevents the typical nutrient-rich cold-water upwellings. The reduction in upwellings results in extensive fish kills off the South American coast. Climatological effects of El Niño are shown in the following figure.

TIP

Questions about El Niño and La Niña are very common on the APES exam. Be sure you know these two processes!

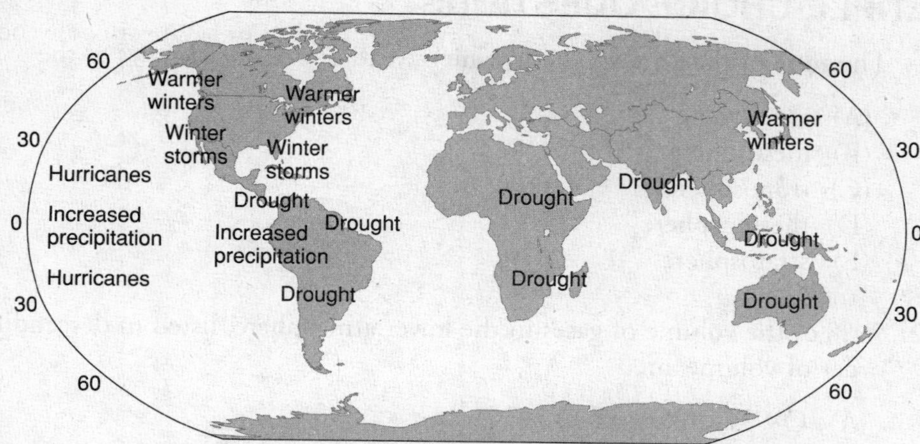

Figure 2.8 Climatological effects of El Niño

La Niña is characterized by unusually cold ocean temperatures in the eastern equatorial Pacific. La Niña tends to bring nearly opposite effects of El Niño to the United States, with wetter-than-normal conditions across the Pacific Northwest and both dryer and warmer-than-normal conditions in the southern states. Winter temperatures are warmer than normal in the southeast United States and cooler than normal in the northwest. The increased temperatures in the southeastern U.S. during La Niña years correlates with the substantial increase in hurricanes that occur during this same time period. La Niña is also responsible for heavier-than-normal monsoons in India and Southeast Asia.

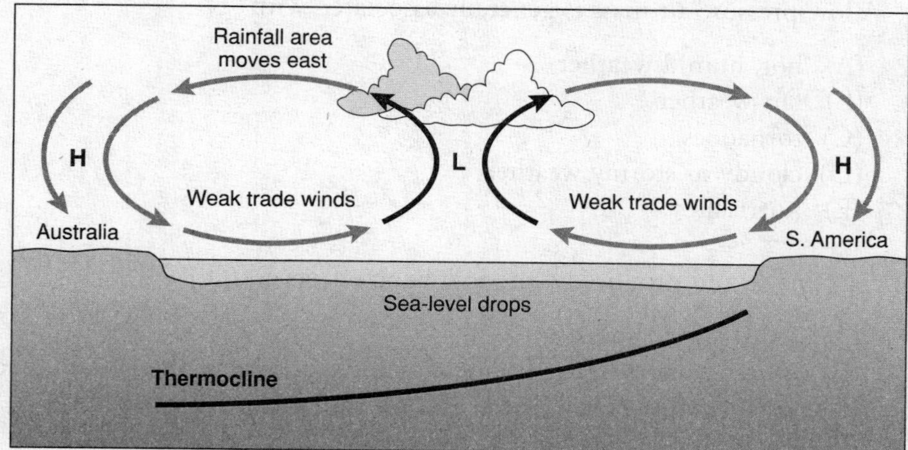

Figure 2.9 The development of El Niño

MULTIPLE-CHOICE QUESTIONS

1. The zone of the atmosphere in which weather occurs is knows as the

   (A) ionosphere
   (B) mesosphere
   (C) troposphere
   (D) thermosphere
   (E) stratosphere

2. 99% of the volume of gases in the lower atmosphere, listed in descending order of volume, are

   (A) O_2, N_2, CO_2, H_2O
   (B) H_2O, N_2, O_2, CO_2
   (C) O_2, CO_2, N_2, H_2O
   (D) O_2, CO_2, N_2, H_2O
   (E) N_2, O_2, H_2O, CO_2

3. Regional climates are most influenced by

   (A) latitude and altitude
   (B) prevailing winds and latitude
   (C) altitude and longitude
   (D) latitude and longitude
   (E) Coriolis effect and trade winds

4. A low-pressure air mass is generally associated with

   (A) hot, humid weather
   (B) fair weather
   (C) tornadoes
   (D) cloudy or stormy weather
   (E) hurricanes

5. La Niña would produce all the following effects EXCEPT

   (A) more rain in southeast Asia
   (B) wetter winters in the Pacific Northwest region of the United States
   (C) warmer winters in Canada and northeast United States
   (D) warmer and drier winters in the southwest and southeast United States
   (E) more Atlantic hurricanes

6. On the leeward side of a mountain range, one would expect

   (A) more clouds and rain than on the windward side
   (B) more clouds but less rain than on the windward side
   (C) colder temperatures
   (D) less clouds and less rain than on the windward side
   (E) no significant difference in climate compared with the windward side

7. The ozone layer exists primarily in what section of the atmosphere?

   (A) Troposphere
   (B) Stratosphere
   (C) Mesosphere
   (D) Thermosphere
   (E) Ionosphere

8. Along the equator,

   (A) warm, moist air rises
   (B) warm, moist air descends
   (C) warm, dry air descends
   (D) cool, dry air descends
   (E) cool, moist air descends

9. The gas that is responsible for trapping most of the heat in the lower atmosphere is

   (A) water vapor
   (B) ozone
   (C) carbon dioxide
   (D) oxygen
   (E) nitrogen

10. Characteristics or requirements of a monsoon include all of the following EXCEPT

   (A) a seasonal reversal of wind patterns
   (B) large land areas cut off from continental air masses by mountain ranges and surrounded by large bodies of water
   (C) different heating and cooling rates between the ocean and the continent
   (D) extremely heavy rainfall
   (E) heating and cooling rates between the oceans and the continents that are equal

11. An atmospheric condition in which the air temperature rises with increasing altitude, holding surface air down and preventing dispersion of pollutants, is known as (a)

   (A) temperature inversion
   (B) cold front
   (C) warm front
   (D) global warming
   (E) upwelling

TIP

Be careful when answering questions that involve placing items in order that you have the order going in the proper direction. A common trick is to have the correct choices in the opposite direction.

12. The surface with the lowest albedo is

    (A) snow
    (B) ocean water
    (C) forest
    (D) desert
    (E) black topsoil

13. Jet streams travel primarily

    (A) north to south
    (B) south to north
    (C) east to west
    (D) west to east
    (E) in many directions

14. The correct arrangement of atmospheric layers, arranged in order from the most distant from Earth's surface to the one closest to Earth's surface, is

    (A) troposphere, stratosphere, mesosphere, thermosphere
    (B) thermosphere, mesosphere, stratosphere, troposphere
    (C) stratosphere, troposphere, mesosphere, thermosphere
    (D) thermosphere, mesosphere, troposphere, stratosphere
    (E) none of the above

15. You notice on a barometer that air pressure has been decreasing rapidly. The type of weather that would normally be expected to occur is

    (A) fog
    (B) drizzle
    (C) snow
    (D) hot and dry
    (E) thunderstorms

16. The global circulation pattern that dominates the tropics is called the

    (A) Ferrel cell
    (B) Polar cell
    (C) Bradley cell
    (D) Hadley cell
    (E) Tropical cell

17. The date with the shortest amount of daylight in the Southern Hemisphere is

    (A) June 22
    (B) January 1
    (C) December 21
    (D) September 1
    (E) March 18

18. Jet streams follow the sun in that as the sun's elevation ________ (increases, decreases) each day in the spring, the jet streams shifts by moving ________ (north, south) during the Northern Hemisphere spring.
    (A) increases, north
    (B) increases, south
    (C) decreases, north
    (D) decreases, south
    (E) none of the above

19. Usually, fair and dry/hot weather is associated with high pressure around ________ latitude with rainy and stormy weather associated with low pressure around ________ latitude.
    (A) 0 degrees N/S, 90 degrees E/W
    (B) 90 degrees E/W, 90 degrees N/S
    (C) 30 degrees N/S, 50–60 degrees N/S
    (D) 50–60 degrees N/S, 30 degrees N/S
    (E) 45 degrees N/S, 45 degrees E/W

20. The three necessary ingredients for thunderstorm formation are
    (A) moisture, lifting mechanism, instability
    (B) lifting mechanism, mountains, oceans
    (C) stability, moisture, heat
    (D) lifting mechanism, fronts, moisture
    (E) deserts, mountains, clouds

FREE-RESPONSE QUESTION

Characteristics of air masses and their worldwide circulation patterns influence the spatial distribution of biomes and the organisms that inhabit them.

(a) Describe a Hadley, Ferrel, or polar cell in terms of what it is, how it develops, and its locations in reference to the equator.
(b) Describe the characteristics of that cell in terms of climatic conditions such as temperature, relative humidity, prevailing winds, and so on.
(c) Describe one biome that would exist at sea level within the specific latitudes of that cell. Give examples of both plants and animals that would exist within that biome.

TIP

Before writing your essays, be sure to map out or brainstorm what you are going to write about. A few minutes planning and organizing your essays will get you a much higher score.

MULTIPLE-CHOICE ANSWERS AND EXPLANATIONS

1. (C) The troposphere is the atmospheric layer closest to Earth and extends for about 11 miles (18 km) above Earth at the equator and about 5 miles (8 km) above Earth at the poles. Temperature declines as altitude increases.
2. (E) Nitrogen (78%), oxygen (21%), water vapor (about 0–4%), and the rest below 1%.
3. (A) Latitude expresses how far north or south of the equator a location is. The equator is 0° latitude, and the poles are at 90°. For every 1,000 feet (300 m) in altitude, there is a 3°F (1.5°C) drop in temperature.
4. (D) A low-pressure air mass (low) occurs when warm air, which is less dense than cooler air, spirals inward toward the center of a low-pressure area. Since the center of the low-pressure area is of even less density and pressure, the air in this section rises and the warm air cools as it expands. The temperature begins to fall and may go below the dew point—the point at which air condenses into water droplets. These water droplets make up clouds. If the droplets begin to coalesce on condensation nuclei, rain follows.
5. (C) During La Niña, large portions of central North America experience increased storminess, increased precipitation, and an increased frequency of significant cold-air outbreaks, while the southern states experience less storminess and precipitation. Also, there tends to be considerable month-to-month variations in temperature, rainfall, and storminess across central North America during the winter and spring seasons.
6. (D) The rain shadow effect occurs on the leeward side of a mountain, the side away from the ocean. Moist air from the ocean rises when it hits mountains, cools, and loses its moisture on the windward side. On the leeward side, air is much drier. For example, the western side of the Sierra Nevada Mountain Range in California is much wetter than the eastern side.
7. (B) 97% of ozone (O_3) is found in the lower stratosphere, which is 9 to 35 miles (15–55 km) above Earth's surface. Temperature increases with altitude in the stratosphere due to absorption of heat energy by ozone molecules.
8. (A) Hadley cells occur between 0° and 25° north and south latitudes (equatorial region). In this area, there is upward air motion, cooling of the air due to uplift, high humidity, high clouds, and heavy rains.
9. (A) Water vapor is present in such abundance throughout the atmosphere that it acts like a blanket of insulation, trapping heat and forcing surface temperatures higher than they would be otherwise. Water vapor is roughly eight times more effective than carbon dioxide as a greenhouse gas.
10. (E) During monsoon season, winds blow from cooler ocean areas (higher pressure) to warmer landmasses (lower pressure). As the air rises over the land masses, it cools and is unable to retain water, producing great amounts of precipitation. In winter, the ocean is now warmer and the cycle reverses. Drier air travels from the land out to the ocean. Monsoons exist in Australia, Africa, and North and South America.
11. (A) Temperature inversions are atmospheric conditions in which the air temperature rises with increasing altitude, holding surface air down and preventing dispersion of pollutants.
12. (E) Albedo is a measure of reflection of sunlight from a surface. Of the choices, dark topsoil absorbs the most energy and therefore reflects the least amount of energy, resulting in the lowest albedo.

13. **(D)** Jet streams are large-scale upper air flows that travel from west to east and are produced by differences in temperature. They can travel as fast as 250 miles per hour (400 kph) and travel between 3 and 8 miles (5–13 km) above Earth's surface.
14. **(B)** Remember, the question required you to place the layers in order from the most distant to the closest.
15. **(E)** Thunderstorms form when air rises rapidly. Rapidly rising air also means a decrease of air mass at the surface, which leads to decreasing pressure.
16. **(D)** Hadley cells dominate the tropics.
17. **(A)** The seasons in the Southern Hemisphere are opposite that of the Northern Hemisphere. For example, spring in the Northern Hemisphere is fall in the Southern Hemisphere.
18. **(A)** The position of the jet stream also determines where the storm track is. As the jet stream moves north during spring, the storm track moves north, leaving the southern plains of Texas and Oklahoma and moving into the northern planes near the Dakotas.
19. **(C)** Except for a few locations, most of the world's deserts are located along 30 degrees N/S with lush forests from abundant rains located around 50–60 degrees N/S.
20. **(A)** Moisture, a lifting mechanism, and instability are all needed for thunderstorms to form. The moisture is needed for rain. The lifting mechanism is needed to get the air moving initially in an upward direction, and the unstable atmosphere insures the upward-moving air continues to do so.

FREE-RESPONSE ANSWER

Let's do this essay together, using it as a teaching tool rather than just providing an answer and rubric. Let's choose the Hadley cell for this example.

The first step is to brainstorm. Write a list of key words that would apply to the question. Remember that the order of the keywords is not important, we will put them into the correct order later. Here is a sample of key words: Hadley, Ferrel, polar, temperature, solar insolation, humidity, biomes, plants, and animals.

Now that we have around 10 key words, let's expand the list by adding details—items that we will discuss in our essay. We can also begin to map out the order in which we will discuss the items.

- Hadley: 0° to 30°, deserts, equatorial regions, tropical rain forests, subtropical areas, savannas
- Location
- Temperature (heat moves from equator to colder areas)
- Relative humidity
- Prevailing winds
- Solar insolation
- Biomes
- Animals
- Plants

Remember, you have 23 minutes to spend on this essay, so do not spend more than 5 minutes in organization. Now look over the list and add anything that you have missed. Now it is time to begin writing.

The first step is to write a thesis statement: **"The world's biomes are primarily determined by climatic conditions. Deserts are characterized as areas of low precipitation, while tropical rain forests are characterized by areas of high precipitation."**

The next step is to begin describing what determines climatic conditions that, in turn, affect the type of life present within that zone. **"Solar insolation, that is the amount of sunlight received on Earth, is greatest at the equator and diminishes toward the poles. Since heat flows from warmer regions to cooler regions, the warmer air produced at the equator moves through major worldwide wind patterns that distribute this energy worldwide. As one moves from the equator to the poles, there are three major air circulation cells—Hadley, Ferrel, and polar."** At this point, a labeled sketch would be helpful.

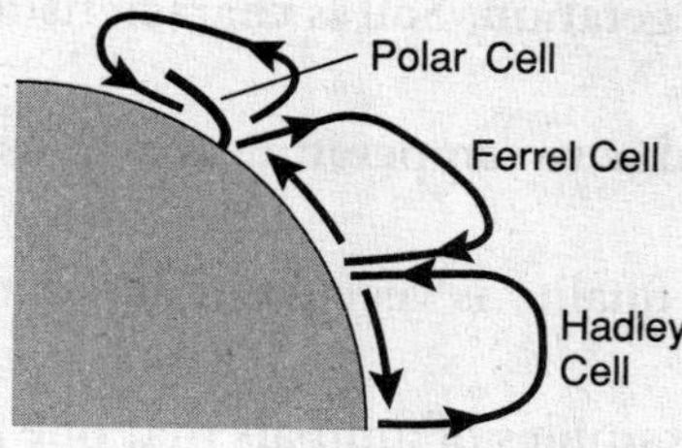

Now, for the remaining time, we can refer to the detailed essay outline and complete the essay, adding details and examples where necessary.

From the equator to 30° north and south latitudes, Hadley cells exist. Since this area of Earth receives the greatest solar radiation due to Earth's axis tilt, this area of Earth is the warmest. Near the equator, this warm, moist air rises. As the warm air rises, it begins to cool and become denser. Since cooler air cannot hold as much water vapor as warmer air, the humidity of the air increases to the point where clouds are produced. This, in turn, causes great amounts of rainfall. Monthly average temperatures are quite high at sea level, and there is no winter. Vegetation near sea level is tropical rain forest. In these tropical systems, temperature variations from day to night are greater than from season to season. Tropical rain forests, which extend about 1,500 miles (2,400 km) north and south of the equator, are found in South and Central America, Africa, and southeast Asia. Climatic conditions can include rain throughout the year, monsoons—a short, dry season followed by a heavy, rainy period, and tropical savanna with characteristic wet and dry seasons.

Tropical rain forests have characteristically high-species plant and animal diversity. Vegetation is dense. Bromeliads, orchids, ferns, and palms are present. Leaves are large in an effort to absorb sunlight, and there is little need to conserve water lost through transpiration. Soils are characteristically low in nutrients with the

nutrients being stored in vegetation. Soil is characteristically acidic. Decomposition of organic material is high due to temperature and moisture. Leaching of soil nutrients is high; therefore, soil quality is very low. Abundant insects and animal biodiversity are characteristic. Examples of animals that one might find in a tropical rain forest biome might include numerous species of butterflies, ants, mosquitoes, millipedes, bats, monkeys, sloths, tarsiers, hippopotamuses, macaws, toucans, parrots, anacondas, alligators, and numerous species of frogs.

At this point, we think we are done. Our last job is to be sure that we have answered all questions. Let's put a check beside each topic that we answered and that we were required to address:

- Describe the cell ✔
- How it develops ✔
- Location of the cell ✔
- Characteristics of the cell
    1. Temperature ✔
    2. Relative humidity ✔
    3. Prevailing winds ✔
    4. Solar insolation ✔
- List examples of plants and animals living within a biome in that cell ✔

Now it is your turn. Take the same question, but this time answer it in terms of either a Ferrel or polar cell. Your teacher may wish to collect your essay and give you pointers on your strengths and weaknesses.

Global Water Resources and Use

CHAPTER 3

All rivers run into the sea, yet the sea is not full: Unto the place from which rivers come, thither they return again.

—Ecclesiastes 1:7

FRESHWATER AND SALTWATER

Over 70% of Earth's surface is covered by water. Oceans hold about 97% of all water on Earth, while freshwater constitutes about 3%. Of the freshwater that is available, most of it is trapped in glaciers and ice caps. The rest is found (in descending order) in groundwater, lakes, soil moisture, atmospheric moisture, rivers, and streams.

Water's has many unique properties:

1. Strong hydrogen bonds hold water molecules to each other.
2. The temperature of water changes slowly due to its high specific heat capacity.
3. Water has a high boiling point.
4. A lot of energy is needed to evaporate water.
5. Water dissolves many compounds.
6. Water filters out harmful UV radiation in aquatic ecosystems.
7. Water adheres to many solid surfaces.
8. Water expands when it freezes.

Most human settlements are determined by the availability of freshwater. The highest per capita supplies of freshwater are in countries with high precipitation and small populations (Iceland, Norway, and so on). Lowest per capita freshwater supplies are in areas with low rainfall and high populations (Egypt, Israel, and so on).

The use of freshwater, a limited resource, is growing at twice the rate of population growth. In the United States, the average amount of freshwater allocated per person for all purposes is approximately 500,000 gallons (190,000 L) per year.

OCEAN CIRCULATION

The Northern Hemisphere is dominated by land and the Southern Hemisphere by oceans. Temperature differences between summer and winter are more extreme in the Northern Hemisphere because the land warms and cools more quickly than water. Heat is transported from the equator to the poles mostly by atmospheric air currents but also by oceanic water currents. The warm waters near the surface and colder waters at deeper levels move by convection. Changes in ocean temperatures have a direct bearing on ocean currents. During summers, a thermocline develops in ocean waters between the warm surface water and the cooler bottom water.

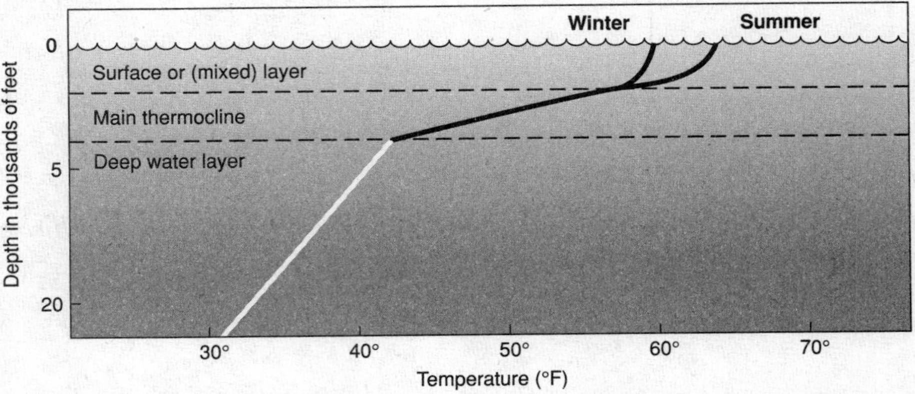

Figure 3.1 Ocean temperature profile

Surface ocean currents are driven by wind patterns that result from the flow of high thermal energy sources generated at the tropics (higher pressure) to low-energy sources in polar areas (lower pressure). They serve to distribute the heat generated near the tropics. Deep-water, density-driven currents are controlled primarily by differences in temperature and salt content. Denser, saltier water sinks, and less-dense water rises. About 90% of the ocean volume circulates due to density differences in temperature and salinities, while the remaining 10% is involved in wind-driven surface currents. In the Northern Hemisphere, north-flowing currents are warm (originating near the equator), and south-flowing currents are colder (originating from the Arctic area).

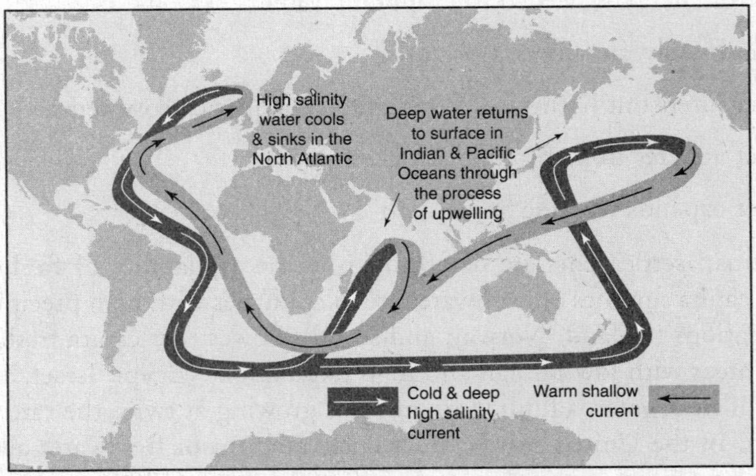

Figure 3.2 Thermohaline circulation

Ocean water has warmed significantly during the past 50 years. The greatest amount of warming has occurred in the top surface layers of the ocean. The temperature of the Antarctic Southern Ocean rose by 0.31°F (0.17°C) between the 1950s and the 1980s—twice the rate for the world's oceans as a whole. Since the 1950s, the California Current that runs southward along the west coast of the United States has risen about 2.7°F (1.5°C) and has resulted in a significant decrease in plankton with resulting rippling effects within the food web. Possible reasons for dramatic increases in ocean temperatures include:

1. Significant slowing of the ocean circulation that transports warm water to the North Atlantic
2. Large reductions in the Greenland and West Antarctic ice sheets
3. Accelerated global warming due to carbon cycle feedbacks in the terrestrial biosphere
4. Decreases in upwelling
5. Releases of terrestrial carbon from permafrost regions and methane from hydrates in coastal sediments.

The Gulf Stream transports warm water from the Caribbean northward. A branch of the Gulf Stream known as the North Atlantic Drift is responsible for bringing warmer temperatures to Europe. Evaporation of ocean water in the North Atlantic results in a cooling effect and a higher salt concentration, both of which increase the density of the water. As the denser water sinks, it creates a southern circulation pattern. As glaciers in Greenland melt due to the effects of global warming, the density of this ocean water decreases due to more freshwater. This, in effect, could stall the North Atlantic Drift and bring colder temperatures and flooding to Europe.

Upwellings occur when prevailing winds produced through the Coriolis effect, and moving clockwise in the Northern Hemisphere, push warmer, nutrient-poor surface waters away from the coastline. This surface water is then replaced by cooler, nutrient-rich deeper waters. The deeper waters contain high levels of nitrates and phosphates, which result from the decomposition and sinking of surface water plankton. When these nutrients are brought to the surface through upwelling, they supply necessary nutrients for phytoplankton, which form the base of the oceanic food chain.

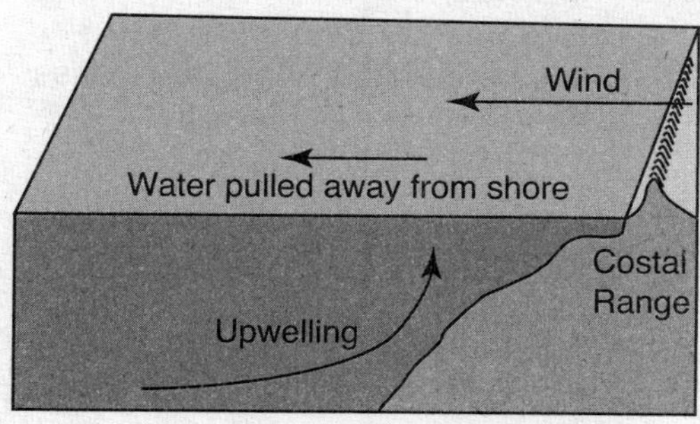

Figure 3.3 Upwelling

AGRICULTURAL, INDUSTRIAL, AND DOMESTIC USE

About 70% of freshwater is used for agriculture. Use of water for agriculture depends upon national wealth, climate, and degree of industrialization. Canada uses about 10% of its freshwater resources for agriculture, whereas India uses about 90%. Up to 70% of water intended for agriculture in developing countries may not reach crops due to seepage, evaporation, and leakage. Drip irrigation, the most efficient type of irrigation, is used on less than 1% of crops worldwide.

Industry uses about 25% of all freshwater, ranging from a high of about 75% in Europe to less than 5% in developing countries. Water used for cooling power plants is the largest sector. Water returns 60 times its economic value when used for industrial purposes rather than for agriculture.

Domestic uses of freshwater include water being used for flushing toilets, bathing, drinking, and so on. People living in developed countries use about 10 times more water for personal use than people living in less-developed countries.

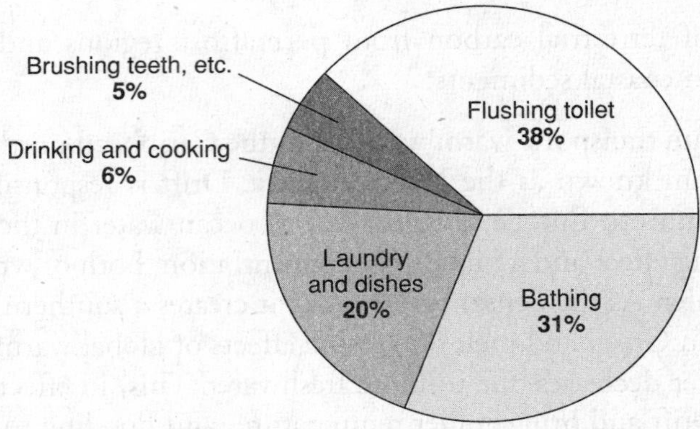

Figure 3.4 Water use

FACTOIDS

- 4,000 gallons (15,000 L) of water produces 1 kilowatt-hour of hydroelectric power (enough to light a 100-watt lightbulb for 10 hours).
- To produce 1 gallon (3.7 L) of milk, dairy cow must drink 4 gallons (15 L) of water. It takes 8 gallons (30 L) of water to grow a tomato.
- 300 million gallons (1 billion L) of water are used to produce one day of newsprint in the United States.
- In the United States, one penny buys about 160 glasses of water.

SURFACE AND GROUNDWATER ISSUES

Surface water infiltrates and percolates through the soil into aquifers—layers of porous rock, sand, and gravel where water is trapped above a nonporous layer or bedrock. The surface area in which water infiltrates into the aquifer is called the recharge zone. If pollution enters an aquifer, the aquifer is no longer a source of safe drinking water. Movement of water through aquifers is very slow. Artesian wells occur where the water breaks through to the surface. Aquifers in the United States hold 30 times more water than all U.S. lakes and rivers combined, with groundwater supplying almost 40% of all freshwater in the United States. When removal of water exceeds the recharge rate, the land sinks (subsidence). Depletion of water in aquifers also leads to sinkholes and saltwater intrusion—a condition in which seawater replaces the freshwater in the aquifer, making it unusable for human use. The region where water-saturated soil meets water-unsaturated soil is called the water table. The water table is unique to a region and can rise and fall with rainfall variations, depletion rates, and so on.

Water-Renewal Rates

Source of Water	Average Renewal Rate
Groundwater (deep)	~ 10,000 years
Groundwater (near surface)	~ 200 years
Lakes	~ 100 years
Glaciers	~ 40 years
Water in the soil	~ 70 days
Rivers	16 days
Atmosphere	8 days

GLOBAL PROBLEMS

Both water shortages and rising sea levels are global problems.

Water Shortages

The rate of water consumption is growing twice as fast as the population growth rate. Freshwater shortages that result from this demand can be due to natural weather patterns that reduce rainfall, rivers changing course, flooding that contaminates existing supplies, competition for available water, overgrazing and the resulting erosion, pollution of existing supplies, and competing interests that reduce water conservation programs. Water is a limiting factor as it limits the amount of food that can be produced in a region. If food cannot be grown locally due to water shortages, then food must be imported as additional costs.

Rising Sea Levels

During the end of the last ice age about 18,000 years ago, when global temperatures were about 10°F (5°C) warmer than they are today, sea level was about 430 feet (130 m) higher than it is today. Much of the rise was due to ice that was on land melting and filling the oceans. When ice that is floating on the water melts, it does not contribute to a rise in sea level. However, it does affect climate. With higher sea levels and more water covering Earth's surface, more heat energy is absorbed by the water and less is reflected back into space, resulting in higher temperatures.

Several other factors contribute to rising sea levels:

1. Land buildup or erosion of mountains (isostatic adjustments)
2. Plate tectonic effects
3. Sedimentation
4. Groundwater and oil extraction
5. Changes in ocean currents and tides
6. Distribution changes in the water cycle.

During the 20th century, sea levels rose 6–8 inches (15–25 cm). Approximately 1–2 inches (2–5 cm) of the rise resulted from the melting of mountain glaciers. Another 1–2 inches (2–5 cm) resulted from the expansion of ocean water that resulted from warmer ocean temperatures. Best scientific estimates indicate that sea levels will rise 7 inches (18 cm) by the year 2030 and 23 inches (58 cm) by the year 2090. Climate models have suggested that temperatures in polar regions will increase more and at a faster pace than in other areas of the world. Since 1995, more than 5,400 square miles (140,000 sq km), an area equal to Connecticut and Rhode Island combined, have broken off the Antarctic ice shelves and melted.

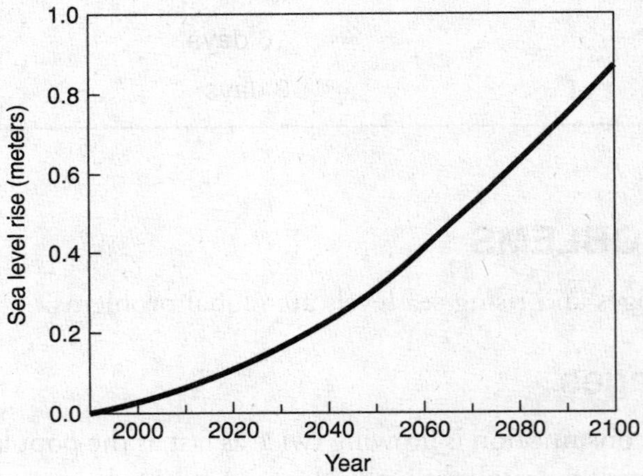

Figure 3.5 Projected rise in sea level

Wetlands are the most-impacted areas affected by rising sea levels. A 1-foot (30 cm) increase in sea level would result in up to 40% of the U.S. wetlands being destroyed. Other impacts would include: erosion of beaches and bluffs, salt intrusion into aquifers and surface waters, inundation of seawater into low-lying areas, and increased flooding and storm damage.

Much of the world's population—20%—lives in coastal regions, and half of the world's population lives within 120 miles (200 km) of the coast. Asia, Latin America, and the Caribbean have the highest percentages of people living near the coast.

CONSERVATION

Several conservation methods can be used to increase the quantity of available freshwater:

Methods to Increase Supplies of Freshwater: Description and Drawbacks

- **Changes in personal habits:** Turn off shower while washing. Wash only full loads of clothes and dishes; turn off water when brushing teeth. Check for and repair leaks around the home. Use a broom when cleaning driveways and patios. Adjust sprinklers to prevent runoff. Water at night or early morning.
- **Construct dams and reservoirs:** Interferes with fish migration and destroys natural rivers. Leakages, earthquakes, evaporation, sediment buildup, and displacement of people are also consequences. Up to 60 million people worldwide (many being indigenous minorities) have been relocated due to dam and reservoir construction.
- **Desalinate water:** Rate of production is low and is expensive (three to four times more expensive than any other process). Issues of brine disposal.
- **Drip irrigation:** Drip irrigation conserves water by reducing evaporation, but it is expensive. Most large agricultural corporations can afford it, but small, independent farmers cannot. Not suitable for annual crops.
- **Education:** Informing and educating the public on water conservation.
- **Encourage the use of recycled products that require less water to produce:** Costs of collecting products may be outweighed by savings.
- **Engineer systems to collect more runoff:** Water may be high in pollutants and expensive to reprocess.
- **Levy taxes or user fees:** Prices would go up on products. International competitiveness would be affected.
- **Line irrigation channels and cover canals:** Prevents loss of irrigation water through seepage and evaporation. Initial and maintenance costs are expensive.
- **Meter all water used:** Municipalities would recoup money on meter installation with paying less to water suppliers.
- **Plant crops that do not require as much water and xeriscaping:** Xeriscaping reduces urban runoff. Issues of market economies, crop prices and demand, weather patterns, and so on.
- **Rebates or legislation of low-flush toilets, shower restrictors, etc.:** Rebates would be offered by water companies, which, over time, would recoup costs by having to buy less water from suppliers.
- **Reduce government subsidies:** Increased water costs would be passed on to consumers and result in greater personal conservation.
- **Reprocess (recycle) water:** The public is not supportive of using reprocessed toilet water. Reprocessed water could be used for irrigation but would require separate pipeline systems. Reuse of gray water is becoming popular in new developments.
- **Seed clouds:** Water availability to other areas would be affected.
- **Tiered price scale:** Would reduce effective family income for larger families. Could be remedied through exemptions or allocated share of water per family member.

TIP

Case studies can be brought into your essay answers to bring a historical connection to the question. Try to bring at least one case study into your essays—your score will go higher. You will also find several multiple-choice questions on the test that focus on case studies. Only the most relevant case studies are included in this book.

- **Use of icebergs:** Expensive and most of it would be lost before it reached final destination.
- **Use more groundwater:** If rate of use exceeds rate of recharge, then subsidence, sinkholes, and saltwater intrusion could occur.

CASE STUDIES

- **Aswan High Dam, Egypt:** Completed in the 1970s, the Aswan High Dam in Egypt was built to supply irrigation water. The water that is available is only half of what was expected due to evaporation and seepage losses in unlined canals. Several other problems were encountered. First, the elimination of nutrients onto farmlands now requires the use of expensive fertilizers. Second, the depletion of nutrients into the Mediterranean caused a decline in certain fish catches. Third, large amounts of standing water caused the proliferation of snails and ultimately resulted in a debilitating disease known as schistosomiasis, with some areas having infection rates of 80%.
- **Bangladesh:** In the 1960s, thousands of wells were dug in Bangladesh by foreign governments and humanitarian organizations in an effort to supply freshwater to the population. Shortly thereafter, arsenic compounds from the soil began to leach into the groundwater. Arsenic poisoning began to appear among the population, with millions of people showing symptoms.
- **Colorado River Basin:** Diversion of water from the Colorado River has led to water right disputes between California, Arizona, and Mexico. Dams on the Colorado River trap large quantities of silt (over 10 million metric tons per year) and reduce nutrient levels in farmlands below the dam. As a result, more fertilizer is required. Farm irrigation has resulted in high levels of sodium chloride in the alkaline soils to become incorporated in agricultural runoff. Millions of acres of once-valuable farmland are now useless due to the salt buildup in soil, a process known as salinization.
- **James Bay, Canada:** Diversion of rivers into Hudson Bay to generate electrical power has resulted in massive flooding. During one flood, up to 10,000 caribou drowned. In addition, mercury has leached out of rocks and into water, with nearby residents showing signs of mercury poisoning. The project also created expensive legal battles and created many issues with indigenous people whose land was flooded.
- **Ogallala Aquifer:** The Ogallala Aquifer underlies eight states from Texas to North Dakota. The Ogallala Aquifer used to hold more freshwater than all freshwater lakes, streams, and rivers on Earth. Due to pumping of this groundwater for agricultural, domestic, and industrial uses, many locations are experiencing water shortages.
- **Three Gorges Dam, China:** In 1949, China had no large reservoirs and only 40 small hydroelectric stations. By 1985, there were 80,000 reservoirs and 70,000 hydroelectric stations. The Three Gorges Dam required relocation of 1.2 million people.

RELEVANT LAWS

- **Water Resources Planning Act (1965):** Provided for plans to formulate and evaluate water and related land resource projects and to maintain a continuing assessment of the adequacy of water supplies in the United States.
- **Coastal Zone Management Act (1972):** Provided funds for state planning and management of coastal areas.
- **Clean Water Act (1972 and 1987 revisions):** Sets objectives for restoring and maintaining the chemical, physical, and biological integrity of the nation's waters. Regulates discharge of pollutants and requires federal agencies to avoid adverse impacts from modification or destruction of navigable streams and associated tributaries, wetlands, or other waters.
- **Safe Drinking Water Act (1974):** Established to protect the quality of drinking water in the United States. This law focuses on ground or underground sources of water.
- **Water Resources Development Act (1986):** Established and maintains dam safety programs.
- **National Estuary Program (1987):** Designed to identify nationally significant estuaries and to restore and protect them.

MULTIPLE-CHOICE QUESTIONS

1. Water vapor returning to the liquid state is called

   (A) evaporation
   (B) transpiration
   (C) boiling
   (D) condensation
   (E) vaporization

2. The temperature at which air becomes saturated and produces liquid is called

   (A) the saturation point
   (B) the dew point
   (C) the condensation point
   (D) relative humidity
   (E) absolute humidity

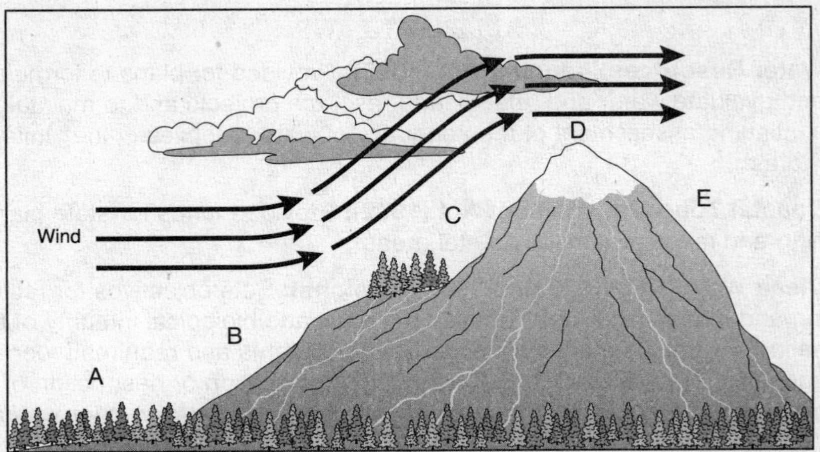

3. The area that would receive the most precipitation would be

   (A) *A*
   (B) *B*
   (C) *C*
   (D) *D*
   (E) *E*

4. The rain shadow effect would be located at point

   (A) *A*
   (B) *B*
   (C) *C*
   (D) *D*
   (E) *E*

5. Of the freshwater on Earth that is not trapped in snow packs or glaciers, most of it (95%) is trapped in

   (A) lakes
   (B) rivers
   (C) aquifers
   (D) dams
   (E) estuaries, marshes, and bogs

6. The primary use of freshwater is for

   (A) industry
   (B) domestic use
   (C) fishing
   (D) agriculture
   (E) landscaping

7. A mixture of freshwater and saltwater is known as

(A) brackish water
(B) gray water
(C) black water
(D) connate water
(E) lentic water

8. A temperate lake is most likely to show thermal stratification and limited mixing of surface and deeper waters during the __________ season.

(A) winter
(B) spring
(C) summer
(D) fall
(E) none of the above

9. Of the following methods of irrigation, the one that currently conserves the most water is

(A) flooding fields
(B) irrigation channels
(C) sprinklers
(D) drip irrigation
(E) misters

10. The largest use for industrial water is for

(A) cooling electrical power plants
(B) automobile manufacturing
(C) mining
(D) the food and beverage industry
(E) aquaculture

11. A country that would represent large per capita water use would be

(A) China
(B) India
(C) Israel
(D) United States
(E) Iceland

12. When compared with the rate of population growth, the worldwide demand rate for water is

(A) about half
(B) about the same
(C) about two times
(D) about three times
(E) about ten times

13. The U.S. per capita use of water on a daily basis is closest to

(A) 50 gallons
(B) 100 gallons
(C) 1,500 gallons
(D) 5,000 gallons
(E) 10,000 gallons

14. Countries that are more likely to suffer from water stress would be located in

(A) North America
(B) South America
(C) western Europe
(D) the Middle East
(E) Asia

15. What fraction of the world's population does not have access to adequate amounts of safe drinking water?

(A) 1/2
(B) 1/3
(C) 1/4
(D) 1/6
(E) 1/10

16. Which of the following conditions may indicate an El Niño?

(A) Sea surface warms, trade winds strengthen
(B) Sea surface warms, trade winds weaken
(C) Sea surface cools, trade winds strengthen
(D) Sea surface cools, trade winds weaken
(E) None of the above

17. Rising sea levels due to global warming would be responsible for all of the following EXCEPT

(A) destruction of coastal wetlands
(B) beach erosion
(C) increased damage due to storms and floods
(D) increased salinity of estuaries and aquifers
(E) all would be the result of rising sea levels

18. What does La Niña bring to the southeastern United States?

(A) Warm winters
(B) Extremely cold winters
(C) Hot summers
(D) Cooler than normal summers
(E) None of the above

19. Which of the following ocean currents flows without obstruction or barriers around Earth?
    (A) Gulf Stream
    (B) California Current
    (C) Antarctic Circumpolar Current
    (D) Aghulas Current
    (E) They all flow unimpeded.

20. A project aimed at producing paleoclimatic maps showing sea surface temperatures in different parts of Earth at various times was
    (A) Global Sun-Temperature Project
    (B) PALEOMAP
    (C) NOAA
    (D) NIMBUS
    (E) CLIMAP

FREE-RESPONSE QUESTION

The oxygen content of water is a limiting abiotic environmental factor that determines the biodiversity of an aquatic environment. An APES class visited a stream near their high school and measured several indicators of water quality.

(a) Describe a technique of determining the oxygen content of water and what it means.
(b) Briefly describe one other test that the students could have conducted to determine the water quality of the stream.
(c) Diagram a typical food web based on a freshwater stream ecosystem.

MULTIPLE-CHOICE ANSWERS AND EXPLANATIONS

1. **(D)** Evaporation is water changing from a liquid state to a gaseous state below the boiling point. Transpiration is water moving through a plant. Vaporization is moving from a liquid state to a gaseous state.
2. **(B)** The saturation point is the maximum amount of water vapor that a particular volume of air at a given temperature can hold. The condensation point is the temperature and pressure at which water vapor turns into liquid water. Absolute humidity is the mass of water vapor in a given volume of air. Relative humidity is the ratio of the actual amount of water vapor held in the atmosphere compared with the maximum amount that the air could hold and is influenced by temperature and atmospheric pressure.
3. **(C)** As the air lifts (orographic lifting), it becomes cooler. Cooler air holds less water vapor. At location *C*, the air is holding the maximum amount of water vapor. Given the fact that the temperature has decreased, it would receive the maximum amount of rain. At the top of the mountain at location *D*, much of the water vapor has been depleted from the air.

4. (E) Point *E* is on the leeward side of the mountain. This side receives little precipitation because most of the rain has been deposited on the windward side. The leeward side is experiencing the rain shadow effect.
5. (C) The oceans hold 98% of all water on Earth. Freshwater only makes up 2%. Of that 2%, 90% of it is trapped in ice and snow, which is rapidly melting due to global warming. Of the freshwater left, the majority is found in groundwater, with the remaining 3% of freshwater found in lakes, rivers, and streams. Of the total amount of water on Earth, only 0.01% is located in lakes, rivers, and streams.
6. (D) Agriculture uses about 70% of all freshwater. Use for agriculture depends upon national wealth, climate, and degree of industrialization. Industry uses about 25% of all freshwater, with Europe using the most and developing countries using the least. Water used for cooling of power plants is the largest sector.
7. (A) Gray water is sewage water that does not contain toilet wastes. Black water is sewage that does contain toilet wastes. Connate water is also known as fossil water and is water that has been trapped within sediment or rock structures at the time the rock was formed. Lentic water is the standing water of lakes, marshes, ponds, and swamps.
8. (C) During the summer, the surface water warms up much faster than the deep water. The warmer surface water is less dense than the cooler, deep water, so it stays on the surface. The wind mixes the surface water but only near the surface. The lake tends to become stratified, with a warmer upper layer or epilimnion and a cooler lower layer or hypolimnion. The boundary between these layers is called the thermocline.
9. (D) Drip irrigation can increase yields and decrease water requirements and labor. It provides the plant with continuous, near-optimal soil moisture by conducting water directly to the plant. It saves water because only the plant's root zone receives moisture.
10. (A) Industry used about 25% of all freshwater, ranging from 75% in Europe to less than 5% in developing countries.
11. (D) Highest per capita supplies of freshwater are in countries with high rainfall and low populations (Iceland, Norway, and so on). These water-rich countries have low water withdrawals. Remember, *per capita* means "per person."
12. (C) Since populations are increasing and increased populations result in higher levels of pollution and since freshwater sources are finite, the amount of freshwater per person is decreasing each year.
13. (C) In the United States, renewable or replacement water averages 2.4 million gallons (9 million L) per person per year. The average amount withdrawn from water supplies in the United States is about 500,000 gallons (1.8 million L) per person per year (1,500 gallons [5,700 L] per day).
14. (D) Areas that do not receive as much precipitation include polar regions (cold air cannot hold as much water as warmer air), midcontinental areas (they are too far from oceans and the clouds have released much of their moisture before they reach inland), subtropical deserts (air masses are subsiding), and the leeward sides of mountains near coastal regions (rain shadow effect).
15. (D) It is estimated that over 1 billion people lack access to safe drinking water. A child dies every 8 seconds worldwide from contaminated water sources (over 5 million children each year).

16. (B) The warm waters of El Niño (ENSO) encompass the central Pacific region. Normally, a warm pool of ocean water builds up in the western equatorial Pacific Ocean. When air pressure patterns weaken, though, the trade winds decrease or even reverse direction. This causes the warmer water to move eastward.
17. (E) Sea levels are rising and are caused by both natural and human factors. A 0.5-inch (1 cm) rise in sea level erodes beaches about 3 feet (1 m) horizontally. It is predicted that within 100 years, there will be a net loss of up to 43% in coastal wetlands due to rising sea levels.
18. (A) La Niña can bring warm winters to the southeast and cooler-than-normal winter temperatures to the northwest United States. It is the cold counterpart of El Niño. La Niña's strong easterly winds bring cold ocean water to the surface in the eastern Pacific and causes increased rainfall in the western Pacific. The jet stream rather than coming through the Pacific Northwest is diverted over Alaska and into the Great Lakes region.
19. (C) The Antarctic Circumpolar Current is the most powerful ocean current system on Earth and exerts a strong influence on climate. It circles Earth in the southern hemisphere and connects the three great ocean basins—Atlantic, Indian, and Pacific. Unlike in the Northern Hemisphere, there are no landmasses to break up this large, continuous stretch of water.
20. (E) CLIMAP was a project developed in the 1970s by the National Oceanic and Atmospheric Administration (NOAA) to develop a detailed climatological map of the prehistoric world.

FREE-RESPONSE ANSWER

(a) Biological oxygen demand (BOD) is a measure of the oxygen used by microorganisms to decompose organic wastes. If there is a large quantity of organic waste in the water supply, bacterial counts will be high. Therefore, the demand for oxygen will be high, resulting in a high BOD level. As the waste is consumed or dispersed through the water, BOD levels will begin to decline. Nitrates and phosphates in a body of water can also contribute to high BOD levels. Nitrates and phosphates are plant nutrients and can cause plant life and algae to grow quickly. When plants grow quickly, they also die quickly. This contributes to the organic waste in the water, which is then decomposed by bacteria—resulting in a high BOD level. When BOD levels are high, dissolved

Notable points:

Using BOD as a measure of oxygen content.

Explaining what BOD means.

Mentioning what factors determine or can affect BOD levels.

Explaining how BOD is determined.

Explaining how BOD levels affect water quality.

oxygen (DO) levels decrease because the oxygen that is available in the water is being consumed by the bacteria. Because less dissolved oxygen is available in the water, fish and other aquatic organisms may not survive.

The BOD test takes five days to complete and is performed using a dissolved oxygen test kit. The BOD level is determined by comparing the DO level of a water sample taken immediately with the DO level of a water sample that has been incubated in the dark for five days. The difference between the two DO levels represents the amount of oxygen required for the decomposition of any organic material in the sample and is a good approximation of the BOD level.

BOD levels of 1–2 ppm are indicative of good water quality and there will not be much organic waste present in the water supply. A water supply with a BOD level of 3–5 ppm is considered moderately clean. In water with a BOD level of 6–9 ppm, the water is considered somewhat polluted because there is usually organic matter present, and bacteria are decomposing this waste. At BOD levels of 10 ppm or greater, the water supply is considered very polluted with organic wastes. At these BOD levels, organisms that are more tolerant of lower dissolved oxygen may appear and become numerous (such as leeches and sludge worms). Organisms that need higher oxygen levels (like caddis fly larvae and mayfly nymphs) will not survive.

Although only 1 test was required, 3 common tests are included here.

(b) (1) Turbidity is a measure of suspended and colloidal particles, including clay, silt, organic and inorganic matter, algae and microorganisms in a water sample. It is caused by soil erosion, excess nutrients, various wastes and pollutants, and the action of bottom-feeding organisms that stir up sediments in the water. Turbidity is measured electronically by quantifying the degree to which light is traveling through a water column. The light is scattered by the suspended organic and inorganic particles. The scattering of light increases with a greater suspended load of particles.

Other suitable tests could have been chosen.

Choosing an appropriate test.

Explaining what the test measures.

Explaining properly how the test is conducted.

Explaining how the results of the test affect water quality.

When turbidity is high, water loses its ability to support a diversity of aquatic organisms. Oxygen levels decrease in turbid water as the result of heat absorption by the particles. Greater concentrations of suspended particles also decrease photosynthesis rates. Suspended solids can clog fish gills, reduce growth rates, reduce disease resistance, and prevent egg and larval development. Settled particles can accumulate and smother fish eggs and aquatic insects on the river bottom, suffocate newly hatched insect larvae, and make river bottom microhabitats unsuitable for mayfly nymphs, stonefly nymphs, caddis fly larvae, and other aquatic insects.

(2) Nitrate (NO_3^-) and nitrite (NO_2^-) ions are inorganic forms of nitrogen in the aquatic environment. Nitrate along with ammonia are the forms of nitrogen used by plants. Nitrates and nitrites are formed through the oxidation of ammonia by nitrifying bacteria, a process known as nitrification. In turn, these ions are converted to other nitrogen forms by denitrification and plant uptake. Nitrogen in its various forms is usually more abundant than phosphorus in the aquatic environment; therefore, nitrogen rarely limits plant growth as phosphorus does. Sources of nitrates include the atmosphere, inadequately treated wastewater from sewage treatment plants, agricultural runoff, storm drains, and poorly functioning septic systems.

(3) pH: pH is measured on a log scale of 0 to 14 and measures the concentration of hydrogen ions (H^+). Pure distilled water is considered neutral with a

pH of 7. The strongest acid has a pH of 0, and the strongest base has a pH of 14. For every 1 unit of pH change, there is a tenfold change. Many species of fish and aquatic plants are sensitive to changes in pH. Acidic lakes and streams can cause leaching of heavy metal ions into the water. pH measurements are determined by using either electronic pH meters or pH paper that contains organic dyes sensitive to narrow pH ranges.

(c)

A sample food web is included here ONLY for reference. The student does NOT need to include any of the organisms listed here to receive full credit. However, the food web that is drawn must be correct.

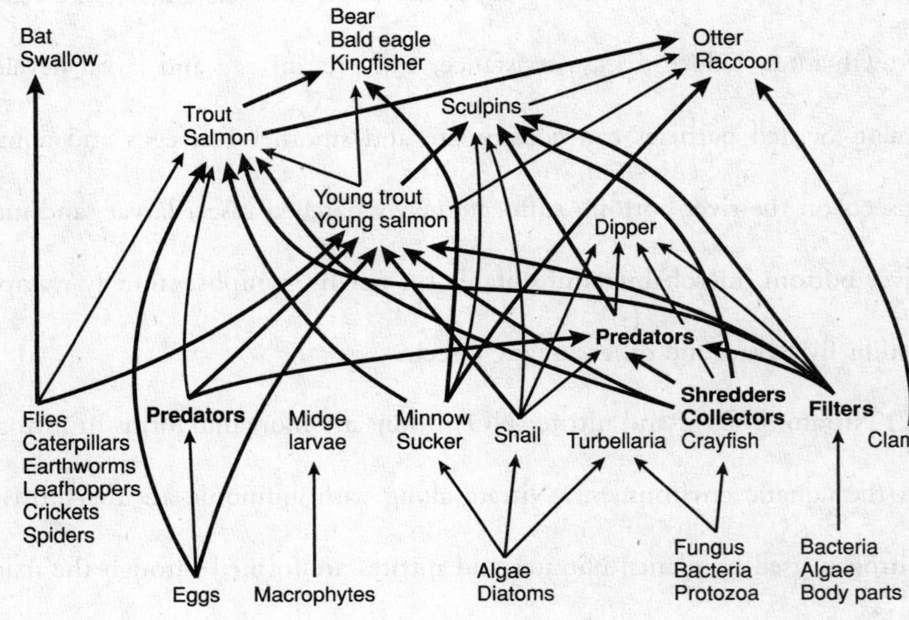

UNIT II: THE LIVING WORLD (10–15%)

Areas on Which You Will Be Tested

A. **Ecosystem Structure**—biological populations and communities, ecological niches, interactions among species, keystone species, species diversity and edge effects, major terrestrial and aquatic biomes.

B. **Energy Flow**—photosynthesis and cellular respiration, food webs and trophic levels, and ecological pyramids.

C. **Ecosystem Diversity**—biodiversity, natural selection, evolution, and ecosystem services.

D. **Natural Ecosystem Change**—climate shifts, species movement, and ecological succession.

E. **Natural Biogeochemical Cycles**—carbon, nitrogen, phosphorus, sulfur, water, and conservation of matter.

Ecosystems

CHAPTER 

Nature encourages no looseness, pardons no errors.
—Ralph Waldo Emerson

ECOLOGY

Understanding ecosystems requires having an understanding of ecology. Several topics, ranging from communities to niches to biomes, will be discussed.

Biological Populations and Communities

Organisms that resemble each other, that are similar in genetic makeup, chemistry, and behavior, and that are able to interbreed and produce fertile offspring belong to the same species. Organisms of the same species (intraspecific) that interact with each other and occupy a specific area form a population. Populations of different species (interspecific) living and interacting within an area create communities. A community is made up of all the populations of different species that live together within a particular area. An ecosystem is a system formed by the interaction of a community of organisms with their physical environment. Organisms make up populations that make up communities that make up ecosystems that make up the biosphere: biosphere → ecosystems → communities → populations → species → organisms.

REMEMBER

"Intra" means within. "Inter" means between.

Members of a population can be dispersed in an area in three ways:

1. **Clumped.** Some areas within the habitat are dense with organisms, while other areas contain few members.
2. **Random.** Little interaction between members of the population leading to random spacing patterns.
3. **Uniform.** Fairly uniform spacing between individuals.

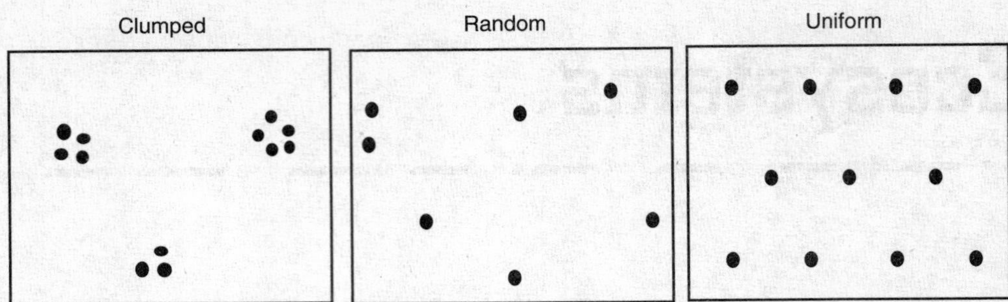

Figure 4.1 Clumped, random, uniform distribution

Ecological Niches

An ecological niche is the particular area within a habitat occupied by an organism and includes the function of that organism within an ecological community. The physical environment influences how organisms affect and are affected by resources and competitors. The niche reflects the specific adaptations that a species has acquired through evolution. To describe an organism's niche involves a description of the organism's adaptive traits, habitat, and place in the food web. A niche also takes into account the types and amounts of resources the species uses and its interactions with both living (biotic) and nonliving (abiotic) factors in its habitat.

Generalist species live in broad niches. They are able to withstand a wide range of environmental conditions. Examples of generalist species include cockroaches, mice, and humans. Specialist species live in narrow niches and are sensitive to environmental changes. Since they cannot tolerate change, they are more prone to extinction. An example of a specialist species is the giant panda, which only eats a certain type of bamboo. When environmental conditions are stable, specialist species have an advantage since there are few competitors as each species occupies its own unique niche (competitive exclusion principle). However, when habitats are subjected to rapid changes, the generalist species usually fare better since they are more adaptable.

Interactions Among Species

Various types of interactions occur among species. They can benefit one or both species, harm one or both species, or not affect one of the species involved.

AMENSALISM (–)(Ø)

Amensalism is the interaction between two species whereby one species suffers and the other species is not affected. Usually this occurs when one organism releases a chemical compound that is detrimental to another organism. Amensalism is common in chaparral and desert communities as it stabilizes the community by reducing competition for scarce nutrients in the water. This chemical interaction is known as allelopathy. Examples of amensalism are the bread mold *Penicillium* secreting penicillin, which is a chemical that kills bacteria, and the black walnut tree releasing a chemical that kills neighboring plants.

COMMENSALISM (+)(Ø)

The interaction between two species whereby one organism benefits and the other species is not affected is commensalism. Forms of commensalism include phoresy, inquilinism, and metabiosis. Phoresy means using another organism for transportation, such as the remora on a shark or mites on dung beetles. Inquilinism means using another organism for housing, such as epiphytic plants like orchids growing on trees or birds living in the holes of trees. Metabiosis means using something that another organism created, such as hermit crabs using the shells of marine snails for protection.

COMPETITION (–)(–)

Competition can be either intraspecific (competition among members of the same species) or interspecific (competition among members of different species). Competition is the driving force of evolution whether it is for competition for food, mating partners, or territory. Intraspecific competition results in organisms best suited for surviving in a changing environment. Competition is prominent in predator-prey relationships, with the predator seeking food and the prey seeking survival. Examples of types of competition include interference, exploitation, and apparent competition. Interference occurs directly among individuals by interfering with foraging, survival, or reproduction or by preventing a species from establishing itself within a habitat. Exploitation occurs indirectly through a common limiting resource that acts as an intermediate. By using the resource, one species depletes the amount available to others. Apparent competition occurs indirectly between two species that are both sought after by the same predator.

MUTUALISM (+)(+)

Mutalism is the interaction between two species whereby both species benefit. Symbiosis is a lifelong positive interaction that involves close physical and/or biochemical contact such as the relationship between trees and mycorrhizal fungi. The relationships can be shorter, as in the case of bees pollinating flowers. Mutualism can be obligatory such as mycorrhizal fungi being totally dependent on their plant hosts. Mutualism can also be nonobligatory (facultative) as seen in *Rhizobia* bacteria reproducing either in the soil or in a symbiotic relationship with legumes.

PARASITISM (+)(–)

Parasitism is the interaction between two species whereby one species is benefited at the expense of the other. If the parasite lives on the host, it is known as an ectoparasite (like a mosquito) and often has elaborate mechanisms and strategies for acquiring a host. For example, leeches locate their host by sensing movement and confirming the identity through skin temperature and chemical cues before attaching. If the parasite lives within the host, it is known as an endoparasite (like a tapeworm) and acquires its hosts by passive mechanisms such as the ingestion of egg cells. Epiparasites feed on other parasites. Biotrophic parasites must keep their hosts alive and represent a successful mode of life. Many viruses are examples of biotrophic parasites as they use the host's genetic and cellular processes to multiply. Necrotrophs are parasites that eventually kill their hosts. Social parasites involve behaviors that benefit the parasite and harm the host. Cuckoo birds using other birds to raise their young or the nectar-robbing behavior of insects and birds are examples of social parasitism. Hosts have evolved defense mechanisms (immune systems, plant toxins) to diminish parasitism.

PREDATION (+)(–)

Predators hunt and kill prey through the act of predation. Whereas predators kill their prey, parasites have evolved mechanisms to keep their host alive since their survival depends on a viable host. Predators can be opportunistic and kill and eat almost anything. They may instead be specialists and prey upon only certain organisms. Predators that eat only meat are called carnivores; those that eat both meat and vegetation are known as omnivores.

SAPROTROPHISM (+)(–)

Saprotrophs obtain their nutrients from dead or decaying plants or animals through absorption of soluble organic compounds. Saprotrophs include many fungi, bacteria, and protozoa. Vultures and dung beetles are also saprotrophs.

Keystone Species

A keystone species is a species whose very presence contributes to a diversity of life and whose extinction would lead to the extinction of other forms of life. Through various interactions, a small number of individuals from a keystone species have a very large and disproportionate impact on how ecosystems function. An ecosystem may experience a dramatic shift if a keystone species is removed even though that keystone species was a small part of the ecosystem as measured by biomass or productivity. A classic keystone species is a small predator that prevents a particular herbivorous species from decimating a dominant plant species. Since the prey numbers are low, the keystone predator numbers could be even lower and still be effective. Yet without the predators, the herbivorous prey would explode in numbers, wipe out the dominant plants, and dramatically alter the character of the ecosystem. Specific examples of keystone species are listed below:

GRIZZLY BEAR

Grizzly bears serve as ecosystem engineers by transferring nutrients from oceanic to forest ecosystems. They do this by first capturing salmon from rivers and streams and then by transporting them onto dry land, where the bears leave nutrient-rich feces and partially eaten carcasses.

SEA STARS

Sea stars prey on sea urchins, mussels, and other shellfish that have no other natural predators. If sea stars are removed from the ecosystem, mussel populations explode and drive out other species. Sea urchin populations rise to the point where they begin to decimate coral reefs

SEA OTTERS IN KELP FORESTS

Sea otters prey on sea urchins. Kelp roots serve as anchors and not as nutrient-absorbing systems as found in land plants. If left unchecked, the sea urchins destroy the kelp forests by foraging on kelp roots.

PRAIRIE DOGS

Prairie dog colonies provide a unique ecological sanctuary for other animals. Where prairie dogs occur, there are greater arthropod and small mammal populations, an increase in bird diversity and density, and greater predator populations. Burrowing owls and reptiles use prairie dog burrows for shelter and hibernation. Prairie dogs are food for many mammalian and avian predators. Approximately 170 vertebrate species are dependent on the prairie dog for survival. Prairie dogs enhance habitat diversity by regulating plant species diversity and by enriching the soil. Burrowing, which aerates and mixes organic material into the soil, alters soil structure and chemistry. The nutrient content of the soil is increased and enhances plant diversity and productivity. Constant grazing near burrows stimulates forb and grass regrowth, and grazed plants have a higher nutritional value than mature plants. Prairie dog grazing and plant regrowth attract livestock, bison, and pronghorn antelope to prairie dog colonies.

Species Diversity

Organisms that live in different environments are specifically adapted to their particular biome.

AQUATIC ORGANISMS

Water provides buoyancy and reduces the need for support structures such as legs and trunks. Water has high thermal capacity, so most aquatic organisms do not spend energy on temperature regulation. Many organisms obtain nutrients directly from the water, thereby reducing energy spent on searching for food. These include filter feeders such as barnacles, clams, and oysters. Water allows dispersal of gametes and larvae to new areas. Water screens out UV radiation. Intertidal organisms have evolved methods of not being swept away by waves and prevent water loss during low tide through shells and exoskeletons, salt-removing mechanisms, and lower metabolic rates due to cooler ambient temperatures.

DESERT ORGANISMS

Desert plants are spaced apart due to limiting factors and consist primarily of succulents (cactus) and short-lived annuals (wildflowers). Succulents store water, have small surface areas exposed to sunlight, have vertical orientation to minimize exposure to the sun, open their stomata at night, have waxy leaves to minimize transpiration, have deep roots to tap groundwater (mesquite and creosote), and have shallow roots to collect water after short rainfalls (prickly pear and saguaro cactus). Sharp spines on cacti reflect sunlight, create shade, and discourage herbivores. Cacti secrete toxins into the soil to prevent interspecific competition (allelopathy). Desert plants store biomass in seeds. Wildflowers have short life spans and are dependent on water for germination.

Desert animals are small and have small surface areas. They spend time in underground burrows where it is cooler, and they are often nocturnal. Aestivation is common. Some animals are able to metabolize dry seeds. Kangaroo rats produce their own water and secrete concentrated urine. Insects and reptiles have thick outer coverings to minimize water loss.

GRASSLAND ORGANISMS

Grasses grow out from the bottom (basal meristem) so they can grow again after being nibbled by grazing animals. Grass species are drought resistant. Deciduous trees and shrubs shed leaves during the dry season to conserve water. Grazers and browsing animals eat vegetation at different heights so as to not compete. Giraffes eat near the top of trees, elephants eat lower down, zebras eat taller grasses, and gazelles eat shorter grasses. Some animals migrate to find water. Others become dormant. Still others survive on seeds during the dry season. Some animals live in burrows to hide and escape predators. In addition their fur color sometimes matches the color of the surroundings to act as camouflage.

FOREST ORGANISMS

In the tropics, some animals live in tree canopies where shelter and available food supplies (leaves, flowers, and fruits) are abundant and where they can escape from predators that live closer to the ground. Epiphytes (orchids and bromeliads) live on trunks and branches of trees, and they catch organic matter falling from the canopy. Epiphytes do not root in the soil. Instead, they obtain their moisture from the air or from dampness on the surface of their host. Some plants have very large leaves to capture scarce light. Roots of trees are shallow and spread out to capture nutrients in poor soil. To compensate for little support by shallow roots, trees may have buttresses. Flowers have elaborate devices to attract pollinators since wind is minimal in dense growth.

In temperate deciduous forests, broadleaf deciduous trees lose their leaves in winter and become dormant to conserve water and energy. Deciduous trees shift their metabolism from a photosynthesis-based system when light and temperature are favorable to one utilizing glucose and amino acids during the winter. This also helps keep the tree from freezing and acts as a kind of antifreeze.

In evergreen coniferous forests, small, waxy-coated needles are able to withstand the cold and drought of winter and have low surface area to reduce transpiration. Furthermore, conifers are always replacing their needles, unlike deciduous trees that replace their leaves only once a year. Decomposed needles make soil acidic, preventing many competing species from surviving in the soil environment. Some animals hibernate to conserve energy during winter when resources are scarce.

TEMPERATE SCRUB FOREST LAND ORGANISMS

Chaparral plants have small, waxy-coated leaves to reduce transpiration. Many plants produce toxins that leach into the soil to prevent competition (allelopathy). Vegetation becomes dormant during the dry season. Leaves do not fall during the dry season due to the stress of replacing leaves without an adequate water supply. Plant thorns are common for protection. Vegetation is adapted to fires and is common due to the high oil content in the brush. Fires reduce competition and allow seeds to germinate. Rodents are common and store seeds in underground burrows.

TUNDRA ORGANISMS

Polar grassland plants are adapted to low sunlight, low amounts of free water, high winds, and low temperatures. Tundra plants primarily grow during summer months. Leaves on plants have waxy outer coatings. Many plants survive winter as roots, stems, bulbs or tubers. Lichens dehydrate during winter to avoid frost damage.

Animals have adapted in several ways. They have extra layers of fat, chemicals in the blood to keep it from freezing, and compact bodies to conserve heat. They also have thick skin, thick fur, and waterproof feathers above downy insulating feathers. Additionally, animals either migrate during the coldest months or live underground (lemmings).

Edge Effects

An edge effect refers to how the local environment changes along some type of boundary or edge. Forest edges are created when trees are harvested, particularly when they are clear-cut. Tree canopies provide the ground below with shade and maintain a cooler, moister environment below. In contrast, a clear-cut allows sunlight to reach the ground, making the ground warmer and drier—environments not suitable for many forest plants. As time passes and a stand of young trees emerges on a clear-cut, the environment in the young stand changes and the edge begins to fade. As the mature forest develops, the edge fades away.

The edge effect is the result of two different conditions influencing the plants and animals that live on the edge. Some animal species (deer, elk, white-tailed deer, and pheasants) survive well in a forest edge since they are able to find food in the clearing, are able to benefit due to various habitats near one another, are able to hide in the nearby trees, or are well-adapted to human disturbances. These animals are called "edge species." Other animals, such as the spotted owl, do not do well in edges. Negative edge effects become most extreme when forest lands share the edge with agriculture or suburbs. Species composition and diversity can change dramatically if organisms adapted to one set of conditions cannot adapt to the change. If the edge effect is gradual or has indistinct boundaries and over which many species cross, it is called an open community. A community that is sharply divided from its neighbors is called a closed community. Preservation of large blocks of habitat and linking smaller habitat blocks together with migration corridors are techniques to protect rare and endangered species.

Major Terrestrial and Aquatic Biomes

Biomes are a major regional or global biotic community characterized by the dominant forms of plant life and the prevailing climate. Temperature and precipitation are the most important determinants of biomes. Biomes are classified by the type of dominant plant and animal life. Species diversity within a biome is directly related to net productivity, availability of moisture, and temperature.

Biomes

ANTARCTIC

Area surrounding south pole. Rainfall <2 inches (5 cm) per year.

BENTHOS (HADAL)

Bottom of oceans. No sunlight, therefore no plant life. Primary input of energy comes from dead organic matter settling and chemosynthesis.

COASTAL ZONES

Includes estuaries, wetlands, and coral reefs. High diversity and counts of animal and plant species due to runoff from land.

CORAL REEFS

Warm, clear, shallow ocean habitats near land and generally in the tropics. There are three types of coral reefs. Fringing reefs grow on continental shelves near the coastline. Barrier reefs are parallel to shoreline but farther from shore. Coral atolls are rings of coral that grow on top of sunken oceanic volcanoes. Coral reefs are disappearing at an alarming rate due to increase in sea temperature, pollution, dredging, and sedimentation.

DESERTS

Occur between 15° and 25° north and south latitude and generally occur in the interior of continents. Occupy about 20% of all landmass. Rainfall less than 20 inches (50 cm) per year. Air currents are descending, which generally diminishes the formation of rain (rain requires ascending air—orographic lifting). Soils often have abundant nutrients but lack organic matter and consist of sand and closely packed boulders. Little humus in the soil profile.

FRESHWATER WETLANDS

Freshwater wetlands include freshwater swamps, marshes, bogs, prairie potholes (important stopping grounds for migrating birds), ponds, peat bogs, and riparian areas (areas near rivers and streams). Ground is saturated with freestanding water, although standing water can be seasonal. Soil is low in oxygen. Important breeding areas rich in insects, amphibians, and reptiles. Prime areas for human development and recreation resulting in large amounts of habitat destruction. Critical for freshwater supplies. Easily polluted. Dominant plants are floating plants (phytoplankton). Animal life is abundant. Estuaries are rich in nutrients and are breeding grounds for fish. Water input includes runoff, groundwater flow, and streams. A lake is a body of water of considerable size surrounded by land. The vast majority of lakes on Earth are freshwater and most, like in the Northern Hemisphere, are at higher altitudes. Lakes can be classified on the basis of their richness of nutrients, which affects plant growth and animal diversity. Oligotrophic lakes are clear and low in nutrients, which results in a small amount of plant life and other forms of biomass. Eutrophic lakes are rich in nitrogen and phosphorus with large and diverse populations of phyto- and zooplankton, which supports a large diversity of fish. During warm weather, the oxygen content of the water may decrease resulting in die-offs. Hypertrophic lakes are excessively enriched with nutrients and subject to algal blooms. These lakes result from human activity such as the heavy use of fertilizers in the lake catchment area.

GRASSLANDS

Found in areas too dry for forests and too wet for deserts. Rainfall is seasonal. Temperatures are moderate. Grasslands occupy approximately 25% of all land area. Few trees and shrubs due to frequent seasonal fires and water availability. Dominant plants are grasses and perennials with extensively developed roots. Soils are rich in organic matter. Upper soil horizons are alkaline, dark, and rich in humus. Extensively used by humans for agriculture.

HYDROTHERMAL VENTS

Occur in the deep ocean where hot-water vents rich in sulfur compounds are found and that provide energy for chemosynthetic bacteria.

INTERTIDAL

The area of the shoreline exposed to water during high tide and air during low tide. Water movement brings in nutrients and removes waste products. Extremely rich in biodiversity. Susceptible and sensitive to pollution from land runoff and ocean pollution.

OCEAN

Oceans occupy 75% of Earth's surface. Areas of low diversity and low productivity except near the shoreline. Low in nitrogen and phosphorus, which limits plant growth and the smaller organisms that feed on them. Large animals occur but in low density.

SAVANNAS

Warm year-round. Scattered trees. Environment is intermediate between grassland and forest. Extended dry season followed by a rainy season. Consists of grasslands with stands of deciduous shrubs and trees. Trees and shrubs generally shed leaves during the dry season, which reduces the need for water. Food is limited during the dry season, requiring many animals to migrate. Soils are rich in nutrients. Contain large herds of grazing animals and browsing animals that provide resources for predators.

TAIGA (CONIFEROUS OR BOREAL FORESTS)

Generally found between 45° and 60° north latitude. Occupies approximately 17% of land surface. Forests of cold climates of high latitudes and altitudes. Two types of taiga: open woodland and dense forest. More precipitation than the tundra. Generally precipitation occurs during the summer months due to midlatitude cyclones. Soils are generally poor in nutrients because of large amounts of leaching caused by rainfall. The soils are acidic due to the decomposition of needles (an adaptation to conserve water). Deep layer of litter on the surface, and decomposition is slow because of low temperatures. Dense stands of small trees cause understory to be low in light. Low biodiversity due to harshness of environment. Disturbances include fires, storms, and insect infestations.

TEMPERATE DECIDUOUS FORESTS

Forests found in milder temperatures than boreal forests of the taiga. Rapid decomposition due to mild temperatures and precipitation, which results in a small amount of litter on surface. Because of the mild climate, this biome has been greatly exploited by humans for agriculture, lumber, and urban development. Soil is generally poor in nutrients. Tall deciduous trees. Rich and diverse understory. Low density of large mammals due to shade that prevents much ground vegetation.

TEMPERATE RAIN FORESTS

Moderate temperatures and rainfall exceeding 100 inches (250 cm) per year. Low biodiversity because of limited light, which limits food for herbivores. Major resource for timber.

TEMPERATE SHRUBLAND (CHAPARRAL)

Hot, dry summers with mild, cool, and rainy winters. During summer and spring, a subtropical high-pressure zone exists over the area. Rain falls during the winter due to midlatitude cyclones. Average rainfall is between 15–40 inches (30–75 cm) per year. Characterized by dense shrub growth. Decomposition is slow during dry months. Few large mammals. Erosion is common after fires. Because of climate, this area is being utilized for urbanization. Found in select coastal regions.

TEMPERATE WOODLANDS

Drier climate than deciduous forests. Dominated by small trees. Stands of trees are open, allowing light to reach ground. Fires are common.

TROPICAL RAIN FORESTS

High and constant temperature (average approximately 80°F [27°C]). Rainfall ranges between 75–100 inches (200–250 cm) per year. Found within Hadley cells. High species diversity for both plants and animals (up to 100 different tree species per square kilometer as opposed to three to five in temperate zones). Vegetation is dense. Soils are low in nutrients, and most nutrients are stored in vegetation. Soil is acidic. Decomposition of organic material is very fast due to temperature and moisture. Leaching is high. Abundant insect and animal biodiversity. Humans are clearing tropical rain forests for agriculture and cattle raising through a technique called slash and burn. Due to poor soil nutrient levels, agricultural activity lasts for only a limited time and can be sustained only by applying expensive fertilizer, which causes farmers to destroy even more forest.

TROPICAL SEASONAL FOREST

Occurs in areas of seasonal rainfall (monsoon) that is followed by long, dry season. Warm temperatures year-round. Contains mixture of deciduous and drought-tolerant evergreen trees.

TUNDRA

60° north latitude and above. Influenced by polar cells. Alpine tundra is located in mountainous areas, above the tree line with well-drained soil. Dominant animals include small rodents and insects. Arctic tundra is frozen, treeless plains, low rainfall, and low average temperatures with poor drainage due to the frozen ground. Growing season lasts about 2 months. Soil has few nutrients because of low vegetation and little decomposition. Permafrost is permanently frozen ground and a barrier for roots.

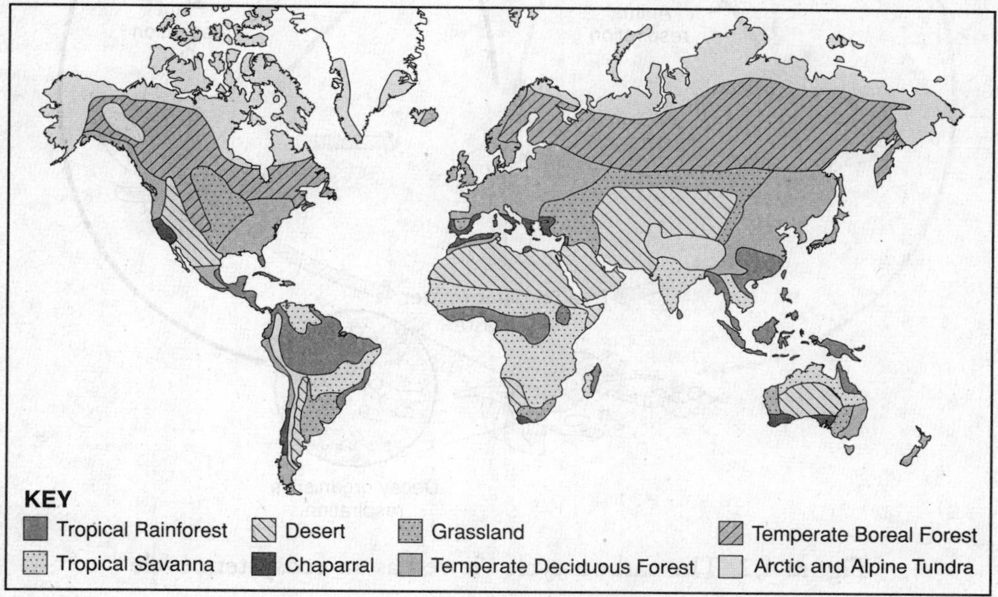

Figure 4.1 Major biomes of the world

ENERGY FLOW

The ultimate source of energy is the sun. Plants are able to use this light energy to create food. The energy in food molecules flows to animals through food webs.

Photosynthesis and Cellular Respiration

$$6CO_2 + 6H_2O + \text{sunlight} \rightarrow C_6H_{12}O_6 + 6O_2$$

Plants remove carbon dioxide from the atmosphere by a chemical process called photosynthesis, which uses light energy to produce carbohydrates and other organic compounds. Plants capture light primarily through the green pigment chlorophyll. Chlorophyll is contained in organelles called chloroplasts. The glucose or the energy derived from its oxidation during cellular respiration is then used to form other organic compounds such as cellulose (for support), lipids (waxes and oils), and amino acids and then proteins. Oxygen gas is released into the atmosphere during photosynthesis. Plants also emit carbon dioxide during respiration. They produce less carbon dioxide than they absorb and therefore become net sinks of carbon.

Organisms that undergo photosynthesis are called photoautotrophs. Factors that affect the rate of photosynthesis are the amount of light and its wavelength, carbon dioxide concentration, the availability of water, and temperature.

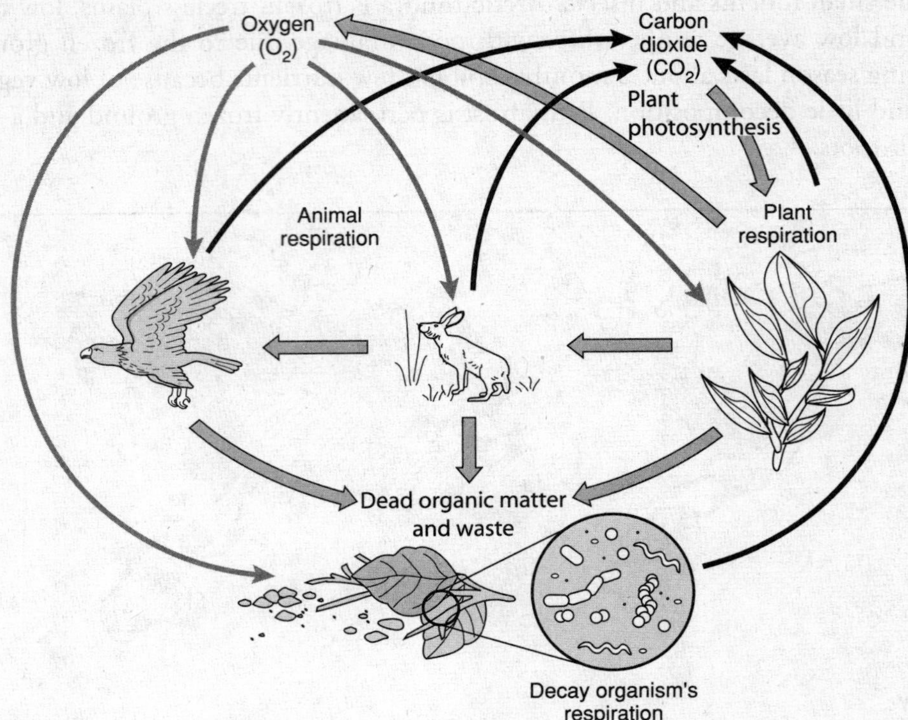

Figure 4.2 The carbon cycle viewed as an ecosystem cycle

Organisms dependent on photosynthetic organisms (autotrophs) are called heterotrophs. In general, cellular respiration is the opposite of photosynthesis. In respiration, glucose is oxidized by the cells to produce carbon dioxide, water and chemical energy. This energy is stored in the molecule adenosine triphosphate (ATP).

$$C_6H_{12}O_6 + 6O_2 \rightarrow 6CO_2 + 6H_2O + \text{energy}$$

Food Webs and Trophic Levels

Primary producers (autotrophs) are plants. They convert solar energy into chemical energy through photosynthesis. Primary consumers are heterotrophs (herbivores—plant eaters) and get their energy by consuming primary producers. Primary consumers have developed defense mechanisms against predation, some of which include:

1. Speed
2. Flight
3. Quills
4. Tough hides
5. Camouflage
6. Horns and antlers.

Secondary (and higher) consumers are also heterotrophs and may be either strictly carnivores (meat eaters) or omnivores (eat both plants and animals).

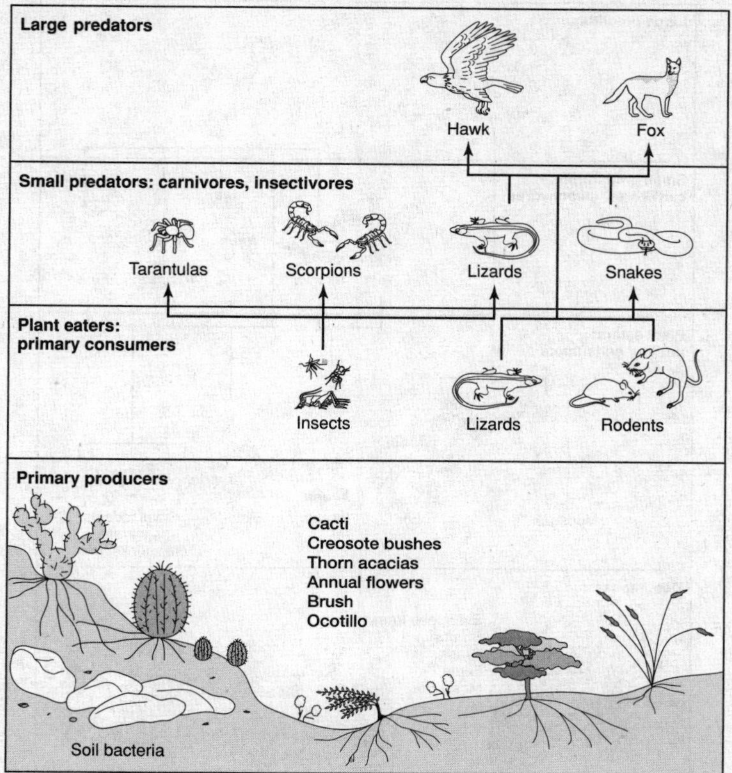

Figure 4.3 Desert food web

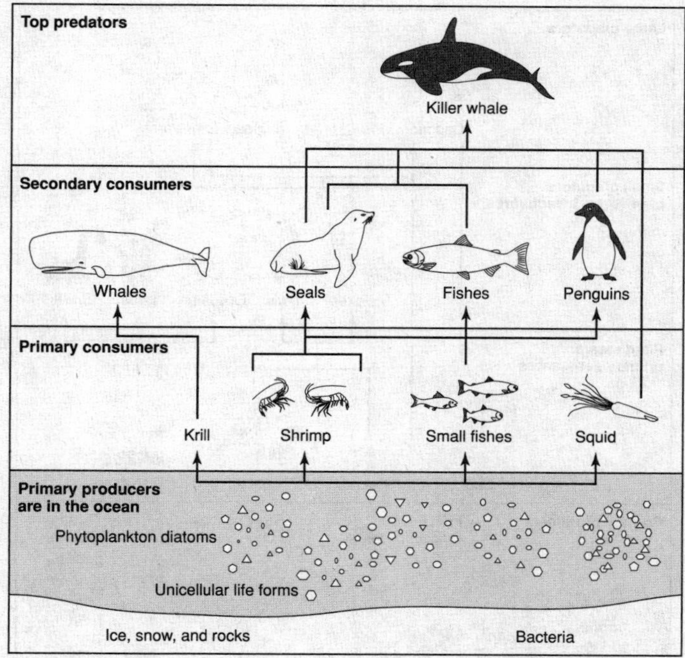

Figure 4.4 Oceanic food web

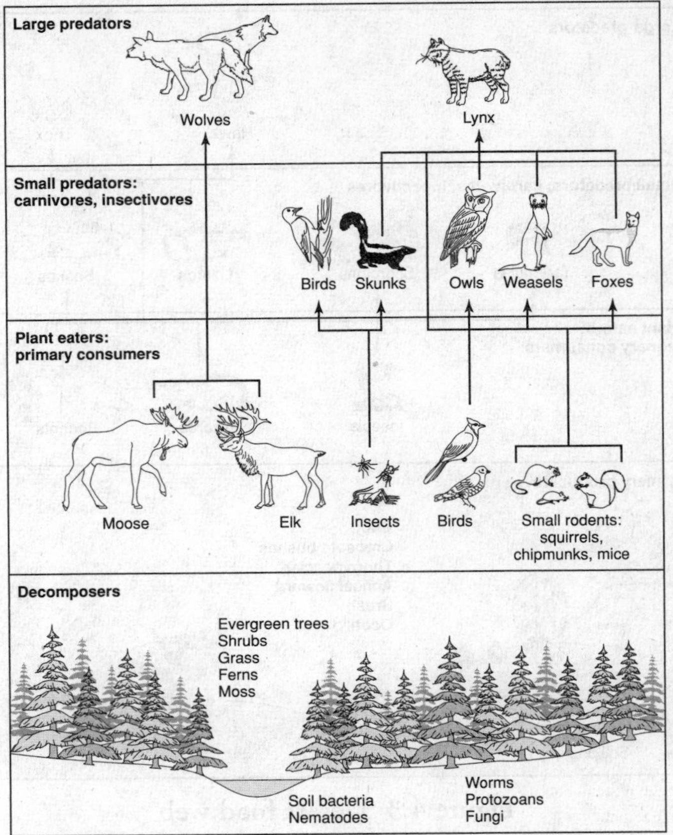

Figure 4.5 Coniferous forest food web

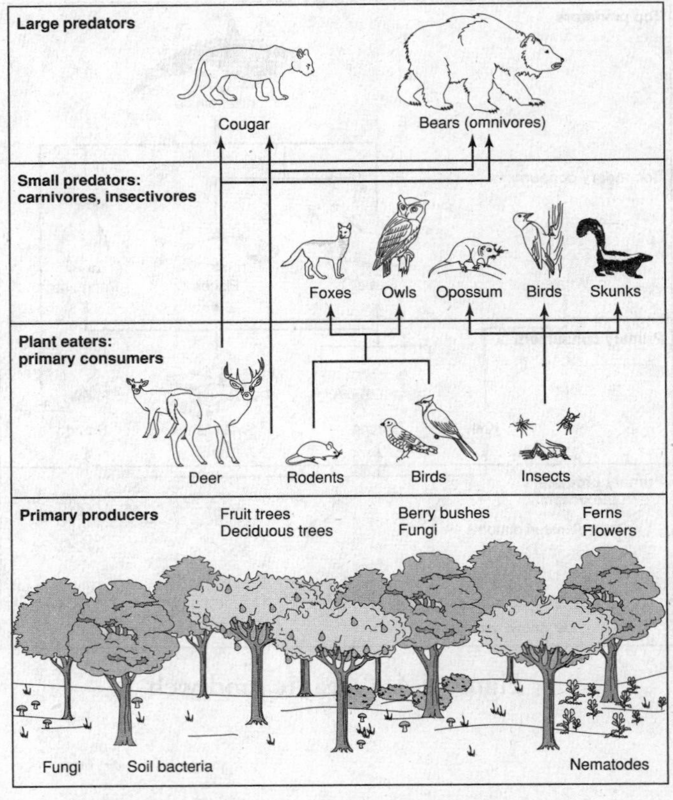

Figure 4.6 Deciduous forest food web

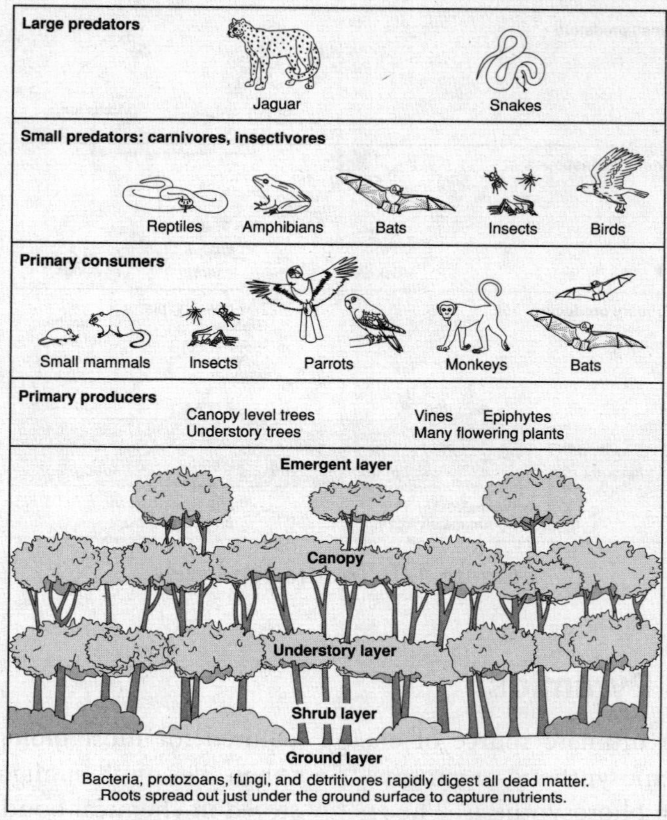

Figure 4.7 Tropical rain forest food web

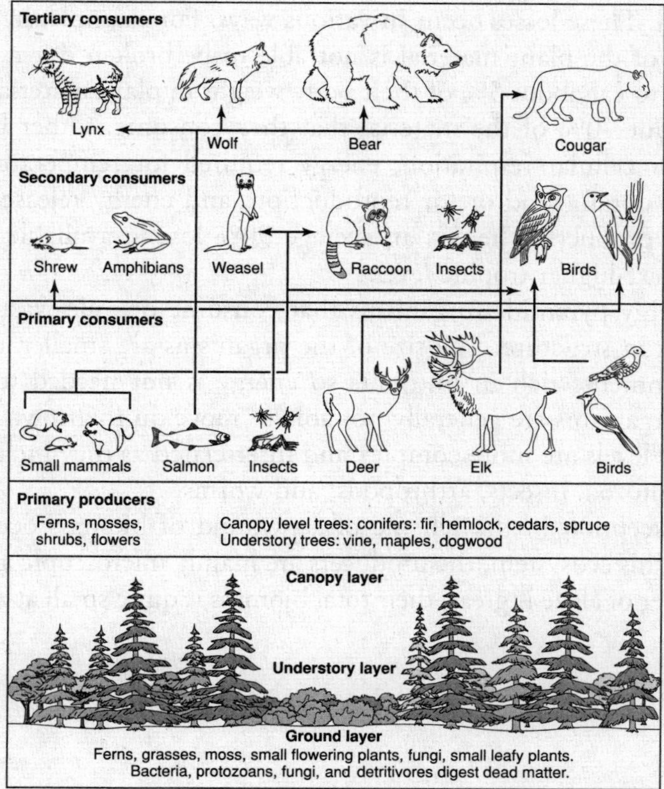

Figure 4.8 Temperate rain forest food web

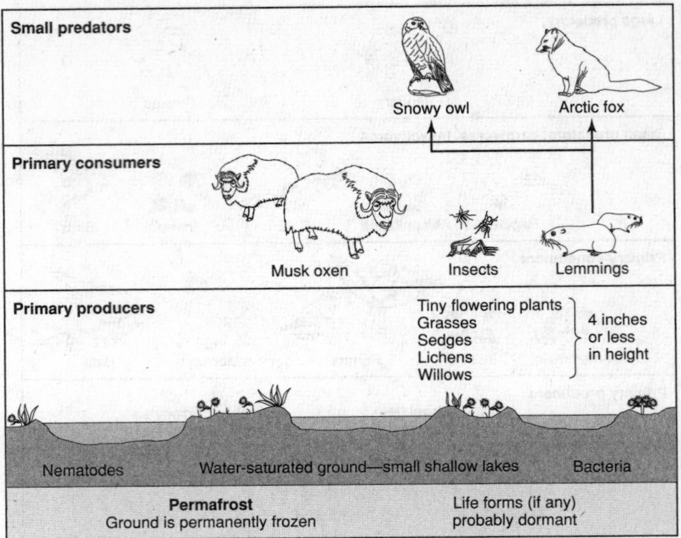

Figure 4.9 Tundra food web

Ecological Pyramids

Sunlight is the ultimate source of energy required for most biological processes (except for chemosynthetic organisms). Less than 3% of all sunlight that reaches Earth is used in photosynthesis. The energy stored in chemical bonds is released to animals and plants through cellular respiration.

> **TIP**
>
> As energy flows through systems, more of it becomes unusable at each step.

Losses of potential energy (in the form of heat energy) occur as one moves up an energy pyramid. These losses occur in various ways. For example, in digestive inefficiency, much of the plant material is not able to be broken down. For example, elephants need to eat about 5% of their body weight in plant material each day but digest only about 40% of the material that they consume. Other losses occur in energy used for cellular respiration, energy required for temperature regulation, energy used to obtain food or for reproduction, and energy released through the decay of waste products. There is an average 90% loss in available energy as one moves to the next-higher trophic level.

Detritus energy pyramids (organisms that consume organic wastes) are significantly different in structure. The size of the organisms are smaller, and organisms exist in environments rich in nutrients so energy is not needed to obtain food. Additionally, organisms are generally not able to move on their own to any degree. Finally, trophic levels are more complex and interrelated as they include algae, bacteria, fungi, protozoa, insects, arthropods, and worms.

A notable exception in the scheme of a pyramid of biomass occurs in aquatic ecosystems. In this ecosystem, the producers are mainly microscopic algae. Although the total number of algae is great, their total biomass is quite small at any given time.

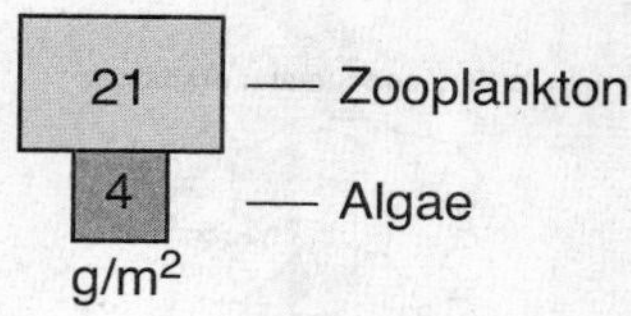

Figure 4.10 Inverted pyramid of biomass—aquatic ecosystem

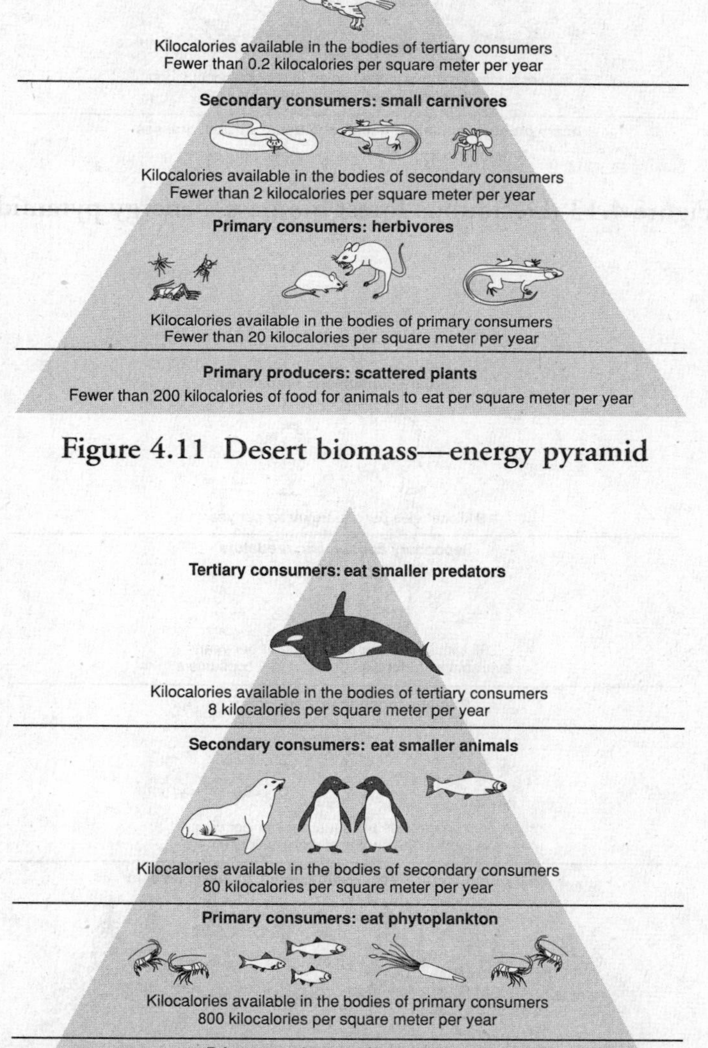

Figure 4.11 Desert biomass—energy pyramid

Figure 4.12 Marine biomass—energy pyramid

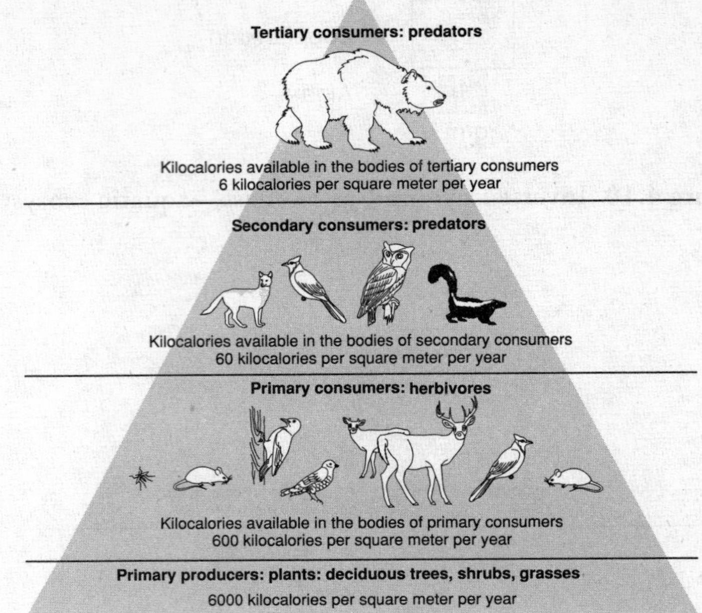

Figure 4.13 Deciduous forest biomass—energy pyramid

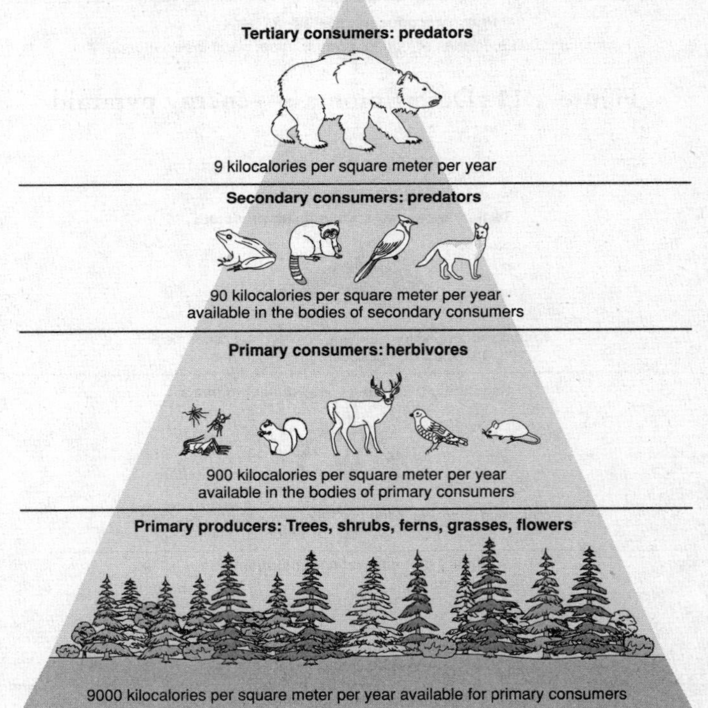

Figure 4.14 Temperate rainforest biomass—energy pyramid

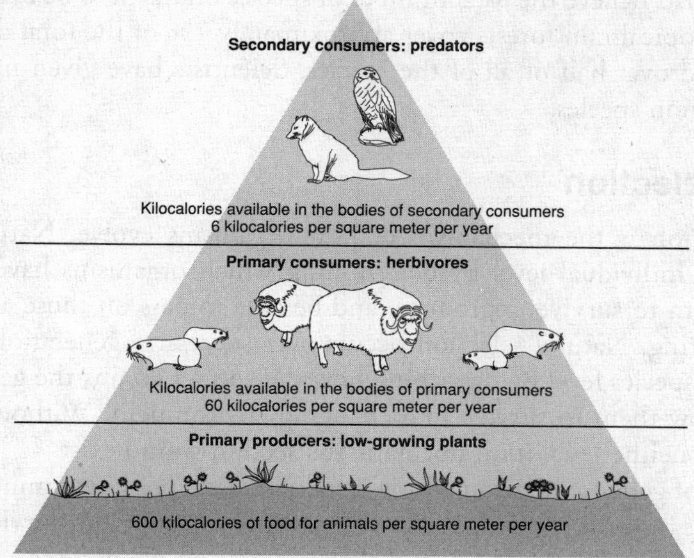

Figure 4.15 Tundra biomass—energy pyramid

ECOSYSTEM DIVERSITY

The topic of ecosystem diversity involves biodiversity, natural selection, evolution, and ecosystem services.

Biodiversity

Biodiversity attempts to describe diversity at three levels: genetic, species, and ecoystem. Genetic diversity involves the range of all genetic traits, both expressed and recessive that makes up the gene pool for a particular species. Species diversity is the number of different species that inhabit a specific area. For example, tropical rain forests have a higher species diversity than extreme deserts. Ecosystem diversity concerns the range of habitats that can be found in a defined area. Ecosystems are composed of both biotic and abiotic components.

Biodiversity

Diversity Increasers	Diversity Decreasers
Diverse habitats	Environmental stress
Disturbance in the habitat (fires, storms, and so on)	Extreme environments
Environmental conditions with low variation	Extreme limitations in the supply of a fundamental resource
Trophic levels with high diversity	Extreme amounts of disturbance
Middle states of succession	Introduction of species from other areas
Evolution	Geographic isolation

Most scientists believe the total number of species on Earth to be between 10 and 30 million. Tropical rain forests cover approximately 7% of the total dry surface of Earth but hold over half of all of the species. Scientists have given names to only about 1.5 million species.

Natural Selection

Natural selection is the mechanism of how organisms evolve. Natural selection works on the individual level by determining which organisms have adaptations that allow them to survive, reproduce, and be able to pass on those adaptive traits to their offspring. Natural selection occurs over successive generations. Evolution works on the species level by describing how the species attains the genetic adaptations that allow them to survive in a changing environment. Without a changing environment, neither evolution nor natural selection would exist.

The range of genetic variation within a species's gene pool determines whether or not the species, not the individual, has the capacity to adapt and survive to changes in the environment. The expression of that variation in phenotypic and behavioral expressions determines whether the individual survives and is commonly referred to as survival of the fittest. "Fittest" means the ability to reproduce and pass on genes to offspring. New genes enter the gene pool through mutation and when combined with a change in gene frequency, results in evolution at the species level. Natural selection operates in three ways: stabilizing, directional, and disruptive.

STABILIZING SELECTION

Stabilizing selection affects the extremes of a population and is the most common form of natural selection. The individuals that deviate too far from the average conditions are removed. The results are a decrease in diversity, maintenance of a stable gene pool, and no evolution. For example, human babies that are too low in weight or too high in weight have survival problems.

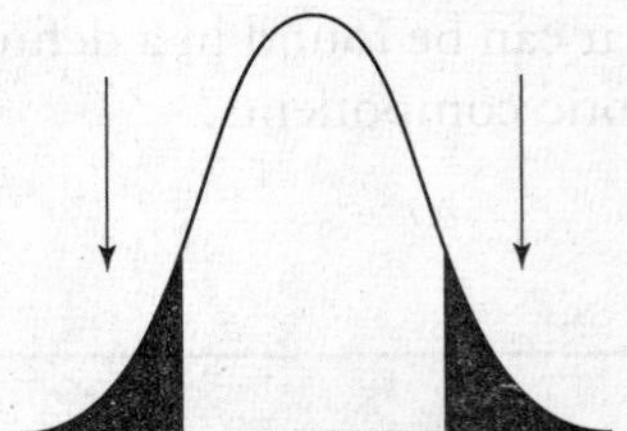

Figure 4.16 Stabilizing selection

DIRECTIONAL SELECTION

Directional selection affects the extremes of a population. Individuals toward one end of the distribution may do especially well, resulting in a frequency distribution toward this advantage in future generations. An example is the industrial melanism of the peppered moth and the evolution of the horse. The early horse present during the early Eocene was a small-bodied creature that moved well through heavy brush and woodlands. Today's horse, with its long legs designed for speed in the open grassland and changes in its dental and toe structure, looks nothing like its ancestor.

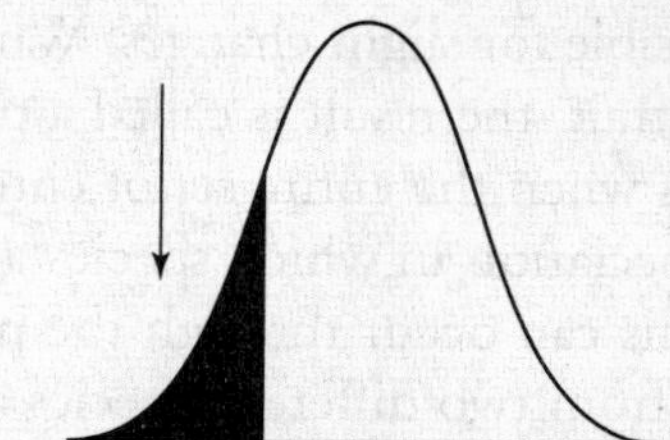

Figure 4.17 Directional selection

DISRUPTIVE SELECTION

Disruptive selection acts against individuals that have the average condition and favors individuals at the extreme ends (bimodal). The population is essentially split into two. Both directional and disruptive selection affects the gene pool. In both cases, the population changes and evolution occurs.

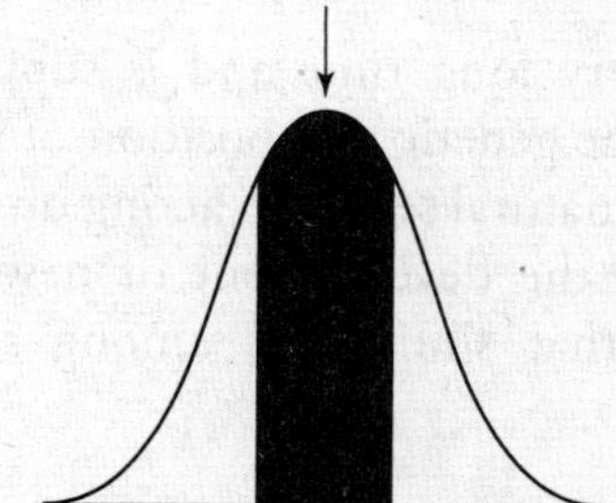

Figure 4.18 Disruptive selection

Natural Selection

Process of Natural Selection	Description
Competition	There is a struggle to survive, and competition exists for limited resources.
Disproportionate increase and persistence in phenotypic adaptation in successive populations	Variations that are advantageous to the individual in terms of survival allow more organisms possessing the trait(s) to survive, reproduce, and pass on the characteristic(s) to future generations.
Geometric (or exponential) increase in population	If all offspring survived, there would be astronomical numbers of individuals.
Individual variations	There is variation in offspring. For natural selection, the variations must be gene expressed and be capable of being inherited.
Limited resources	Earth has finite resources.

More often than not, natural selection is based upon the cumulative effects of numerous genes, each responsible for slight changes. When genes at more than one locus contribute to the same trait, the result is called a polygenic effect.

Polyploidy occurs in plants when the entire set of chromosomes is multiplied. It is an example of sympatric speciation in which species arise within the same, overlapping geographic range. This can occur through the process of hybridization. In hybridization, chromosomes from two different species are artificially combined to form a new species (hybrid). In another type of hybridization, chromosomes naturally fail to segregate during meiosis, producing diploid gametes. If the hybrid has adaptive traits to survive in the new environment, a new species of plant has been produced. Although the plant may not be able to reproduce sexually, it may be able to reproduce through vegetative means. Examples of polyploidy include cotton, tobacco, sugarcane, bananas, potatoes, and wheat. More than half of all known species of plants today (260,000 species) may have originated through polyploidy.

Evolution

Change generally takes a very long time and is supported by the fossil record. Evolution is the change in the genetic composition of a population during successive generations as a result of natural selection acting on the genetic variation among individuals and resulting in the development of new species. The concept of a "common ancestor" states that similarities among species can be traced back through a phylogenetic tree.

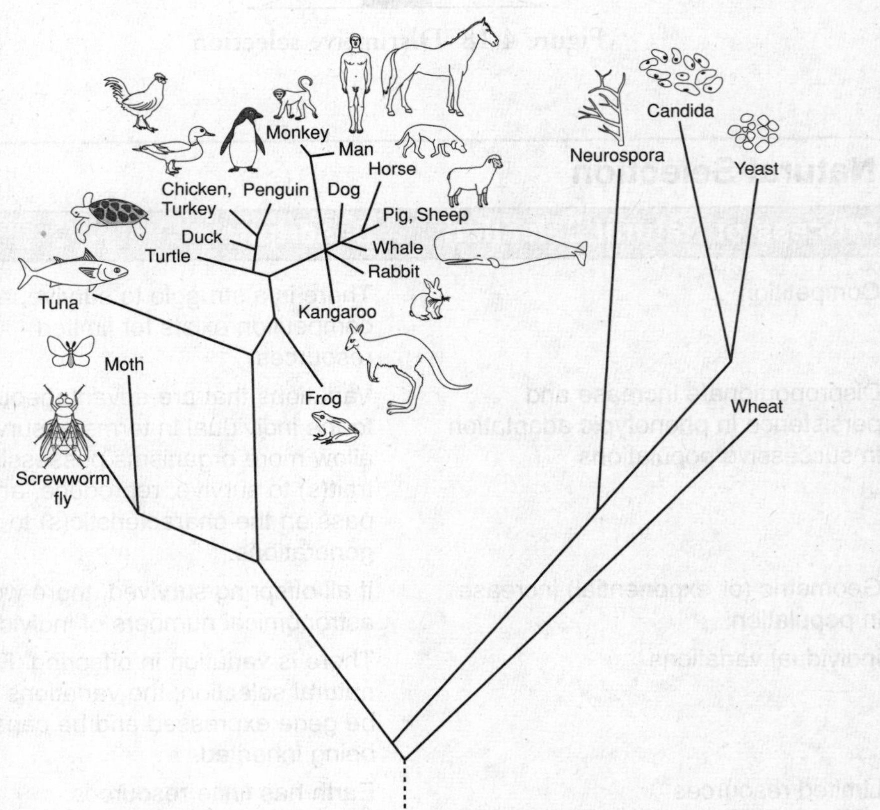

Figure 4.19 Phylogenetic tree traced through analysis of cytochrome C.

SPECIATION

Speciation results when segments of a population of one species become so isolated that gene flow stops. Adaptive radiation describes rapid speciation to fill ecological niches and is driven by either mutation and/or natural selection. Three basic types of adaptive radiation are: general adaptation, environmental change, and geographic isolation. In general adaptation, a species develops a radical new ability to reach new parts of its environment, for example, bird flight. With environmental change, a species that can, in contrast with the other species in the ecosystem, successfully survive in a radically changed environment and branch into new species that covers the new niches. An example is the rapid spread and development of mammalian species after the extinction of the dinosaurs. In geographic isolation, isolated ecosystems, such as archipelagos and mountain areas, can be colonized by a species that upon establishing itself undergoes rapid divergent evolution. A classic example of geographic isolation is the Darwin finches of the Galapagos Islands in which ancestral groups of finches from the mainland of Ecuador were blown to the Galapagos Islands. The isolated groups of finches were subjected to different selective pressures on different islands that, over time, resulted in separate species.

CONVERGENT EVOLUTION

Convergent evolution describes the process whereby organisms not closely related to each other independently acquire similar (analogous) characteristics while evolving in separate and sometimes varying ecosystems. An example of convergent evolution is the development of wings in birds, insects, and bats even though the organisms are not closely related.

EVOLUTIONARY RELAY

Evolutionary relay occurs when independent species acquire similar characteristics through their evolution in similar ecosystems but not at the same time. An example of evolutionary relay is the development of a dorsal fin in both sharks and the prehistoric ichthyosaurs (extinct marine reptiles).

PARALLEL EVOLUTION

Parallel evolution occurs when two independent species evolve together at the same time and in the same ecosystem and acquire similar characteristics. A classic example of parallel evolution is the evolution of the placentals and the marsupials. Placental animals bear their young fully developed. Marsupials give birth prematurely and nurture their young in a pouch. In the plant kingdom, parallel evolution is seen in the similar forms of leaves, where similar patterns have appeared over and over again in separate genera and families.

GRADUALISM AND PUNCTUATED EQUILIBRIUM

Gradualism views evolution as a slow, stepwise development of a species over long periods of time (millions of years). Punctuated equilibrium proposes that some species arose suddenly in a short period of time (thousands of years) after long periods of stability. These bursts of rapid evolution are thought to have been triggered by changes in the physical or biological environment—perhaps a period of drought or the appearance of a new, more challenging predator. A classic example of punctuated equilibrium is the abrupt appearance of flowering plants without a fossil record.

Ecosystem Services

Ecosystem services are the processes by which the environment produces resources. Examples include clean water, timber, habitat for fisheries, and pollination of native and agricultural plants. Ecosystems provide the following services:

- Moderate weather extremes and their impacts
- Disperse seeds
- Mitigate droughts and floods
- Protect people from the sun's harmful ultraviolet rays
- Cycle and move nutrients
- Protect stream and river channels and coastal shores from erosion
- Detoxify and decompose wastes
- Control agricultural pests
- Maintain biodiversity
- Generate and preserve soils and renew their fertility
- Contribute to climate stability
- Purify the air and water
- Regulate disease-carrying organisms
- Pollinate crops and natural vegetation

NATURAL ECOSYSTEM CHANGES

An ecosystem can change for various natural reasons. These include climate shifts, species movement, and ecological succession.

Climate Shifts

Evidence for changes in the climate come from data used to measure climate (which is available for only a few hundred years), written accounts (subjective), and data from material present at the time. These materials consists of tree rings, fossilized plants, insect and pollen samples, gas bubbles trapped in glaciers, deep ice core samples, lake sediments, stalactites and stalagmites, marine fossils including coral analysis, sediments including rafted debris, dust analysis, and isotope ratios in fossilized remains. The bottom line is that Earth's climate has gone through many cycles of warming and cooling trends. Many different factors influence the climate.

ALBEDO

Albedo is reflectivity. Materials like ocean water have low albedo, whereas land masses have moderate albedo. The highest albedo is snow and ice. Hence, periods when polar ice is highly extended will promote further cooling. This is a positive-feedback mechanism. Dust in the atmosphere has the same effect. It forms a high albedo veil around Earth so that a significant amount of solar radiation is reflected before it reaches the surface. The dust may come from dry climate periods, volcanic eruptions, meteor impacts, and so on.

CARBON CYCLE

The consumption of carbon in the form of carbon dioxide (CO_2) results in cooling. Two different processes consume carbon dioxide: carbonate rock weathering and silicate rock weathering.

Carbonate rock weathering: $\mathbf{CO_2} + H_2O + CaCO_3 \rightarrow Ca^{2+} + 2HCO_3^-$

Silicate rock weathering: $\mathbf{2CO_2} + H_2O + CaSiO_3 \rightarrow Ca^{2+} + 2HCO_3^- + SiO_2$

The production of carbon in the form of carbon dioxide results in warming. Both carbonate formation in the oceans and metamorphic breakdown of carbonate yield carbon dioxide.

Carbonate formation in the oceans: $Ca^{2+} + 2HCO_3^- \rightarrow \mathbf{CO_2} + H_2O + CaCO_3$

Metamorphic breakdown of carbonate: $SiO_2 + CaCO_3 \rightarrow \mathbf{CO_2} + CaSiO_3$

GREENHOUSE EFFECT

The most important greenhouse gases are water (H_2O), carbon dioxide (CO_2), and methanae (CH_4). Without this effect, Earth would be cold and inhospitable. If taken too far, Earth could evolve into a hothouse.

LANDMASS DISTRIBUTION

Materials absorb and reflect solar radiation to different extents. Ocean water is much more absorbent than landmasses so that continents reflect a lot more solar energy back into space than the oceans. Earth receives more solar radiation at low latitudes (near the equator) than near the poles. An Earth with landmasses clustered at low latitudes would reflect more solar energy into space, resulting in a cooler planet than one with more equatorial ocean area. Approximately 600–800 million years ago, there were significant glacial deposits in North America, Australia, and Africa. Paleomagnetism of rocks suggests that these continents were near the south pole and that the equatorial Earth was largely ocean.

PLATE TECTONICS AND VOLCANOES

Plate tectonics affect atmospheric carbon dioxide, which factors into climate changes through the greenhouse effect. Volcanoes produce carbon dioxide. If global volcanism slows, as would be the case when supercontinents stabilize, less atmospheric carbon dioxide would trigger global cooling. Increased volcanism puts more carbon dioxide in the atmosphere and results in more greenhouse warming.

PRECESSION

The wobble of Earth on its axis changes the amount of energy received by the sun. Changes in the orientation of Earth in space (tilt and obliquity) also have an effect on climate.

SOLAR OUTPUT

Changes in solar output of only 1% per 100 years would change Earth's temperature by up to 1°F (0.5°C). Times of sunspot activity (every 11, 90, and 180 years) correspond to decreases in solar radiation reaching Earth. The sun's magnetic field reverses every 22 years.

Major Climatic Periods

Several major climatic periods have occurred. They are described below.

2,000,000 B.C.E. TO 12,000 B.C.E. (PLEISTOCENE ICE AGE)

Large glacial ice sheets covered much of North America, Europe, and Asia. The Pleistocene had periods when the glaciers retreated (interglacial) because of warmer temperatures and advanced because of colder temperatures (glacial). During the coldest periods of the Pleistocene Ice Age, average global temperatures were probably 7°F–9°F (4–5°C) colder than they are today.

12,000 B.C.E. TO 3000 B.C.E.

This warming of Earth and subsequent glacial retreat began about 14,000 years ago. The warming was shortly interrupted by a sudden cooling period between 10,000–8500 B.C.E. Scientists speculate that this cooling may have been caused by the release of fresh water trapped behind ice on North America draining into the North Atlantic Ocean. The release altered vertical currents in the ocean, which exchange heat energy with the atmosphere. The warming resumed by 8500 B.C.E. By 5000 to 3000 B.C.E., average global temperatures reached their maximum level and were 2°F–4°F (1–2°C) warmer than they are today, a period known as the Climatic Optimum. During the Climatic Optimum, many of Earth's great ancient civilizations began and flourished. In Africa, the Nile River had three times its present volume, indicating a much larger tropical region.

3000 B.C.E. TO 750 B.C.E.

From 3000 to 2000 B.C.E., a cooling trend occurred. This cooling caused large drops in sea levels and the emergence of many islands (Bahamas) and coastal areas that are still above sea level today. A short warming trend took place from 2000 to 1500 B.C.E., followed once again by colder conditions. Colder temperatures from 1500 to 750 B.C.E. caused renewed ice growth in continental glaciers and alpine glaciers and a sea level drop of between 6 to 10 feet (2–3 m) below present-day levels.

750 B.C.E. TO 900 C.E.

The period from 750 B.C.E. to 900 C.E. saw warming up to 150 B.C.E. During the time of the Roman Empire (150 B.C.E. to 300 C.E.) a cooling began that lasted until about 900 C.E. At its height, the cooling caused the Nile River and the Black Sea to freeze.

900 C.E. TO 1200 C.E. (LITTLE CLIMATIC OPTIMUM)

During this warm period, the Vikings established settlements on Greenland and Iceland. The snow line in the Rocky Mountains was about 400 yards (370 m) above current levels. A period of cool and more extreme weather followed the Little Climatic Optimum. There are records of floods, great droughts, and extreme seasonal climate fluctuations up to the 1400s.

1550 C.E. TO 1850 C.E. (LITTLE ICE AGE)

From 1550 to 1850 C.E., global temperatures were at their coldest since the beginning of the Holocene. During the Little Ice Age, the average annual temperature of the Northern Hemisphere was about 2°F (1.0°C) lower than today.

1850 C.E. TO PRESENT

The period from1850 to the present is one of general warming.

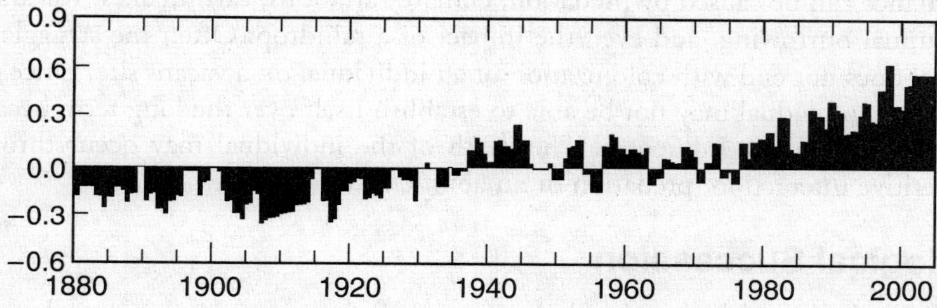

Figure 4.20 Changes in ocean and land temperatures (°C) from 1880 to 2000.

Species Movements

Many of the different types of organisms that inhabit Earth have the ability to move. This movement can be accomplished either passively or actively. Examples of active movement include walking, running, flying, or swimming. In passive movement, the organism uses some external force to cause transit. For example, plants can use wind for seed dispersal while oyster larvae can passively travel great distances by sea currents.

One common factor of why organisms move is to disperse to new habitats to reduce intraspecific competition. By finding new suitable habitats, individuals can increase the range of their species. A larger range makes the species better off in terms of evolution.

The geographic distributions of plant and animal species are never fixed over time. Geographic ranges of organisms shift, expand, and contract. These changes are the result of two contrasting processes: colonization and establishment, and localized extinction. Colonization and establishment takes place when populations expand into new areas. A number of processes can initiate this process, including disturbance and abiotic environmental change. Localized extinction results in the elimination of populations from all or part of their former range. It can be caused by biotic interactions or abiotic environmental change.

Plants have developed a number of different mechanisms for dispersing their offspring:

1. The use of specialized structures to aid the transport of an individual by wind
2. The employment of particular structures to transport the individual by moving water
3. The production of fruit-encased seeds that other organisms consume and disperse
4. Adhesion mechanisms
5. The physical ejection of seeds

Once dispersed, an individual can colonize a new site only if it is devoid of other organisms and if the necessary abiotic requirements and conditions exist for its survival. Sites within ecosystems become devoid of organisms through disturbance. A disturbance can be caused by predation, climate variations, earthquakes, volcanoes, fire, animal burrowing, and even the impact of a raindrop. Often the struggle for survival does not end with colonization of an individual on a vacant site. Once colonized, an individual may not be able to establish itself over the long term because of abiotic and biotic influences. The death of the individual may occur through competitive interaction, predation or an abiotic factor like fire.

Ecological Succession

Succession is the gradual and orderly process of ecosystem development brought about by changes in community composition and the production of a climax community characteristic of a particular geographic region. It describes the changes in an ecosystem through time and disturbance. Rates of succession are affected by several factors.

- Facilitation is when one species modifies an environment to the extent it meets the needs of another species.
- Inhibition is when one species modifies the environment to an extent that it is not suitable for another species.
- Tolerance is when species are not affected by the presence of other species.

Earlier successional species frequently called pioneer species are generalists. Pioneer plants have short reproductive times (annuals), and pioneer animals have small biomass and fast reproductive rates. Later successional species include larger perennial plants and animals with greater biomass, longer generational times, and higher parental care.

Types of Succession

Type	Description
Allogenic	Changes in the environmental conditions create conditions beneficial to new plant communities.
Primary	The colonization and establishment of pioneer plant species on bare ground. (Example: lichens on bare rocks)
Progressive	Communities become more complex over time by having a higher species diversity and greater biomass.
Retrogressive	The environment deteriorates and results in less biodiversity and less biomass.
Secondary	Begins in an area where the natural community has been disturbed but topsoil remains. (Example: forest fire)

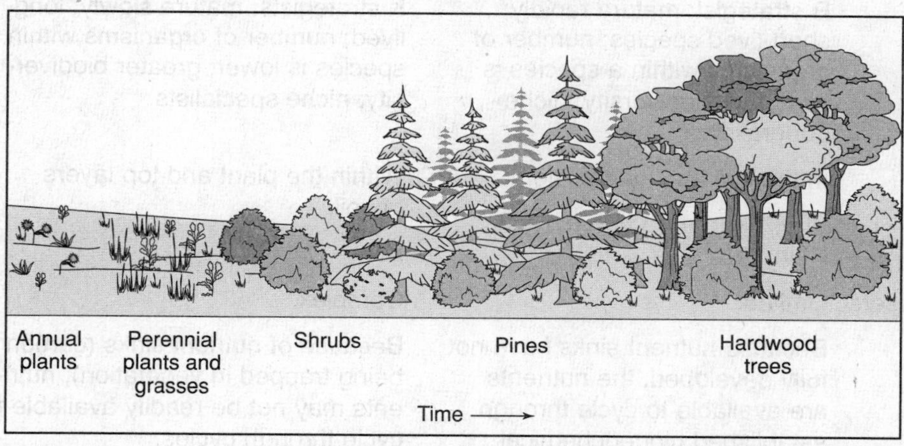

Figure 4.21 Stages of succession in a temperate deciduous forest. The time span from annual plants to hardwood trees is over 100 years.

The following table lists the characteristics of succession within plant communities:

Characteristics of Succession-Plant Communities

Characteristic	Early Successional State	Late Successional Stage
Biomass	Limited.	High in tropics and wetlands; limited in deserts.
Consumption of soil nutrients	Nutrients are quickly absorbed by simpler plants.	Since biomass is greater and more nutrients are contained within plant structures, nutrient cycling between the plant and soil tends to be slower.
Impact of macroenvironment	Early plants depend primarily on conditions created by macro-environmental changes (fires, floods, etc.).	These plant species appear only after pioneer plant communities have adequately prepared the soil.
Life span of seed	Long. Seeds may become dormant and able to withstand wide environmental fluctuations.	Short. Not able to withstand wide environmental fluctuations.
Life strategy	R strategist: mature rapidly; short-lived species; number of organisms within a species is high; low biodiversity; niche generalists.	K strategists: mature slowly; long-lived; number of organisms within a species is lower; greater biodiver-sity; niche specialists.
Location of nutrients	In the soil and in leaf litter.	Within the plant and top layers of soil.
Net primary productivity	High.	Low.
Nutrient cycling by decomposers	Limited.	Complex.
Nutrient cycling through biogeochemical cycles	Because nutrient sinks have not fully developed, the nutrients are available to cycle through established biogeochemical fairly rapidly.	Because of nutrient sinks (carbon being trapped in vegetation), nutri-ents may not be readily available to cycle through cycles.
Photosynthesis efficiency	Low.	High.
Plant structure complexity	Simple.	More complex.
Recovery rate of plants from environmental stress	Plants quickly and easily come back.	Recovery is slow.
Seed dispersal	Widespread	Limited in range.
Species diversity	Limited	High.
Stability of ecosystem	Since diversity is limited, eco-system is subject to instability.	Due to high diversity, ecosystem can withstand stress.

MULTIPLE-CHOICE QUESTIONS

For questions 1–3, choose from the following items:

(A) Tropical rain forest
(B) Temperate deciduous forest
(C) Savanna
(D) Taiga
(E) Tundra

1. Forests of cold climates of high latitudes and high altitudes.

2. Warm year-round; prolonged dry seasons; scattered trees.

3. Low biodiversity due to lots of shade, which limits food for herbivores. Major resource for timber.

4. The annual productivity of any ecosystem is greater than the annual increase in biomass of the herbivores in the ecosystem because

   (A) plants convert energy input into biomass more efficiently than animals
   (B) there are always more animals than plants in any ecosystem
   (C) plants have a greater longevity than animals
   (D) during each energy transformation, some energy is lost
   (E) animals convert energy input into biomass more efficiently than plants do

5. A Type I survivorship curve would apply to

   (A) humans
   (B) redwoods
   (C) bacteria
   (D) flies
   (E) tapeworms

6. All of the following are factors that increase population size EXCEPT

   (A) ability to adapt
   (B) specialized niche
   (C) few competitors
   (D) generalized niche
   (E) high birthrate

7. A specialist faces _____ competition for resources and has _____ ability to adapt to environmental changes. A generalist faces _____ competition for resources and has _____ ability to adapt to environmental changes.

   (A) less, greater, greater, less
   (B) greater, less, less, greater
   (C) less, less, greater, greater
   (D) greater, greater, less, less
   (E) none of the above

8. Whether a land area supports a deciduous forest or grassland depends primarily on

   (A) changes in temperature
   (B) latitude north or south of the equator
   (C) consistency of rainfall from year to year and the effect that it has on fires
   (D) changes in length of the growing season
   (E) none of the above

9. The main difference between primary and secondary succession is that

   (A) primary succession occurs in the year before secondary succession
   (B) primary succession occurs on barren, rocky areas and secondary succession does not
   (C) secondary succession ends in a climax species and primary succession ends in a pioneer species
   (D) secondary succession occurs on barren, rocky areas and primary succession does not
   (E) all of the above statements are true

10. The biggest threat to species is

   (A) low reproductive rates
   (B) disease
   (C) alien, invasive species
   (D) collecting, hunting, and poaching
   (E) loss of habitat

11. Darwin noted that the Patagonian hare was similar in appearance and had a niche similar to the European hare. However, the Patagonian hare is not a rabbit. It is a rodent related to the guinea pig. This example illustrates the principle known as

   (A) allopatric speciation
   (B) adaptive radiation
   (C) divergent evolution
   (D) coevolution
   (E) convergent evolution

For questions 12–15, choose from the following items:

(A) Adaptive radiation
(B) Isolation
(C) Natural selection
(D) Stable gene pool
(E) Convergent evolution

12. Members of the same species of moths are prevented from interbreeding because they live on opposites sides of a mountain range.

13. Darwin's finches are a good example of this biological principle.

14. Members of a large population mate at random.

15. In the evolutionary history of the horse, the early horse (*Eohippus*) was replaced by the modern one-toed horse.

16. Species that serve as early warnings of environmental damage are called

(A) keystone species
(B) native species
(C) specialist species
(D) indicator species
(E) generalist species

17. Which one of the following statements is false?

(A) When environmental conditions are changing rapidly, a generalist is usually better off than a specialist.
(B) The fundamental niche of a species is the full range of physical, chemical, and biological factors it could use if there were no competition.
(C) The competitive exclusion principle states that no two species with the same fundamental niche can indefinitely occupy the same habitat.
(D) Interspecific competition is competition between two members of the same species.
(E) Resource partitioning limits competition by two species using the same scarce resource at different times, in different ways, or in different places.

18. Which of the following best describes a nonanthropogenic secondary succession?

(A) Plants and other vegetation die gradually due to drought
(B) Wildflowers grow in an area that was previously destroyed by fire
(C) A farmer removes weeds using a herbicide
(D) Lichens and mosses secrete acids that allow other plants to grow
(E) None of the above

19. The location of where an organism lives would be best described as its

(A) niche
(B) habitat
(C) range
(D) biome
(E) ecosystem

20. Mount San Jacinto at almost 11,000 feet elevation is only a few miles from Palm Springs, California, located in the lower Mojave-Colorado Desert in southern California. Temperatures in Palm Springs have been recorded as high as 120°F. If you were to take the Palm Springs Aerial Tramway to the very top of Mount San Jacinto, what type of biome would you find?

(A) Desert
(B) Temperate forest
(C) Taiga
(D) Tundra
(E) Grassland

FREE-RESPONSE QUESTION

TIP

Be sure to explain technical terms when you use them in an answer. Dropping in terms like "bioaccumulation" without demonstrating an understanding of the term will not earn you any credit.

There are over 25,000 species of bees in the world today. Africanized honey bees (AHB), *Apis mellifera L. scutellata* (Lepetelier), sometimes referred to as killer bees, are the same species as European honeybees (EHB) but a different subspecies. Only through careful examination under a microscope, or through DNA tests, can AHBs be distinguished from EHBs. AHBs are called "Africanized" honeybees as a result of interbreeding experiments. The thought was that AHBs might be better suited for the tropical climates of South America than EHBs. In the 1950s, AHBs were inadvertently released in Brazil. AHBs began migrating (about 300 miles per year), arriving in southern Texas in 1990, Arizona and New Mexico in 1993, and southern California in 1994. Since 1994, migration has appeared to slow down significantly. Although the two subspecies are virtually indistinguishable, it is the behavior of the two subspecies that sets them apart. AHBs defend their colonies much more vigorously than do EHBs. When AHBs sting, more of them participate in extremely aggressive behavior. AHBs nest where EHBs do not, such as small, confined spaces near the ground such as flowerpots, abandoned tires, and cracks in building foundations.

(a) Discuss the influx of AHBs in the southwest United States in terms of the processes of natural selection. In what areas do the processes of natural selection apply to the colonization of AHBs, and in what areas do they not? What issues warrant further study?
(b) Provide possible explanations of why the expansion of AHBs may have stopped.
(c) What adaptations have honeybees in general developed to be able to survive?
(d) Provide some possible methods to control AHBs.
(e) What are possible economic implications of AHBs to agriculture?

MULTIPLE-CHOICE ANSWERS AND EXPLANATIONS

1. **(D)** Refer to the page of biome descriptions (pages 76–79).
2. **(C)** Refer to the page of biome descriptions (pages 76–79).
3. **(B)** Refer to the page of biome descriptions (pages 76–79).
4. **(D)** Less energy is available at each trophic level because energy is lost by organisms through cellular respiration and incomplete digestion of food sources.
5. **(A)** Type I survivorship curves are for species that have a high survival rate of the young, live out most of their expected life span, and die in old age. Humans are a good example of a species with a Type I survivorship curve. Type II survivorship curves are for species that have a relatively constant death rate throughout their life span. Death could be due to hunting or diseases. Examples of species exhibiting a Type II survivorship curve are corals, squirrels, honeybees, and many reptiles. Type III survivorship curves are found in species that have many young, most of which die very early in their life. Plants, oysters, and sea urchins are examples of species that have Type III survivorship curves.

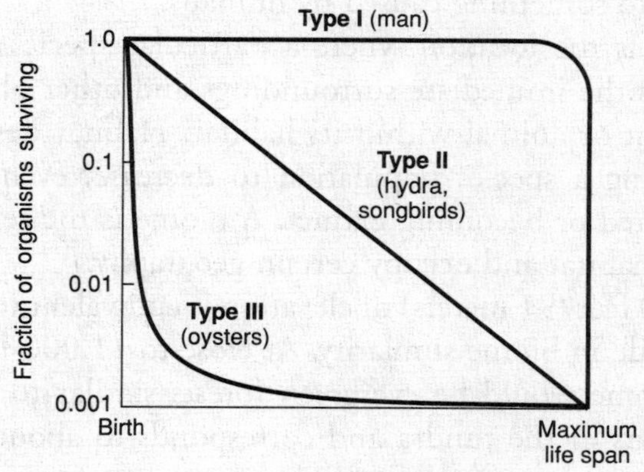

6. **(B)** Specialized niches are more susceptible to environmental changes and have a direct effect on the stability of populations (population size).
7. **(C)** Specialist species are adapted to a narrow range of habitats and conditions, while generalist species are able to live in a variety of habitats.

8. (C) The question of determining whether it is a deciduous forest or grassland is dependent on yearly patterns of rainfall since both biomes can exist over similar temperature ranges. Frequent fires are an important factor in determining grasslands.
9. (B) Primary succession occurs on bare rocks and starts with lichens. Secondary succession occurs in areas where there is intact topsoil.
10. (E) Scientists warn that human activities may be bringing about the sixth mass extinction of species in the world's history.
11. (E) Convergent evolution describes the process whereby organisms not closely related independently acquire similar characteristics while evolving in separate and sometimes varying environments.
12. (B) "Species" is defined as a group of organisms that look similar, have the ability to interbreed, and produce fertile offspring. Two forms of isolation that prevent interbreeding are geographic isolation and reproductive isolation.
13. (A) Adaptive radiation is the development of many species that are derived from a single, ancestral population.
14. (D) This question relates to the Hardy-Weinberg theory and the concept of a stable gene pool. It assumes a large population, random mating, no mutations, no migration between populations, and no selection.
15. (C) Natural selection is the process by which only the organisms that are best adapted to their environment tend to survive and transmit their genes to successive generations.
16. (D) An indicator species is a species whose presence, absence, or relative well-being in a given environment is indicative of the health of the ecosystem as a whole.
17. (D) Interspecific competition is competition among members of different species.
18. (B) Ecosystems undergo secondary succession following some artificial or natural disturbance such as a forest fire or farming. The question included the word "nonanthropogenic" which would rule out choice (C) since anthropogenic refers to something caused by humans.
19. (B) A habitat is the location where a particular species lives and grows. A microhabitat is the immediate surroundings and other physical factors of an individual plant or animal within its habitat. Habitat destruction is a major factor in causing a species population to decrease, eventually leading to it being endangered or becoming extinct. A biome is the set of flora and fauna that lives in a habitat and occupy certain geography.
20. (D) Every 300 feet (91 meters) of elevation is equivalent to 62 miles (100 km) north in latitude in biome similarity. At close to 11,000 feet (3.3 km) in elevation, the biome would be evergreen forests similar to those found in the southern regions of the tundra and corresponds to about 52° north latitude (southern Canada).

FREE-RESPONSE ANSWER

Part (a): 2 points maximum
Part (b): 2 points maximum
Part (c): 2 points maximum
Part (d): 2 points maximum
Part (e): 2 points maximum

Ten points possible for full credit.

Essay

(a) The range of genetic variation within a species's gene pool determines whether or not the species, not the individual, has the capacity to adapt and survive to changes in the environment. The colonization of the American southwest by AHBs differs from classical theories of natural selection in that AHBs and EHBs belong to the same species and therefore are presumably able to interbreed and produce fertile offspring. The two subspecies seem to differ only in behavioral patterns. As to whether the behavioral patterns of the two subspecies are so radically different as to cause a prezygotic reproductive barrier or not is worth investigation.

Another issue that differs from classical natural selection theory is that, in this case, the environment, for all practical purposes, is not changing. In effect, it is not a variable involved in survival rates. Both subspecies can and do survive in the environment. Additionally, the environment is not affecting disproportionate natality rates between the two subspecies based upon one subspecies having a survival advantage over the other.

In terms of following classic natural selection theory, what exists in this scenario is competition between the two subspecies for limited resources. Both subspecies are competing for space and food supplies. As to whether or not the more aggressive behavior of protecting the nest allows more AHBs to survive disproportionately, thereby increasing the ratio of AHBs to EHBs is another factor that warrants further study.

Notable points:

(a) Recognition of genetic variation as a determinant in the process of natural selection.

Recognition that subspecies are able to interbreed.

Recognition that behavior may be a reproductive barrier.

Recognition that the environment is remaining constant and not acting to select one subspecies over another.

Recognition that the two subspecies are competing for the same limited natural resources—a basic tenet of natural selection.

Recognition that more study is needed in the advantage of aggressive behavior and its effect on survivability of the colony.

(b) Specific adaptations of AHBs to tropical climates.

Temperature tolerances may be narrower in AHBs.

Geographic barriers.

Mountain ranges.

Low humidity in deserts.

Lack of sufficient food resources in deserts.

Variation in seasonal photoperiod.

Timing of forage availability (seasons) as opposed to no season in the tropics.

Parasites.

Pathogenic organisms.

(b) It appears that since 1994, the expansion of AHBs into the southwestern United States has slowed down. Possible reasons might include the following.

(1) AHBs are adapted to tropical climates (Africa and Brazil). Their ability to survive may be restricted to climatic zones that have fairly warm temperatures during most of the year (similar to the American Southwest). Extremely cold temperatures (common in the Midwest and northeast United States) may influence the viability of offspring and, consequently, their limited expansion into these areas.

(2) Geographic barriers may be another reason for a decline in the expansion of AHBs. Mountain ranges, dry deserts with low humidity, and the lack of adequate food resources may also be slowing expansion. Finally, variation in seasonal photoperiod, timing of forage availability, and/or the existence of parasites such as mites, pathogenic bacteria, or viruses may be having an effect on the newly introduced subspecies.

(c) Clustering behavior.

Predetermined roles.

Temperature regulation of hive to create stable environment for young.

(c) Three adaptations of honeybees which have been essential to their evolution and biology. First is clustering behavior, which is working as a social unit with specific predetermined behavioral patterns and duties of each class in the hierarchy. The second is the ability to cool the hive through the process of evaporation of water collected outside the hive, ensuring a stable internal temperature. The ability to ensure temperature stability within the nest allows honeybees to colonize a wide variety of environments, as opposed to bees that lack this trait and are therefore restricted to thermally stable environments (tropics). The third is the ability to communicate information about food sources, direction, and distance through dance behavior. The ability to communicate such a wide variety of complex information is unparalleled in the animal kingdom (except for humans).

Internal stability of nest temperature allows colonization into more diverse areas.

Ability to communicate.

(d) Since AHBs and EHBs are coexisting within the same areas, any attempt to control the AHBs through pesticides or other extermination techniques would probably have an effect on the EHBs as well. For the most part, EHBs and their less-aggressive behavior are more suited to coexist with humans for agricultural pollination purposes than are AHBs. Introduction of any bacterial agent, parasite, or other biological controlling mechanism would also probably affect the EHBs. One method that might work to control the domination of an area by AHBs would be to ensure that there are sufficient EHBs in the area. First, this would provide competition for resources between the two subspecies, thereby reducing the carrying capacity of the area. In turn, this would limit expansion of the AHB population. Second, ensuring sufficient EHBs in the area would provide sufficiently high enough EHBs in an area to ensure interbreeding of an AHB queen with EHB males. The goal of interbreeding would be to possibly dampen the highly aggressive behavioral characteristics of the AHBs. This would be an area in need of more research. Third, introducing large numbers of sterile, male AHBs into hives might result in smaller populations. Physically disrupting hives would only enhance recolonization.

(d) Recognition that wide use of pesticides would not be effective.

EHBs are more suitable for agriculture.

Bacterial agents or viruses would affect both subspecies.

Only way to control AHBs would be to ensure competitive population numbers of EHBs.

Reduce carrying capacity of environment.

Interbreeding and its role to diminish aggressive behavior.

Introduction of sterile males.

(e) Bees are necessary for agriculture.

People are closer today to agricultural areas and are thereby more impacted.

Public fear and hysteria of AHBs.

Availability of suitable sites for EHBs.

Public pressure.

Liability issues (insurance).

Legislative mandates.

Fewer beekeepers.

Cost of maintaining EHBs would increase.

As cost of bees goes up, so would cost of agricultural products.

(e) Agriculture as practiced today requires the use of honeybees for pollination. Honeybees are rented and placed into fields for this purpose by beekeepers. Urban expansion, human population increases, and sophisticated transportation systems have all made rural areas more accessible. AHBs strike terror in many people, and yet documented cases of deaths due to AHBs are rather low and exaggerated. Public pressure to eradicate "killer bees" will certainly involve economics. The cost of renting EHBs would certainly increase due to issues of suitability and availability of sites, public concerns, legislative mandates, and destroying EHBs in an attempt to control AHBs. As costs of renting EHBs increase, profit margins could decrease, forcing many beekeepers out of business. The result would be fewer bees for the needs of agriculture. This, in turn, could result in less agricultural production and, consequently, through supply and demand, higher prices to the consumer.

NOTE TO STUDENT . . .

It is your turn now. Take the same question above, and write your own answer. Do not worry if you do not get all of the points above. Your chance of getting a free-response question on killer bees is practically zero. The point of you answering this same question in your own words is to practice your writing and organizational skills from initial brainstorming, organization, topic sentences, paragraph development, evidence to support your statements, and finally writing a conclusion. You may wish to refer to the rubric initially and use that as your guide as you write. Have your teacher review your essay for style and suggestions about ways for you to improve.

Natural Biogeochemical Cycles

CHAPTER 5

In all things there is a law of cycles.

—Publius Cornelius Tacitus
(Roman historian)

CARBON CYCLE

Carbon is the basic building block of life and the fundamental element found in carbohydrates, fats, proteins, and nucleic acids (DNA and RNA). Carbon is exchanged among the biosphere, geosphere, hydrosphere, and atmosphere. Although carbon is found in rocks, it is a minor component when compared with the mass of either oxygen or silicon atoms in rocks. Carbon is found in carbon dioxide (CO_2), which makes up less than 1% of the atmosphere.

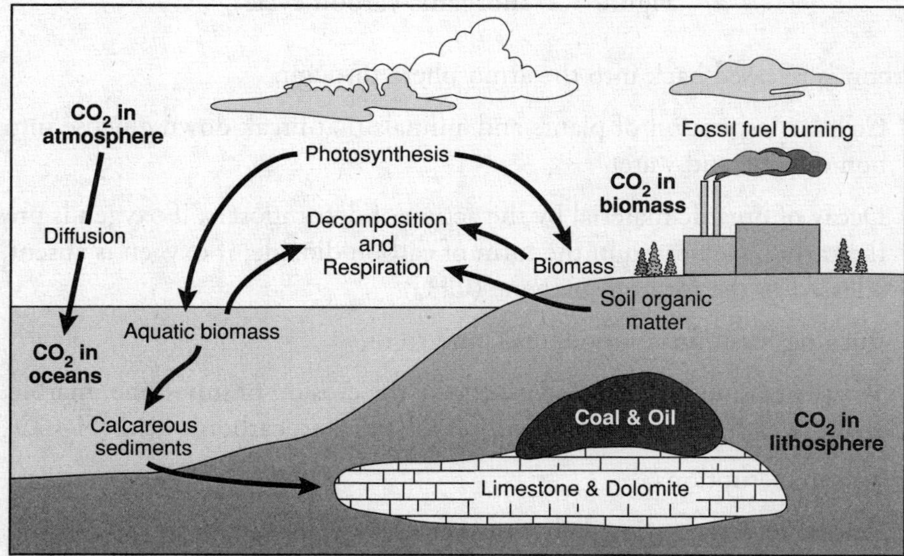

Figure 5.1 The carbon cycle viewed as chemical processes.

A portion of of atmospheric carbon—15%—is removed through photosynthesis, where the carbon is incorporated into plant structures and compounds; and dissolving into cooler ocean waters. The oceans are gaining approximately 2 gigatons (4×10^{12} kg) of carbon each year.

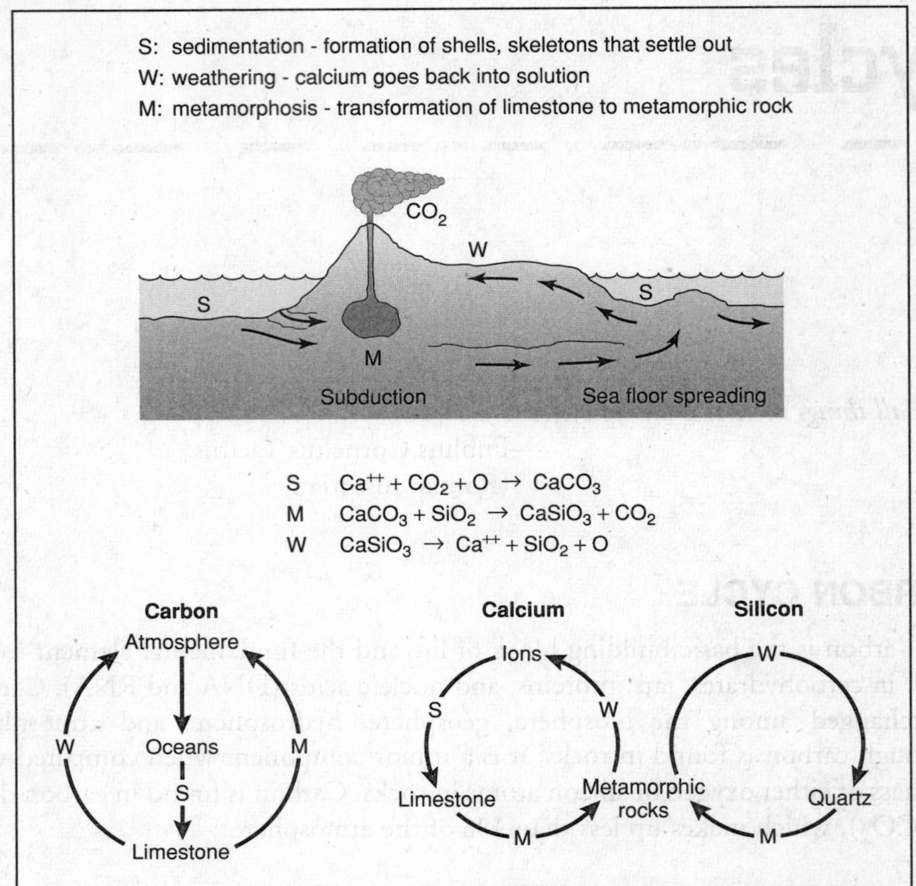

Figure 5.2 Inorganic carbon cycle

Carbon is released back into the atmosphere through:

1. Cellular respiration of plants and animals that break down glucose into carbon dioxide and water
2. Decay of organic material by the action of decomposers; if oxygen is present, the carbon is released in the form of carbon dioxide; if oxygen is absent, it is released in the form of methane (CH_4)
3. Burning fossil fuels, wood, coal, and so on
4. Weatherization of rocks and especially the erosion of limestone, marble, and chalk, which break down to carbon dioxide and carbonic acid (H_2CO_3)
5. Volcanic eruptions
6. Release of carbon dioxide by warmer ocean waters.

The oceans contain the largest amount of dissolved carbon dioxide. However, most of it is not involved with rapid exchanges with the atmosphere. Removing carbon dioxide from water raises the pH making the water more basic. The oceans are gaining a net 2 gigatons of carbon each year (1 Gt = 2×10^{12} kg), the lithosphere 7 Gt, and the atmosphere 5 Gt.

Carbon Sink

Carbon Sink	Amount (Billions of Metric Tons)
Marine sediments and sedimentary rocks	~ 75,000,000
Ocean	~ 40,000
Fossil fuel deposits	~ 4000
Soil organic matter	~ 1500
Atmosphere	578 (in 1700 C.E.) to 766 (in 2000 C.E.)
Terrestrial plants	~ 580

NITROGEN CYCLE

Nitrogen is an essential element needed to make amino acids, proteins, and nucleic acids. Nitrogen makes up 78% of the atmosphere. Other nitrogen stores include organic matter in the soil and the ocean (1 million times more nitrogen is found in the atmosphere than is contained in either land or ocean waters).

The nitrogen cycle involves:

- Nitrogen fixation
- Nitrification
- Assimilation
- Ammonification
- Dentrification

TIP

For the test, be sure you can describe spatial patterns of environmental change and the stepwise sequence of ecological events that accompany such changes.

Nitrogen Fixation

Nitrogen fixation is the conversion of atmospheric nitrogen (N_2) to ammonia (NH_3) or nitrate (NO_3^-) ions. Nitrate is the product of high-energy fixation by lightning, cosmic radiation, or meteorite trails. In high-energy fixation, atmospheric nitrogen and oxygen combine to form nitrates, which are carried to Earth's surface in rainfall as nitric acid (HNO_3). High-energy fixation accounts for about 10% of the nitrate entering the nitrogen cycle.

In contrast, biological fixation accounts for 90% of the fixed nitrogen in the cycle. In biological fixation, molecular nitrogen (N_2) is split into two free nitrogen atoms ($N_2 \rightarrow N + N$). The nitrogen atoms combine with hydrogen to yield ammonia (NH_3).

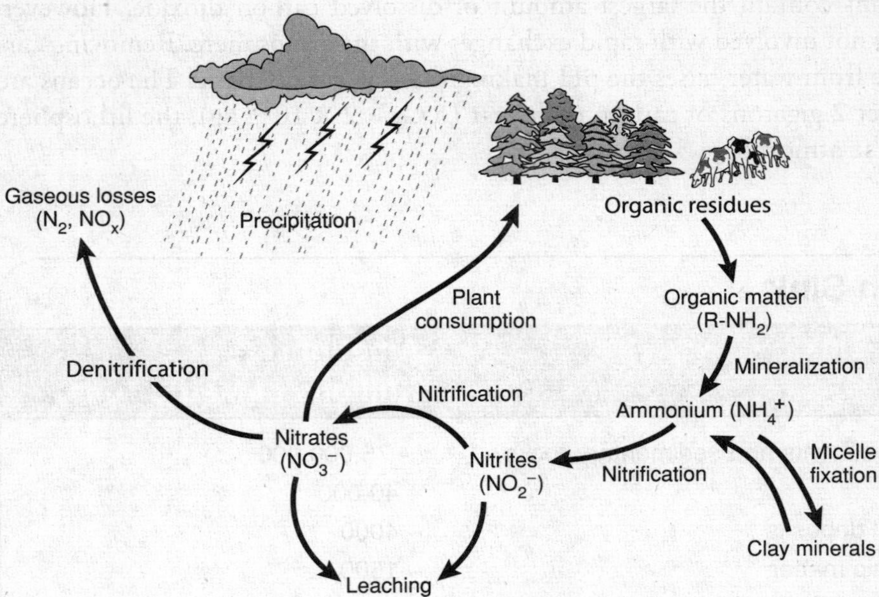

Figure 5.3 The nitrogen cycle

The fixation process is accomplished by a series of different microorganisms. The symbiotic bacteria *Rhizobium* is associated with the roots of legumes. To a lesser extent, some nonleguminous plants also exhibit symbiotic relationships with bacteria. Some free-living aerobic bacteria, such as *Azobacter* and *Clostridium*, freely fix nitrogen in the soil. Finally, blue-green algae (cyanobacteria) such as *Nostoc* and *Calothrix* can also fix nitrogen both in the soil and in water, yielding ammonia as the stable end product.

Nitrification

Nitrification is the process in which ammonia is oxidized to nitrite (NO_2^-) and nitrate (NO_3^-), the forms most usable by plants. Two groups of microorganisms are involved in nitrification. *Nitrosomonas* oxidize ammonia to nitrite and water. *Nitrobacter* oxidize the nitrite to nitrate.

Assimilation

Nitrates are the form of nitrogen most commonly assimilated by plants through their root hairs. Nitrogen is used by plants to synthesize amino acids, oils, and nucleic acids. Rains and extensive irrigation can leach soluble nitrates and nitrites into groundwater, which, in high amounts, can interfere with blood-oxygen levels in human infants. Soluble nitrates that run off the land and enter aquatic and wetland habitats result in cultural eutrophication and the destruction of these habitats. Animals assimilate nitrogen-based compounds by consuming plants and other organisms that consume plants.

Ammonification

Ammonification is a one-way reaction. In it, organisms break down amino acids and produce ammonia (NH_3).

Denitrification

Denitrification is the process in which nitrates are reduced to gaseous nitrogen. This process is used by facultative anaerobes. These organisms flourish in an aerobic environment but are also capable of breaking down oxygen-containing compounds (NO_3^-) to obtain oxygen in anaerobic environments. Examples include fungi and the bacteria *Pseudomonas.*

PHOSPHORUS CYCLE

Phosphorus is essential for the production of nucleotides, production of ATP, fats in cell membranes, bones, teeth, and shells. Phosphorus is not found in the atmosphere but, rather, in sedimentary rocks and does not depend on the action of bacteria. Generally, phosphorus is found in the form of the phosphate ion (PO_4^{3-}) or the hydrogen phosphate ion (HPO_4^{2-}). Phosphorus is slowly released from terrestrial rocks by weathering and the action of acid rain. It then dissolves into the soil and is taken up by plants. It is often a limiting factor for soils due to its low concentration and solubility. Phosphorus is a key element in fertilizer. A fertilizer labeled 6-24-26 contains 6% nitrogen, 24% phosphorus, and 26% potassium.

Humans have impacted the phosphorus cycle in several ways: First, humans have mined large quantities of rocks containing phosphorus for inorganic fertilizers and detergents. Second, clear-cutting tropical habitats for agriculture decreases the amount of available phosphorus as it is contained in the vegetation. Third, humans allow runoff from feedlots, from fertilizers, and from the discharge of municipal sewage plants. The runoff collects in lakes, streams, and ponds. It causes an increase in the growth of cyanobacteria (blue-green bacteria), green algae, and aquatic plants. In turn, this growth results in decreased oxygen content in the water, which then kills other aquatic organism in the food web. Fourth, humans apply phosphorus-rich guano and other phosphate-containing fertilizers to fields.

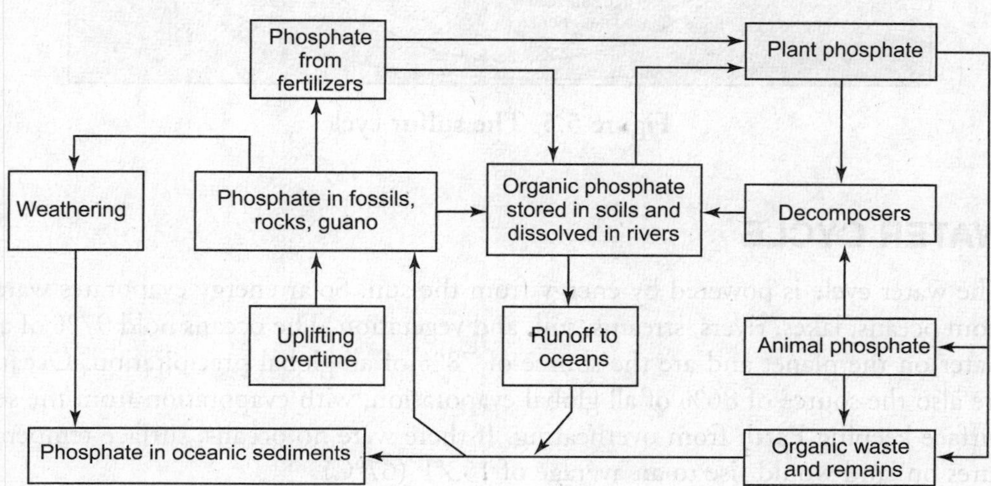

Figure 5.4 The phosphorus cycle

SULFUR CYCLE

Most sulfur is found in underground rocks and deep oceanic deposits. The natural release of sulfur into the atmosphere comes from the weathering of rock and gases released from seafloor vents and volcanic eruptions. Sulfur is released mostly in the form of hydrogen sulfide (H_2S) and sulfur dioxide (SO_2) from these volcanic eruptions. This sulfur dioxide is converted to sulfur trioxide (SO_3) and eventually to tiny droplets of sulfuric acid (H_2SO_4). Sulfuric acid mixes with rain to fall back to Earth and is known as acid rain, acid precipitation, or acid deposition. Sulfate (SO_4^{2-}) salt particles enter the atmosphere from sea sprays, dust storms, and forest fires. These sulfate ions are eventually absorbed by plants, which then incorporate them into proteins. Certain marine algae produce gaseous dimethyl sulfide (DMS), which in the atmosphere also forms tiny condensation nuclei for rain. Humans contribute to adding sulfur to the cycle through refining and burning fossil fuels and by converting (smelting) sulfur-containing metallic mineral ores into free metals such as copper, lead, and zinc.

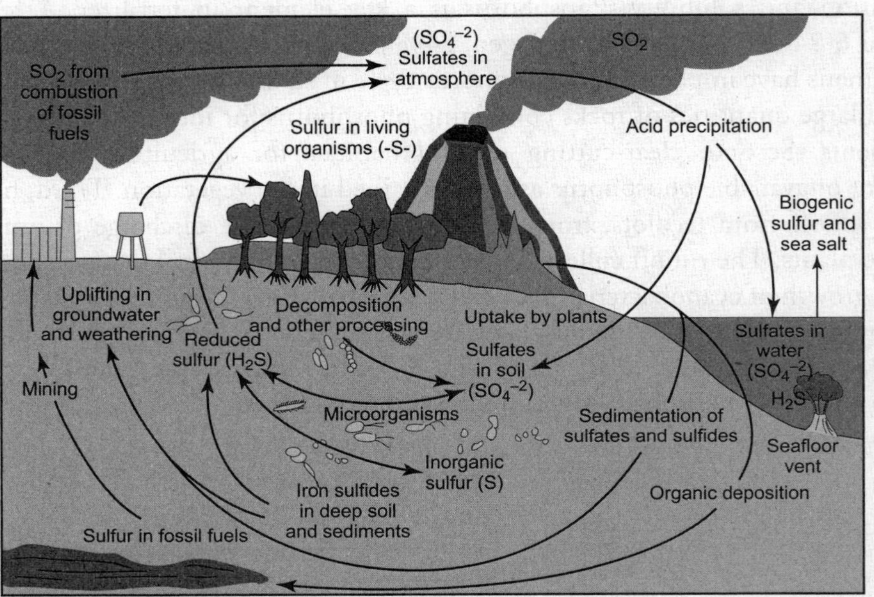

Figure 5.5 The sulfur cycle

WATER CYCLE

The water cycle is powered by energy from the sun. Solar energy evaporates water from oceans, lakes, rivers, streams, soil, and vegetation. The oceans hold 97% of all water on the planet and are the source of 78% of all global precipitation. Oceans are also the source of 86% of all global evaporation, with evaporation from the sea surface keeping Earth from overheating. If there were no oceans, surface temperatures on land would rise to an average of 153°F (67°C).

The water cycle is in a state of dynamic equilibrium according to the law of conservation of mass. The rate of evaporation equals the rate of precipitation. Warm air holds more water vapor than cold air. Processes involved in the water cycle include evaporation, evapotranspiration, condensation, infiltration, runoff, and precipitation.

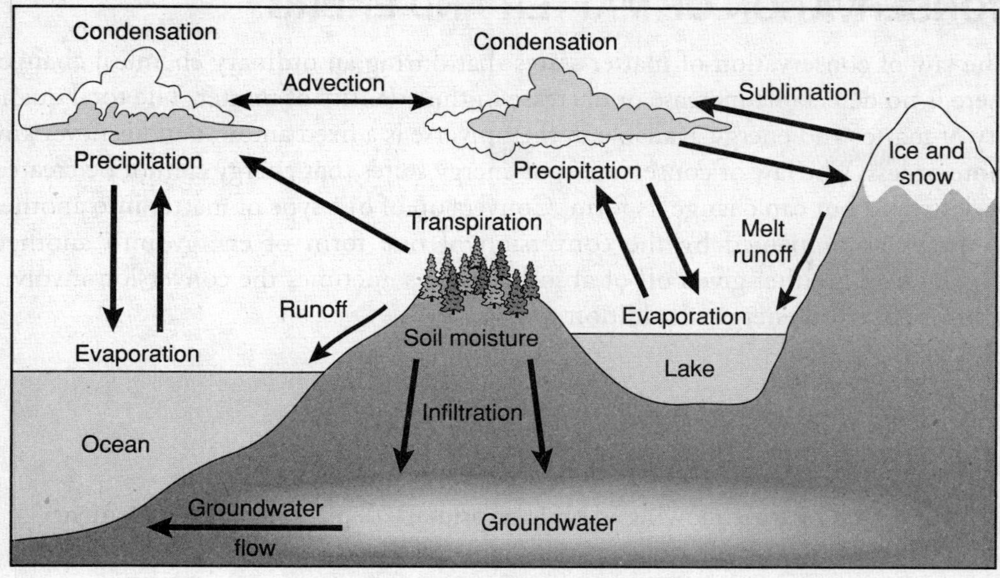

Figure 5.6 The water cycle

Human Impact on Water Cycle

Human Activity	Impact on Water Cycle
Withdrawing water from lakes, aquifers, and rivers	Groundwater depletion and saltwater intrusion
Clearing of land for agriculture and urbanization	Increased runoff Decreased infiltration Increased flood risks Accelerated soil erosion
Agriculture	Runoff contains nitrates, phosphates, ammonia, etc.
Destruction of wetlands	Disturbing natural processes that purify water
Pollution of water sources	Increased occurrences of infectious agents such as cholera, dysentery, etc.
Sewage runoff, feedlot runoff	Cultural eutrophication
Building power plants	Increased thermal pollution

CONSERVATION OF MATTER AND ENERGY

The law of conservation of matter states that during an ordinary chemical change, there is no detectable increase or decrease in the quantity of matter. The total quantity of matter and energy available in the universe is a fixed amount and is never any more or less. The law of conservation of energy states that energy cannot be created or destroyed but can change its form. Conversion of one type of matter into another is always accompanied by the conversion of one form of energy into another. Usually heat is either given off or absorbed, but sometimes the conversion involves light or electrical energy in addition to heat.

MULTIPLE-CHOICE QUESTIONS

1. Which of the following is NOT a primary depository for the element listed?

   (A) Carbon—coal
   (B) Nitrogen—nitrogen gas in the atmosphere
   (C) Phosphorus—marble and limestone
   (D) Sulfur—deep ocean deposits
   (E) All are correct

2. Burning fossil fuels coupled with deforestation increases the amount of __________ in the atmosphere.

   (A) NO_2
   (B) CO_2
   (C) SO_2
   (D) O_3
   (E) all of the above are correct

3. In the nitrogen fixation cycle, cyanobacteria in the soil and water and *Rhizobium* bacteria in root systems are responsible for converting

   (A) organic material to ammonia and ammonium ions
   (B) ammonia, ammonium ions, and nitrate ions to DNA, amino acids, and proteins
   (C) ammonia and nitrite ions to nitrate ions
   (D) nitrogen and hydrogen gas to ammonia
   (E) ammonia to nitrite and nitrate ions

4. The cycle listed that has the most immediate effect on acid precipitation would be the

   (A) carbon cycle
   (B) sulfur cycle
   (C) water cycle
   (D) phosphorous cycle
   (E) rock cycle

5. Nitrogen is assimilated in plants in what form?

   (A) Nitrite, NO_2^-
   (B) Ammonia, NH_3
   (C) Ammonium, NH_4^+
   (D) Nitrate, NO_3^-
   (E) Choices B, C, and D

6. Plants assimilate sulfur primarily in what form?

   (A) Sulfates, SO_4^{2-}
   (B) Sulfites, SO_3^{2-}
   (C) Hydrogen sulfide, H_2S
   (D) Sulfur dioxide, SO_2
   (E) Elemental sulfur, S

7. Humans increase sulfur into the atmosphere and thereby increase acid deposition by all of the following activities EXCEPT

   (A) industrial processing
   (B) processing (smelting) ores to produce metals
   (C) burning coal
   (D) refining petroleum
   (E) clear-cutting

8. Phosphorus is being added to the environment by all of the following activities EXCEPT

   (A) runoff from feedlots
   (B) slashing and burning in tropical areas
   (C) stream runoff
   (D) burning coal and fossil fuels
   (E) mining to produce inorganic fertilizer

9. Carbon dioxide is a reactant in

   (A) photosynthesis
   (B) cellular respiration
   (C) the Haber-Bosch process
   (D) nitrogen fixation
   (E) none of the above

10. Human activity adds significant amounts of carbon dioxide to the atmosphere by all of the following EXCEPT
    (A) brush clearing
    (B) burning wood
    (C) burning fossil fuels
    (D) clear-cutting
    (E) agricultural runoff

11. Clearing of land for either habitation or agriculture does all of the following EXCEPT
    (A) increases runoff
    (B) increases flood risks
    (C) increases potential for landslides
    (D) increases infiltration
    (E) accelerates soil erosion

12. All of the following have an impact on the nitrogen cycle EXCEPT
    (A) the application of inorganic fertilizers applied to the soil
    (B) the action of aerobic bacteria acting on livestock wastes
    (C) the overplanting of nitrogen-rich crops
    (D) the discharge of municipal sewage
    (E) the burning of fossil fuels

13. Which of the following is a macronutrient essential for the formation of proteins?
    (A) Sulfur
    (B) Nitrogen
    (C) Iron
    (D) Cobalt
    (E) Molybdenum

14. Which of the following bacteria are able to convert soil nitrites to nitrates?
    (A) *Nitrosomonas*
    (B) *Nitrobacter*
    (C) *Rhizobium*
    (D) *Penicillium*
    (E) *Clostridium*

15. Which of the following cycles would be considered sedimentary?
    (A) Carbon cycle
    (B) Water cycle
    (C) Phosphorus cycle
    (D) Nitrogen cycle
    (E) Sulfur cycle

16. The cycle that is most common to all other cycles is the
   (A) nitrogen cycle
   (B) carbon cycle
   (C) hydrologic cycle
   (D) life cycle
   (E) the rock cycle

17. The energy that drives the hydrologic cycle comes primarily from
   (A) trade winds
   (B) solar energy and gravity
   (C) Earth's rotation on its axis
   (D) ocean currents and wind patterns
   (E) solar radiation

18. Which one of the following processes is working against gravity?
   (A) Precipitation
   (B) Percolation
   (C) Runoff
   (D) Transpiration
   (E) Infiltration

19. Ammonium ions (NH_4^+) are converted to nitrate (NO_3^-) and nitrite ions (NO_2^-) through which process?
   (A) Assimilation
   (B) Denitrification
   (C) Nitrogen fixation
   (D) Nitrification
   (E) Ammonification

20. Which form of nitrogen is most usable by plants?
   (A) Nitrate
   (B) Nitrite
   (C) Nitrogen gas
   (D) Ammonia
   (E) Atomic nitrogen

FREE-RESPONSE QUESTION

By: Sarah E. Utley
College Board APES Reader; Princeton, New Jersey
Environmental Content Specialist
Center for Digital Innovation
AP Environmental Science Division
University of California at Los Angeles

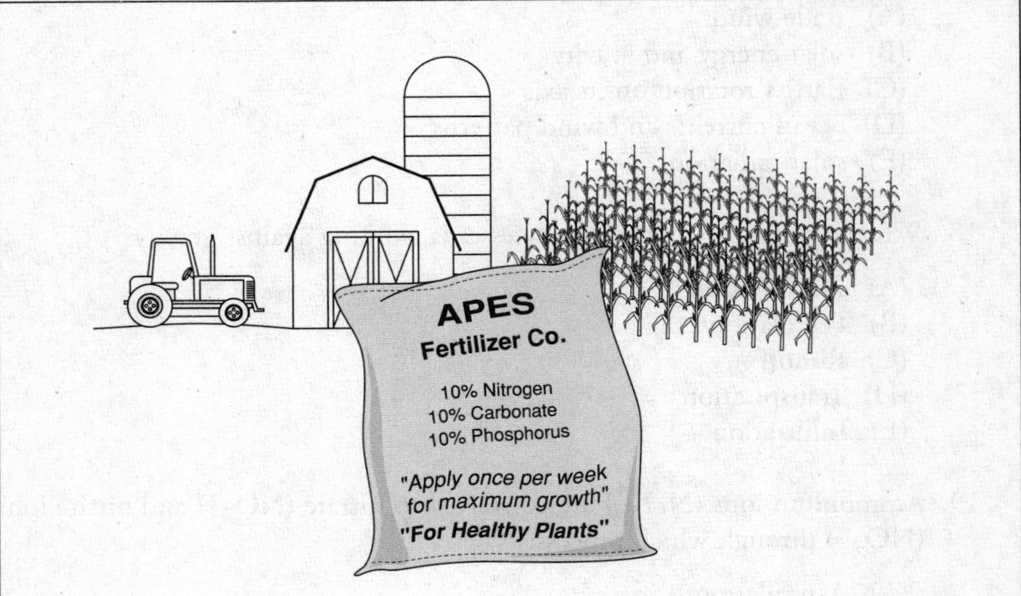

The APES Fertilizer Company manufactures a soil additive that contains 10% (w/v) of nitrogen, carbonate, and phosphorus. Use this information to answer the following questions.

(a) Choose either carbon or nitrogen and clearly explain the biogeochemical cycle of that nutrient. Include all of the important steps in the cycle and indicate movement from the atmosphere, water, and soil where necessary.

(b) Agricultural runoff from overapplication of fertilizers is a major source of water pollution. Choose either nitrogen or phosphorus and fully explain the cause and effect of this excess on the aquatic ecosystem.

(c) Rather than rely on the application of fertilizers, current agricultural research is encouraging the practice of more conservation-minded farming. Name and explain one such environmental practice that can decrease dependence on human-applied nutrients without causing a significant decrease in crop yield.

MULTIPLE-CHOICE ANSWERS AND EXPLANATIONS

1. **(C)** The primary sinks for phosphorus are ocean sediments and certain islands rich in guano.
2. **(B)** Burning fossil fuels releases sulfur oxides (SO_x), carbon oxides (carbon dioxide on complete combustion, or carbon monoxide on incomplete combustion) and nitrogen oxides (NO_x). Ozone is not produced by burning fossil fuels. Deforestation releases carbon dioxide. Since the question said "coupled," the gas that is common to both processes is carbon dioxide.
3. **(D)** This is the first step in the nitrogen cycle and is called nitrogen fixation.
4. **(B)** Sulfur dioxide produced by industry enters the atmosphere and returns to Earth as sulfuric acid.
5. **(E)** The nitrite ion is toxic to plants. In the nitrogen cycle, during assimilation, plant roots absorb nitrate.
6. **(A)** Hydrogen sulfide and sulfur dioxide are toxic to living organisms. Some sulfate compounds are soluble in water, which allows the sulfate ion to be able to be absorbed by plants. Elemental sulfur is not soluble in water and therefore cannot be absorbed.
7. **(E)** Clear-cutting produces carbon dioxide, not sulfur dioxide.
8. **(D)** Animal manure and guano are rich in phosphate. In the tropics, most of the nutrients are contained within the vegetation. Little is being retained in the soil since much of it leaches due to high rainfall. Phosphorus would therefore be released back into the environment by cutting down vegetation and then burning it—thereby releasing it and being subjected to runoff. Mining phosphates for fertilizer and industrial products takes phosphorus out of sinks and puts it into the environment. Burning coal and petroleum does not add appreciable amounts of phosphorus to the environment.
9. **(A)** The reaction for photosynthesis is $6CO_2 + 6H_2O + \text{sunlight} \rightarrow C_6H_{12}O_6 + 6O_2$. Cellular respiration is the reverse of photosynthesis.
10. **(E)** Agricultural runoff, primarily from fertilizers and feedlots, adds nitrates and phosphates to streams and results in cultural eutrophication. All other choices involve combustion, which produces carbon dioxide.
11. **(D)** Infiltration is the movement of water into the soil. Removing vegetation decreases infiltration by not allowing water to percolate through the soil slowly. Since runoff is increased, the potential for flooding increases. As the soil structure loses its integrity, the chances of landslides increase. Runoff also carries with it topsoil and nutrients, thus accelerating soil erosion.
12. **(B)** The bacteria that normally work to decompose livestock wastes are anaerobic, operate only in environments with little or no oxygen, and produce nitrous oxide (N_2O). Digesters can be constructed to reduce livestock wastes to methane gas (CH_4), which can then be burned as a fuel.
13. **(B)** The growth of all organisms depends on the availability of macro- and micronutrients, and none is more important than nitrogen, which is required in large amounts as an essential component of proteins, nucleic acids, and other cellular constituents.
14. **(B)** *Nitrosomonas* bacteria oxidize ammonia to nitrite: $NH_3 \rightarrow NO_2^-$. *Nitrobacter* then oxidize nitrite to nitrate: $NO_2^- \rightarrow NO_3^-$. Denitrifying bacteria anaerobically reduce nitrate to nitrogen gas: $NO_3^- \rightarrow N_2$.

15. (C) The largest reservoir of phosphorus is sedimentary rock. The phosphorus cycle originates with the introduction of phosphate (PO_4^{3-}) into soils from the weathering of rocks. Phosphate enters living ecosystems when plants take up phosphate ions from the soil.
16. (C) Water plays a part somewhere in every cycle.
17. (B) Solar energy allows water to change phase (water to gas, liquid to ice, and so on) and gravity causes rain to fall and rivers to flow.
18. (D) Transpiration is the process by which water moves upward (against gravity) from the soil through the roots and out through the leaves of plants.
19. (D) Nitrification is the biological oxidation of ammonia into nitrite ions followed by the oxidation of nitrites into nitrates. The oxidation of ammonia into nitrites is done by bacteria belonging to the genera *Nitrosomonas* and *Nitrosococcus*. The oxidation of nitrite ions into nitrate ions is done by bacteria belonging to the genus *Nitrobacter*.
20. (A) Nitrite and ammonia are more toxic to plants than nitrate. Plants cannot use nitrogen from the atmosphere directly.

FREE-RESPONSE ANSWER

Part (a): Maximum 5 points
Part (b): Maximum 3 points
Part (c): Maximum 2 points

Total: maximum 10 points

Notable points:

Movement through Earth's life support systems (atmosphere, lithosphere, biomass).

Introduction to nitrogen cycle.

Definition of nitrogen fixation.

Sources of fixation.

(a) Earth is a closed system. A finite amount of matter (such as life-supporting nutrients) is found in the biosphere. These nutrients cycle through Earth's varied support systems in different forms. Nitrogen, an essential nutrient for life, is found primarily in the atmosphere (making up 78% by volume). However, it also cycles through the lithosphere and the biomass of organisms. Unfortunately, the producers cannot absorb nitrogen in its most commonly found form, N_2. In order for this gaseous element to be used by living organisms, it has to be converted in a series of steps called the nitrogen cycle.

The first step in the cycle is called nitrogen fixation. This is the process by which atmospheric nitrogen is converted to ammonia (NH_3). There are some natural sources of nitrogen fixing such as lightning and volcanoes. However, most of the biological fixing is done by nitrogen-fixing bacteria that live in the root nodules of certain plants called legumes, such as peas and clover. The nitrogen (N_2) is converted to ammonia (NH_3) using the enzyme nitrogenase.

Next step—nitrification.

After the nitrogen is fixed into ammonia, it goes through a process called nitrification. This is a two-step process where special soil bacteria first convert the ammonia to nitrite (NO_2^-) and then to nitrate (NO_3^-), the form that plants can most easily absorb through their roots. Plants can also take up a limited amount of ammonia. The process by which the roots absorb the nitrate and ammonia is called assimilation. During this step, the plants incorporate nitrogen into their tissues. This absorbed nitrogen is used by plants to form nitrogen-containing molecules such as proteins and nucleic acids.

Definition of nitrification (step 1).

Definition of nitrification (step 2).

Next step in cycle—assimilation.

Since animals cannot directly absorb nitrogen, they assimilate it into their body tissues when they eat plants or other animals. After animals absorb the nitrogen into their tissues, it is excreted as urea or uric acid. Uric acid is the end product of nitrogen metabolism in birds and reptiles. In humans and most other higher animals, the main product of nitrogen detoxification is urea. In fish, bacteria, and protozoa, it is ammonia. In those animals that do produce uric acid in high quantities, it is excreted in the feces as a dry mass. Although this compound is produced through a complex and energetically costly metabolic pathway (in comparison with the production of other nitrogenous wastes), its elimination minimizes water loss and is therefore commonly found in the excretions of animals that live in desert habitats.

Next step in cycle—ammonification.

Next step in cycle—denitrification.

Final step in cycle—conversion into gaseous N_2.

In addition, when the organism dies, the nitrogen in the tissues is released back into the cycle. There nitrogen-containing substances are decomposed into ammonia in a process called ammonification. Then during denitrification, other bacteria convert the ammonia found in the soil into nitrate and nitrite. The nitrite and nitrate are further converted into gaseous nitrogen, which is then available to be refixed and used by other organisms.

Cause of the aquatic pollution.

Initial effect of excessive fertilizer.

Effect after initial algal bloom.

Result of overapplication of fertilizer.

Final result of excessive application.

(b) When fertilizers such as phosphorus are overapplied, rain or irrigation causes the excess fertilizer to run off into local waterways. Since phosphorus is often a limiting factor in aquatic ecosystems, this excess amount results in proliferation of algae and is known as an algal bloom. When these algal populations die, they are decomposed by large amounts of bacteria that consume much of the dissolved oxygen in the water. This, in turn, causes aquatic organisms to die from lack of oxygen. The result is a sterile lake or stream that lacks the nutrients necessary to support life.

Name of the conservation method.

Initial description of the method.

Further explanation of the conservation-minded agricultural method.

Reason why the crop yield will not decrease.

Further explanation of the yield maintenance.

(c) Conservation-minded agriculture can lead to a decreased dependence on human-applied nutrients. One of the most common, environmentally conscious agricultural practices is conservation or no-till farming. In this practice, the fields are not plowed under at the end of the growing season. The old stalks and vegetation are left on the fields over the winter. Then, in the spring, special machinery is used to plant the new crop on the untilled field. By leaving the previous season's biomass on the untilled land, the natural decomposition process can occur. The nutrients that would normally have been removed and lost are allowed to decompose back into the soil. This leads to a decreased need for artificial fertilizers without a loss of crop yield.

UNIT III: POPULATION (10–15%)

Areas on Which You Will Be Tested

A. **Population Biology Concepts**—population ecology, carrying capacity, reproductive strategies, and survivorship.

B. **Human Population**—

1. Human population dynamics—historical population sizes, distribution, fertility rates, growth rates and doubling times, demographic transition, and age-structure diagrams.
2. Population size—strategies for sustainability, case studies, and national policies.
3. Impacts of population growth—hunger, disease, economic effects, resource use, habitat destruction.

Populations

CHAPTER 6

Bring out every kind of living creature that is with you—the birds, the animals, and all the creatures that move along the ground—so they can multiply on the Earth and be fruitful and increase in number upon it.

—Genesis 8:16–18

POPULATION BIOLOGY CONCEPTS

Several topics must be considered when discussing population biology:

- Population ecology
- Carrying capacity (K)
- Reproductive strategies
- Survivorship

Population Ecology

Population ecology studies the dynamics of species' populations and how these populations interact with the environment. Population ecology plays an important role in the development of the field of conservation biology, especially in the development of population viability analysis (PVA). PVA allows ecologists to predict the long-term probability of a species persisting in a given habitat. The following table lists those factors that affect population viability.

Most organisms live in groups (flocks, schools, nests, and so on). Living in groups provides several advantages: increased protection from predators, increased chances for mating, and division of labor.

Factors That Affect Population Viability

Increase (+) Viability	Decrease (–) Viability
Favorable environmental conditions (light, temperature, and nutrients)	Unfavorable environmental conditions (insufficient light, temperature extremes, and/or poor supply of nutrients)
High natality	Low natality
Generalized niche	Specialized niche
Satisfactory habitat	Habitat not satisfactory or has been seriously impacted
Few competitors	Too many competitors
Suitable predatory defense mechanism(s)	Unsuitable predatory defense mechanism(s)
Adequate resistance to diseases and parasites	Little or no suitable defense mechanisms against diseases or parasites
Able to migrate	Unable to migrate
Flexible—able to adapt	Inflexible—unable to adapt
Sufficient food supply	Deficient food supply

Carrying Capacity (K)

Carrying capacity refers to the number of organisms that can be supported in a given area sustainably. It varies from species to species and is subject to changes over time. As an environment degrades, the carrying capacity decreases. Factors that keep population sizes in balance with the carrying capacity are called regulating factors. They include food availability, space, oxygen content in aquatic ecosystems, nutrient levels in soil profiles, and the amount of sunlight. Below the carrying capacity, populations tend to increase in size. Populations cannot live indefinitely above the carrying capacity; eventually the population will crash.

A population's biotic potential is the maximum rate at which a population can grow. Its biotic potential, given optimal conditions (food, water, and space), occurs when resources are unlimited. Factors that influence biotic potential include age at reproduction, frequency of reproduction, number of offspring produced, reproductive life span, and average death rate under ideal conditions. If a population in a community is left unchecked, the maximum population growth rate can increase exponentially and takes on a J-shaped curve.

Predation not only removes the very old, the very young, and the weaker members from the population, it may also reduce the population of the prey. If the predators do not keep the prey population in balance, the carrying capacity is exceeded and the prey may starve. Predator and prey populations are closely interdependent. Populations of algae, annual plants, and insects with short life spans are controlled by seasonal and nutritional environmental changes. A J-shaped growth curve may show a plunge in population as the reproductive potential declines because of environmental changes. The following year, there may be an exponential increase in the population.

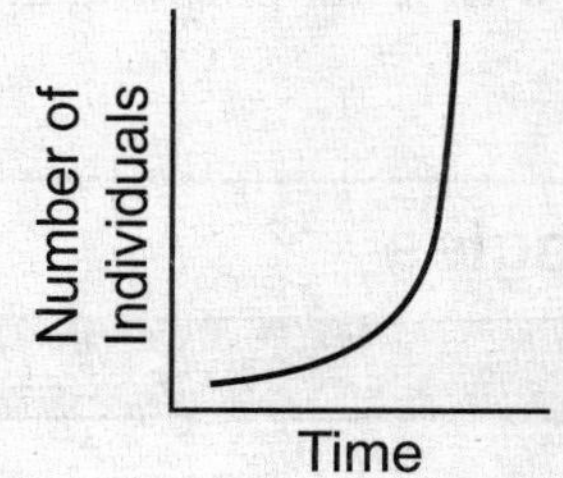

Figure 6.1 J-shaped curve

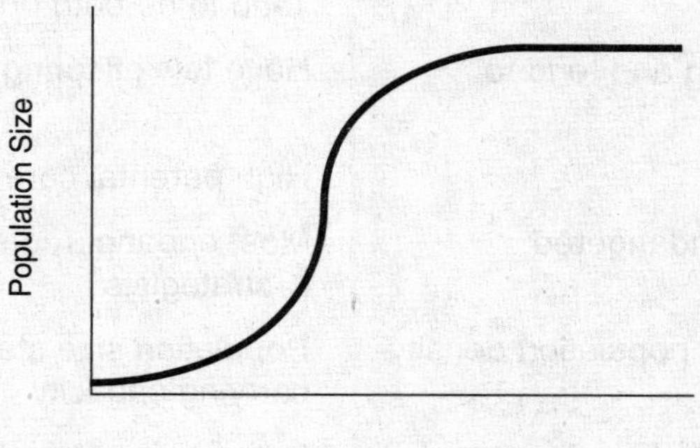

Figure 6.2 S-shaped curve

An S-shaped curve is used to describe the pattern of growth over extended periods of time when organisms move into an empty niche. Growth rates are density-dependent. They are characterized by maximum population growth rate and the carrying capacity. In an S-shaped curve, population size initially increases due to unlimited resources for a small population. However, as resources become limited, the population growth rate slows down and stabilizes around the limits of the carrying capacity at the point where the birth rate equals the death rate. The average American's ecological footprint (the demand of an individual with an average amount of resources) is about 12 acres.

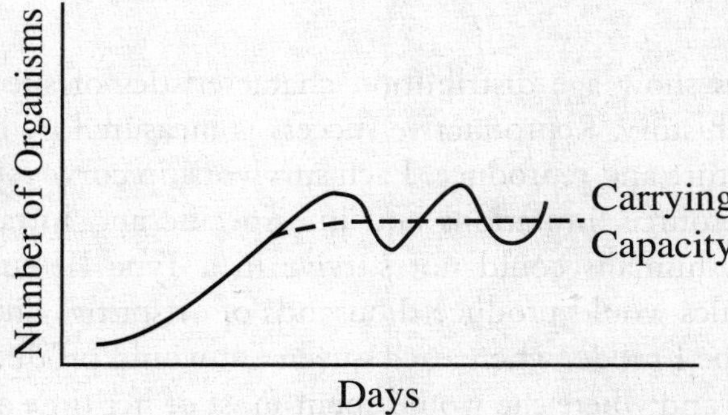

Figure 6.3 Flucations around the carrying capacity

Reproductive Strategies

Organisms have adapted either to maximize growth rates in environments that lack limits or to maintain population size at close to the carrying capacity in stable environments. Species that have high reproductive rates are known as r-strategists. Species that reproduce later in life and with fewer offspring are known as K-strategists.

Reproductive Strategies

r-Strategists	K-strategists
Mature rapidly	Mature slowly
Short lived	Long lived
Tend to be prey	Tend to be both predator and prey
Have many offspring and tend to overproduce	Have few offspring
Low parental care	High parental care
Are generally not endangered	Most endangered species are K-strategists
Wide fluctuations in population density (booms and busts)	Population size stabilizes near the carrying capacity
Population size limited by density-independent limiting factors, including climate, weather, natural disasters, and requirements for growth	Density-dependent limiting factors to population growth stem from intra-specific competition and include competition, predation, parasitism, competition, and migration
Tend to be small	Tend to be larger
Type III survivorship curve	Type I or II survivorship curve
Examples: most insects, algae, bacteria, rodents, and annual plants	Examples: humans, elephants, cacti, and sharks

Survivorship

Survivorship curves show age distribution characteristics of species, reproductive strategies, and life history. Reproductive success is measured by how many organisms are able to mature and reproduce. Each survivorship curve represents a balance between natural resource limitations and interspecific and intraspecific competition. For example, humans could not survive in a Type III survivorship mode, where human females would produce thousands of offspring. Likewise, ants could not survive in a Type I mode, where each queen ant would produce only a few eggs during her lifetime and where she would spent most of her time and energy raising offspring.

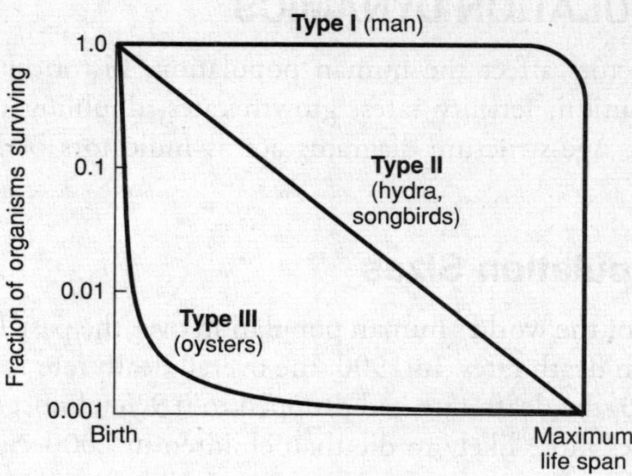

Figure 6.4

Survivorship Curves

Type	Description
I Late Loss	Reproduction occurs fairly early in life. Most deaths occur at the limit of biological life span. Low mortality at birth; high probability of surviving to advanced age. Death rates increase during old age. Advances in prenatal care, nutrition, disease prevention, and cures including immunization have meant longer life spans for humans. Examples: humans, annual plants, sheep, and elephants.
II Constant Loss	Individuals in all age categories have fairly uniform death rates. Predation affecting all age categories is primary means of death. Typical of organisms that reach adult stages quickly. Examples: rodents, perennial plants, and songbirds.
III Early Loss	Typical of species that have great numbers of offspring and reproduce for most of their lifetime. Death is prevalent for younger members of the species (environmental loss and predation) and declines with age. Examples: sea turtles, trees, internal parasites, fish, and oysters.

HUMAN POPULATION DYNAMICS

Many different factors affect the human population: historical population sizes, population distribution, fertility rates, growth rates, doubling times, and demographic transition. Age-structure diagrams act as indicators of future population trends.

Historical Population Sizes

The rapid growth of the world's human population over the past 100 years has been due to a decrease in death rates. In 1900, the overall death rate in the United States was 1.7%. In 2000, the death rate had dropped to 0.9% (almost half). Children in 1900 were 10 times more likely to die than children in 2000. Several factors have reduced death rates:

1. Increased food and more efficient distribution that result in better nutrition
2. Improvements in medical and public health technology
3. Improvements in sanitation and personal hygiene
4. Safer water supplies

Human population has had three surges in growth. These surges in population have been attributed to three factors: The first was the use of tools and fire. The second was the agricultural revolution, when humans stopped being hunter-gatherers and began to raise crops. The third was the industrial and medical revolutions within the last 200 years. Population change can be calculated from the following formula:

Population Change = (Crude birth rate + Immigration) – (Crude death rate + Emigration)

EXAMPLE

In 1950, the population of a small suburb in Los Angeles, California, was 20,000. The birth rate was measured at 25 per 1,000 population per year, while the death rate was measured at 7 per 1,000 population per year. Immigration was measured at 600 per year, while emigration was measured at 200 per year. By how much did the population increase (or decrease) in that year?

Answer:

Population change = (Crude birth rate + Immigration) – (Crude death rate + Emigration)

= (25(20) + 600) – (7(20) + 200)

= +760

The population grew from 20,000 to 20,760 in one year.

A current world growth rate of approximately 1.3% when applied to the about 6 billion people on Earth yields an annual increase of about 85 million people. Because of the large and increasing population size, the number of people added to the global population will remain high for several decades, even if growth rates decline.

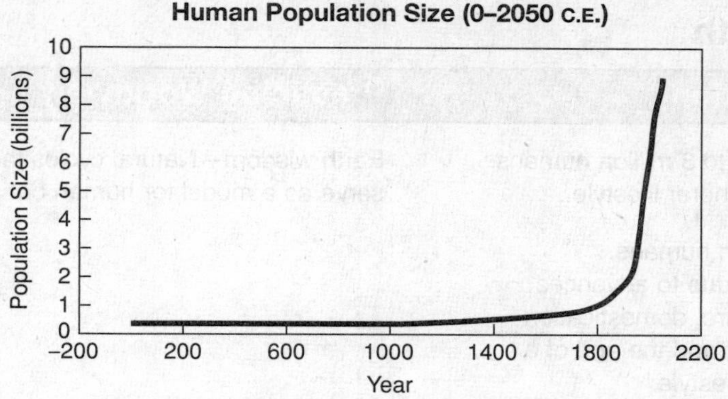

Figure 6.5 Estimated human population growth from 200 B.C.E. to 2200 C.E.

Distribution

In 1800, the vast majority of the world's population resided in Asia (65%) and Europe. By 1900, 25% of the human population lived in Europe largely due to the Industrial Revolution. Between 2000 and 2030, most of the growth will occur in the less-developed countries in Africa, Asia, and Latin America whose growth rates are much higher than those in more-developed countries. The more-developed countries in Europe and North America will have growth rates less than 1%. Some countries such as Russia, Germany, Italy, and Japan will even experience negative growth rates.

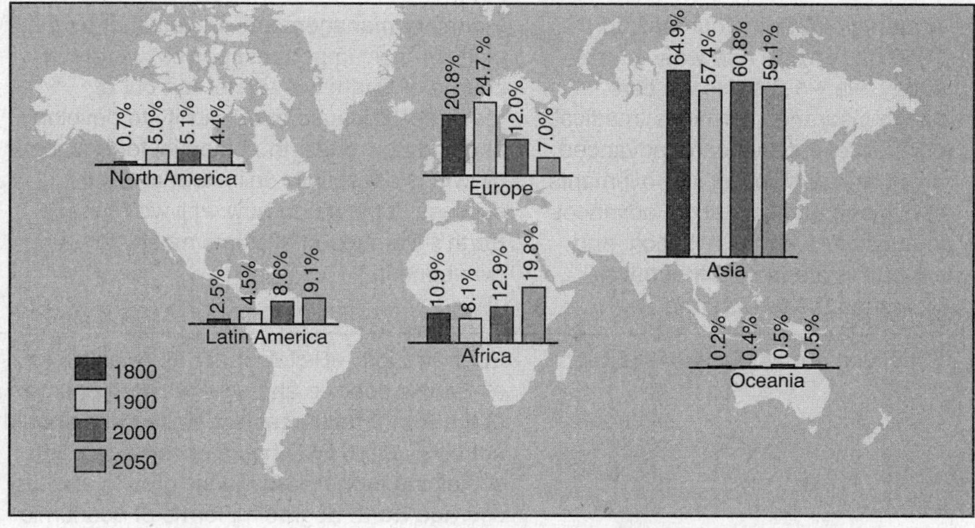

Figure 6.6 Human distribution patterns, 1800 C.E. to 2050 C.E.
(Source: United Nations Population Division, Briefing Packet, 1998 Revision of World Population Prospects)

Human Population Growth

Time Period	Description	Practicing Worldview
Before Agricultural Revolution	~ 1 million to 3 million humans. Hunter-gatherer lifestyle.	Earth wisdom—Natural cycles that can serve as a model for human behavior.
8000 B.C.E. to 5000 B.C.E.	~ 50 million humans. Increases due to advances in agriculture, domestication of animals, and the end of a nomadic lifestyle.	
5000 B.C.E. to 1 B.C.E.	~ 200 million humans. Rate of population growth during this period was about 0.03 to 0.05%, compared with today's growth rate of 1.3%.	Frontier worldview—Viewed undeveloped land as a hostile wilderness to be cleared and planted, then exploited for its resources as quickly as possible.
0 C.E. to 1300 C.E.	~ 500 million humans. Population rate increased during the Middle Ages because new habitats were discovered. Factors that reduced population growth rate during this time were famines, wars, and disease (density-dependent factors).	
1300 C.E. to 1650 C.E.	~ 600 million humans. Plagues reduced population growth rate. Up to 25% mortality rates are attributed to the plagues that reached their peak in the mid-1600s.	
1650 C.E. to present	Currently ~ 6 billion humans. In 1650 C.E. growth rate was ~ 0.1%. Today it is ~ 1.3%. Health care, health insurance, vaccines, medical cures, preventative care, advanced drugs and antibiotics, improvements in hygiene and sanitation, advances in agriculture and distribution, and education are factors that have increased the growth rate.	Planetary management—Beliefs that as the planet's most important species, we are in charge of Earth; we will not run out of resources because of our ability to develop and find new ones; the potential for economic growth is essentially unlimited; and our success depends on how well we manage Earth's life-support systems mostly for our own benefit.
Present to 2050 C.E.	Estimates are as high as ~ 15 billion.	Earth wisdom—Beliefs that nature exists for all Earth's species and we are not in charge of the Earth; resources are limited and should not be wasted. We should encourage Earth-sustaining forms of economic growth and discourage Earth-degrading forms of economic growth; and our success depends on learning how Earth sustains itself and integrating such lessons from nature into the ways we think and act.

Fertility Rates

Replacement level fertility (RLF) is the level of fertility at which a couple has only enough children to replace themselves, or about 2 children per couple. It takes a RLF of 2.1 to replace each generation since some children will die before they grow up to have their own two children. RLF rates are lower in moderately developed countries (MDC) and higher in less-developed countries (LDC) due to higher infant mortality rates in LDCs. Infant mortality rates are good indicators of comparative standards of living. The total fertility rate (TFR) is the average number of children that each woman will have during her lifetime. The country of Niger in Africa leads the world's TFR at 7.46.

Worldwide Total Fertility Rate

Country	TFR
Niger	7.46
India	2.73
Mexico	2.42
Israel	2.41
United States	2.09
China	1.73
European Union	1.47
Japan	1.40
Russia	1.28
Hong Kong	0.95
World average	**2.59**

Despite half of the world's population having subreplacement fertility rates, the world's population is still growing quickly. This growth is due to nations with above-replacement TFRs and a population momentum caused by large numbers of younger females who have as yet not had children.

Declines in fertility rates can be attributed to several factors. Urbanization results in a higher cost of living. Urbanization reduces the need for extra children to work on farms. There is a greater personal acceptance and government encouragement of contraception and abortion. The numbers of females in the workforce and female educational opportunities are increasing. More individuals desire to increase their standard of living by having less children. Many people are postponing marriage until their careers are established.

The two main effects of TFRs less than 2.1 without additions through immigration are population decline and population aging. In the United States the TFR has been inconsistent. The greatest TFR occurred during the post-World War II years (baby boomers). New immigrants and their descendants are projected to contribute 66% of the expected growth by 2050.

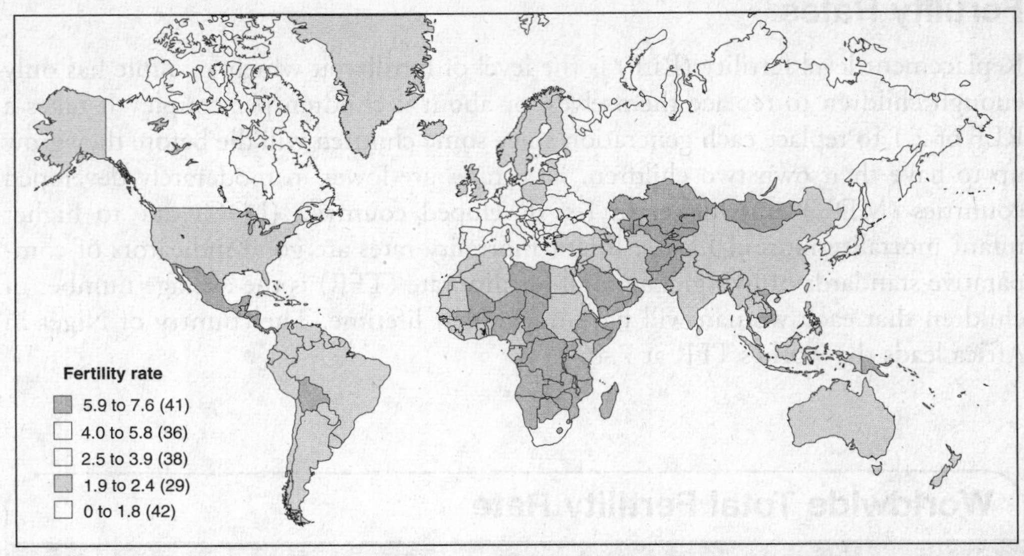

Figure 6.7 Worldwide total fertility rates

Growth Rates and Doubling Times

The 20th century saw the largest increase in the world's population in human history. The following chart shows the doubling times of the world's population:

950 C.E. to 1600	650 years
1600 to 1800	200 years
1800 to 1925	125 years
1925 to 1975	50 years
1975 to 2025	50 years

Note that the doubling times from 1925 to 1975 and projected from 1975 to 2025 remain constant. This does NOT mean that the world is not increasing in population—it means that the growth *rate* has decreased as shown in the following figure.

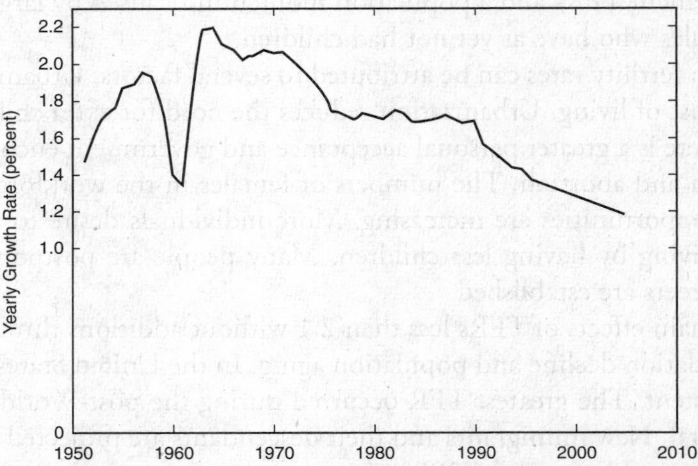

Figure 6.8 Yearly growth rate of human population (1950–2010)

Demographic Transition

Demographic transition is the name given to the process that has occurred during the past century. It leads to a stabilization of population growth in the more highly developed countries and is generally characterized as having four separate stages: preindustrial, transitional, industrial, and postindustrial.

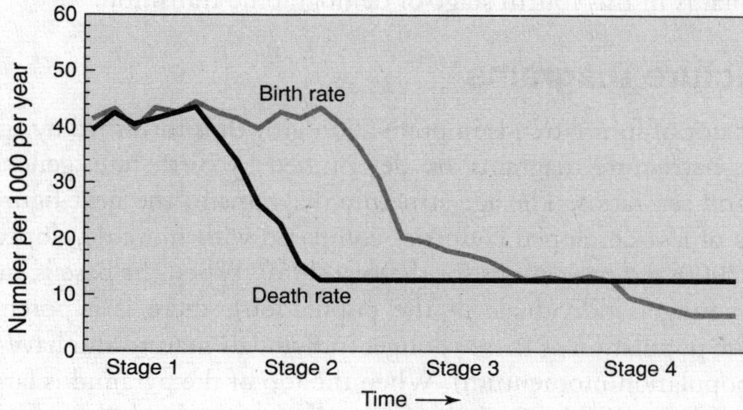

Figure 6.9 Demographic transitions occuring in human populations.

STAGE 1: PRE-INDUSTRIAL

Living conditions are severe, medical care is poor or nonexistent, and the food supply is limited due to poor agricultural techniques, preservation, and pestilence. Birth rates are high to replace individuals lost through high mortality rates. The net result is little population growth. Many countries in sub-Saharan Africa have reverted back to this stage due to the increase of AIDS.

STAGE 2: TRANSITIONAL

This stage occurs after the start of industrialization. Standards of hygiene and more modern medical techniques begin to drive down the death rate, leading to a significant upward trend in population size. Mortality rates drop as a result of advances in medical care, improved sanitation, cleaner water supplies, vaccination, and higher levels of education. The net result is a rapid increase in population. Examples include India, Pakistan, and Mexico.

STAGE 3: INDUSTRIAL

Urbanization decreases the economic incentives for large families. The cost of supporting an urban family grows, and parents are more actively discouraged from having large families. Educational and work opportunities for women decrease birth rates. Obtaining food is not a major focus of the day. Leisure time is available. Retirement safety nets are in place, reducing the need for extra children to support parents. In response to these economic pressures, the birth rate starts to drop, ultimately coming close to the death rate. An example is China.

STAGE 4: POST-INDUSTRIAL

Birth rates equal mortality rates, and zero population growth is achieved. Birth and death rates are both relatively low, and the standard of living is much higher than during the earlier periods. In some countries, birth rates may actually fall below mortality rates and result in net losses in population. Examples of declining populations include Russia, Japan, and many European countries. The developed world currently remains in this fourth stage of demographic transition.

Age-Structure Diagrams

A good indicator of future trends in population growth is furnished by age-structure diagrams. Age-structure diagrams are determined by birth rate, generation time, death rate, and sex ratios. The age-structure diagrams in the next figure show age distributions of less-developed countries compared with more developed countries for the year 2000 and projections for the year 2050. When the base is large (greater number of younger individuals in the population), there is a potential for an increase in the population as these younger individuals mature and have children of their own (population momentum). When the top of the pyramid is larger, it indicates a large segment of the population is past their reproductive years and indicates a future slowdown in population growth. Age-structure diagrams reflect demographic transitions.

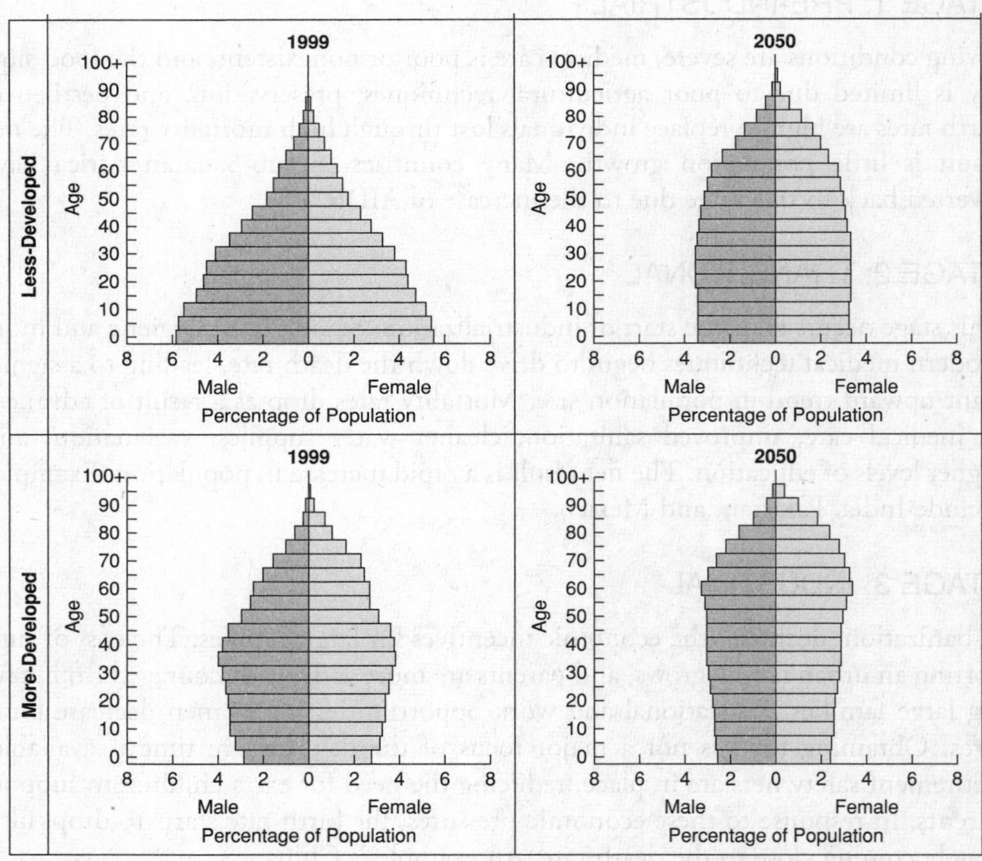

Figure 6.10 Age structure diagrams comparing less-developed countries to more-developed countries in the next 50 years. *(Source: United Nations)*

In Mexico, large family size is due to the necessity for farm labor, the need to support parents when they no longer work, a need to increase family income, and cultural and religious beliefs. The death rate has declined due to social and medical programs. However, the birth rate continues to remain high.

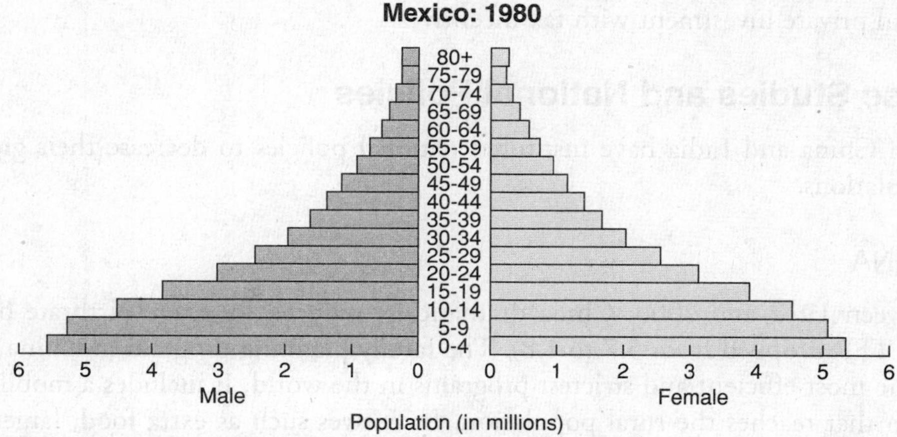

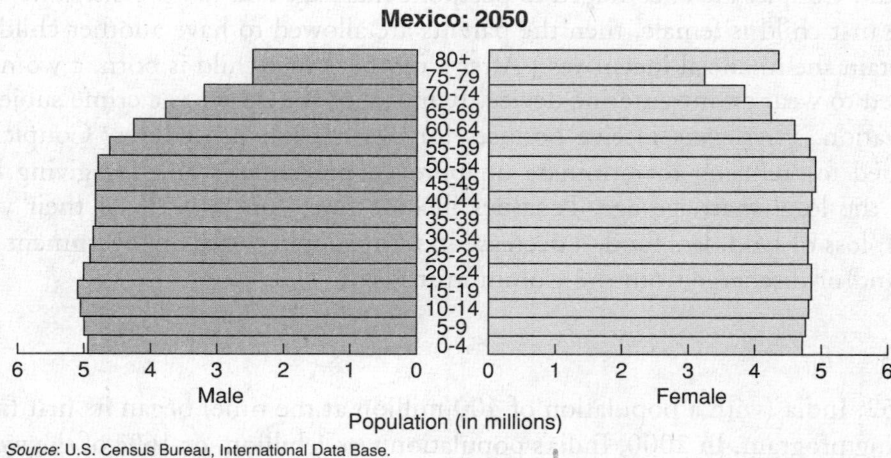

Figure 6.11 Age structure in Mexico, 1980 and 2050 (estimated)

POPULATION SIZE

This section discusses several strategies to sustain population size. It then provides a few case studies.

Strategies for Sustainability

- Provide economic incentives for having fewer children.
- Empower and educate women.
- Education usually leads to higher incomes. Higher incomes decrease the need for having extra children to take care of older parents.
- Higher education usually results in having children later in life.

- Provide government family planning services. These practices have ranged from education and birth control to sterilization and abortion.
- Improve prenatal and infant health care. Women would not need more children if the ones they had survived.
- Increase economic development in less-developed countries through free trade and private investment with tax incentives.

Case Studies and National Policies

Both China and India have instituted national policies to decrease their growing populations.

CHINA

Between 1972 and 2000, China dramatically reduced its crude birthrate by half (the TFR dropped from 5.7 to 1.8). The family planning program in China is one of the most efficient and strictest programs in the world. It includes a mobile program that reaches the rural population. Incentives such as extra food, larger pensions, better housing, free medical care, free school tuition, and salary bonuses for parents who limit their number of children are responsible for the success of this program. Couples are encouraged to postpone marriage and to have only one child. (If the first child is female, then the parents are allowed to have another child and still retain the financial incentives.) After a mother's first child is born, a woman is required to wear an intrauterine device. Removal of the device is a crime subject to sterilization. Physicians receive bonuses for sterilization procedures. Couples are punished for refusing to terminate unapproved pregnancies and for giving birth under the legal marriage age. Penalties include fines (up to 50% of their yearly salary), loss of land, less food, a decrease in farm supplies, loss of government benefits, and/or discharge from the Communist Party.

INDIA

In 1952, India (with a population of 400 million at the time) began its first family planning program. In 2000, India's population was 1 billion, or 16% of the world's population. Each day there are 50,000 live births in India. One-third of the population of India earns less than 40 cents per day, and cropland has decreased 50% per capita since 1960. In the 1970s, India instituted a mandatory sterilization program involving vasectomies. Some of the reasons for India's failures were poor planning, low status of women, favoring male children, and insensitivity to cultures and religion. Tubal ligation is the preferred method of family planning in India today. Condoms are free from the Indian government but have less than 10% use. Other birth control methods are usually accepted by only the upper/educated class.

IMPACTS OF POPULATION GROWTH

When left unchecked, population growth often results in hunger. Diseases and economics also affect population growth.

Hunger

A large portion of the world's population—25%—is malnourished. Areas of greatest malnutrition are Africa, Asia, and parts of Latin America. Consequently, these are areas with the highest TFRs. Several factors cause malnutrition.

1. Poverty
2. Droughts, which will only increase as the impact of global warming becomes more severe
3. Populations that have surpassed their carrying capacity
4. Political instability and wars, which cause mass migrations
5. Pestilence
6. Foreign investors who own large landholdings and whose sole motivation is profit (selling the food to the highest bidder, which often means exporting it)

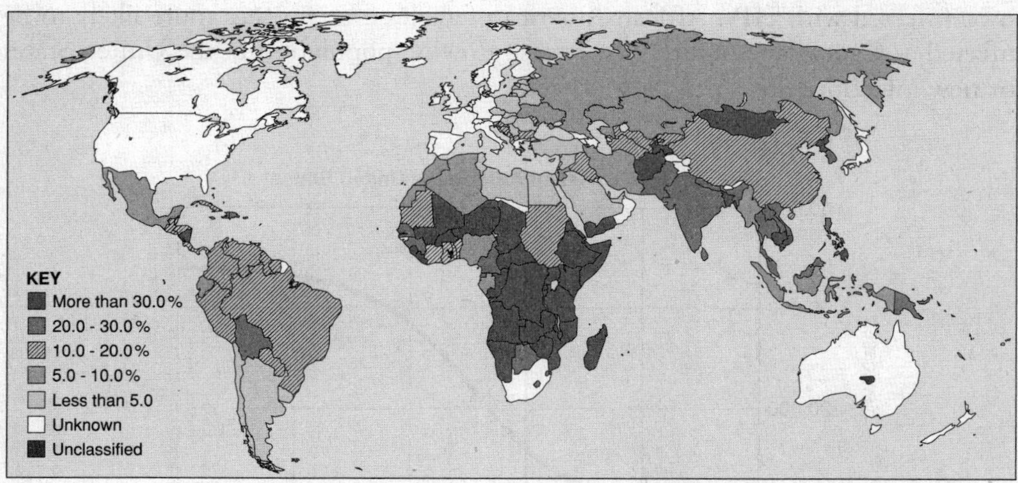

Figure 6.12 Worldwide malnutrition

Advances made during the first and second green revolutions (which focused on food production) are not ending world famine. In India, Mexico, and the Philippines, where large advances were made in crop production, famine is still common. Exports from these and other poor countries caused by distributors receiving higher profits from richer countries is actually making famine worse. Profit motives result in removing food from countries that grow it and sending the food to other countries that are able to pay higher prices.

Large-scale agribusiness or corporate ownership of farmland will not solve world hunger. Land reform in Japan, Zimbabwe, Taiwan, and Brazil has redistributed land into smaller holdings. It has also raised agricultural output an average of 80%.

The issue of malnutrition is not that the world does not produce enough food. The issue is that too many people cannot afford it or that it is not distributed efficiently. Enough wheat, rice, and grain are produced on Earth each day to provide each human with 3,500 calories per day (2,500 calories is the minimum daily requirement for men, and 2,000 calories is the minimum daily requirement for women). This does not even include vegetables, beans, nuts, meats, and fish processed each day. If all food grown and raised on Earth was distributed equally, it would result in 4.3 pounds (2 kg) per person per day.

Disease and Economic Impact

This section discusses how disease and economic realities affect population growth.

AIDS

In the United States, more than half a million people have died from complications arising from AIDS since 1981. An estimated 15,000 currently die annually according to the Federal Centers for Disease Control and Prevention. More than 1 million people in the United States are living with the virus, and 40,000 become infected each year. African-Americans, who make up approximately 13% of the U.S. population, account for half of new U.S. infections and a third of all AIDS-related deaths. African-American males are 7 times more likely as Caucasian males to be infected with HIV. African-American females are 20 times more likely to be infected as Caucasian females. The fastest growing population in the United States for new infections are 15 to 24 years old.

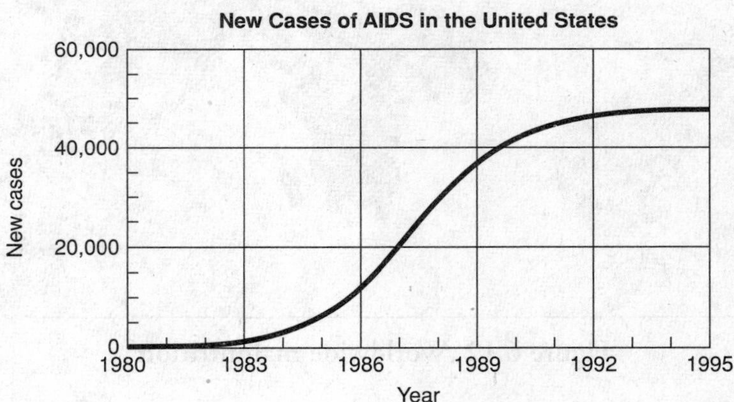

Figure 6.13 Growth curve for new cases of AIDS in the U.S. between 1980 and 1995. Notice how it follows the classic logistic growth model.

Worldwide, at least 25 million people have died from AIDS, and 2.8 million will have died in 2006, according to the World Health Organization. AIDS could kill 31 million people in India and 18 million in China by 2025, according to projections by U.N. population researchers. The toll from AIDS over the next 25 years will go far beyond the 34 million people thought to have died from the bubonic plague in the 14th century or the 20 to 40 million people who perished in the 1918 Spanish flu pandemic. AIDS is the leading cause of death in Africa, which has accounted for nearly half of all global AIDS deaths.

The epidemic is still growing. Its peak could be a decade or more away. In at least seven countries, the U.N. estimates that AIDS has reduced life expectancy to 40 years or less. Botswana has the highest HIV prevalence in the world; 36% of the country's 1.6 million people are HIV positive. AIDS experts predict that by 2010, more than 50% of Botswana's children will be AIDS orphans and the average life expectancy will have fallen from 47 years to 27 years.

PANDEMICS

The Spanish influenza pandemic of 1918 to 1919 killed somewhere between 20 and 40 million people worldwide, with 28% of all Americans infected. It has been cited as the most devastating pandemic in recorded world history. More people died of Spanish influenza in a single year than in the four years of the plague from 1347 to 1351. The flu was most deadly for people ages 20 to 40. This was unusual for influenza, which is normally a killer of the elderly and of young children.

According to the World Health Organization, if a mutant form of bird flu virus (H5N1) develops that could pass from person to person, it could kill as many as 7 million people worldwide and infect nearly one-third of the world's population. Infectious disease experts fear the virus may mutate within a pig that harbors both human and avian forms of the flu virus.

OTHER DISEASES

The World Health Organization estimates that by 2020, tobacco-related illnesses, including heart disease, cancer, and respiratory disorders, will be the world's leading killer. They will be responsible for more deaths than AIDS, tuberculosis, road accidents, murder, and suicide combined. In 2005, estimated losses in national income from heart disease, strokes, and diabetes was $18 billion in China, $11 billion in the Russian Federation, and $9 billion in India.

Tuberculosis is the leading cause of death in many poorer countries. Its economic impact on global economies is estimated to be $12 billion.

Each year, approximately 500 million malaria infections lead to over 1 million deaths, 75% of which occur among children living in Africa. Mortality rates due to malaria are rising among young children, and drug therapies that were once fairly effective in treating malaria are becoming less effective as newer strains become more drug-resistant.

Resource Use and Habitat Destruction

Three methods are used to estimate the effects of humans on patterns of resource utilization: measure net primary productivity, estimate how much impact humans have had on Earth, and examine finite resources and from that draw conclusions on increasing productivity. Net primary productivity (NPP) is the total amount of solar energy converted into biochemical energy through photosynthesis minus the energy needed by those plants for their own metabolic requirements. It is a quantifiable measure of resources available on Earth. The NPP without human activity has been estimated to be 150 billion tons of organic matter per year. Human activity has caused a 12% decline in the NPP due to deforestation. Humans utilize about 27% of the NPP for their own purposes (food, building material, energy, and so on) or by converting productive land to nonproductive purposes. When added up, humans utilize about 40% of the NPP and leave all other life on Earth with 60%. The difference grows with the ever-increasing human population growth.

Of the NPP of the oceans, about 8% is utilized for human purposes. In nutrient-rich upwelling areas, humans utilize about 25%. In temperate continental shelf waters, humans utilize approximately 35% of the NPP.

If humans utilized 100% of the NPP (leaving nothing for all other life-forms on Earth), the theoretical maximum sustainable human population at 100% of the carrying capacity would be 15 billion people—which could be achieved during the 21st century. In this scenario, every square centimeter of Earth would be utilized for human needs. The following table lists factors that affect resource utilization.

Factors That Affect Resource Utilization

Factor	Description
Carrying capacity	See the section "Carrying Capacity (K)" in this chapter.
Energy resources	One average American consumes as much energy as 500 Ethiopians or 35 people from India.
Environmental degradation	Due to increased population size, erosion, desertification, pollution, impact on the ozone layer, and gases that contribute to global warming all increase.
Exploitation of natural resources as a function of gross domestic product	The richest 20% of the world's population contribute to resource depletion of energy and raw materials through overconsumption. This in turn, leads to disproportionate amounts of pollution. The poorest 20% of the world's population are forced to deplete resources by being forced to cut down forests, clear land, and farm marginal land.
Extinction of animal and plant species	Close to 50% of all species of animals and plants on Earth could be on a path toward extinction within 100 years. 11% of all bird species and 13% of all plant species are at risk for extinction.
Famine	See the section "Hunger" in this chapter.
Political unrest	Affects employment, food distribution and standard of living that, in turn, affect utilization of resources.
Population density	Density, more than population size, has a greater effect on the amount of pollution and use of energy.
Population size	Large numbers of people lead to high rates of habitat loss and natural resource depletion.
Poverty	20% of the world's richest countries control 80% of the world's wealth. The poorest 20% of the world's population controls just 1.5% of the world's economic resources.
Technological development	More-developed countries consume more resources than less-developed countries. The United States represents about 5% of the world's population but consumes 25% of the world's resources and generates 25% of the world's waste.

FACTOIDS

- Between 39% and 50% of Earth's surface has been converted for either agricultural or urban purposes.
- Carbon dioxide concentrations have increased 30% due to human activity such as fossil fuel burning and deforestation.
- More than 50% of all easily accessible freshwater resources are utilized by humans.
- More than 50% of all nitrogen fixation is caused by human activity. This includes the use of fertilizers, raising nitrogen-fixing crops, and releasing nitrogen pollutants into the environment through the combustion of fossil fuels.
- Humans have disrupted and introduced more than 25% of all plant species into nonnative continental areas. In some areas (especially islands), this number is greater than 50%.

CASE STUDY

Conference on Population and Development, Cairo (1994): The plan calls for improved healthcare and family planning services for women, children, and families throughout the world. It also emphasizes the importance of education for girls as a factor in the shift to smaller families.

TIP

Do NOT start preparing for the APES exam a few weeks ahead of time. The APES exam is actually one of the toughest AP Exams given. In 2006, half of all students taking the exam received a 1 or 2. Pace yourself and be sure to get lots of rest before the exam day. Don't just read the practice essays in this book—write them out.

MULTIPLE-CHOICE QUESTIONS

1. Most populations in nature
   (A) show characteristic J-curves
   (B) show characteristic S-curves
   (C) show characteristics of dynamic equilibrium
   (D) show wide fluctuations in population growth
   (E) cannot be characterized

2. A population showing a growth rate of 20, 40, 60, 80 . . . would be characteristic of
   (A) logarithmic growth
   (B) exponential growth
   (C) static growth
   (D) arithmetic (linear) growth
   (E) power curve growth

3. If a population doubles in about 70 years, it is showing a ___ % growth rate.
   (A) 1
   (B) 5
   (C) 35
   (D) 140
   (E) 200

Questions 4 and 5 refer to an island off the coast of Costa Rica where 500 birds of a particular species live. Population biologists determined that this bird population was isolated with no immigration or emigration. After one year, the scientists were able to count 60 births and 10 deaths.

4. The net growth for this population was
   (A) 0.5
   (B) 0.9
   (C) 1.0
   (D) 1.1
   (E) 1.5

5. The doubling time for this population would be
   (A) 10 years
   (B) 13.3 years
   (C) 24.2 years
   (D) 63.6 years
   (E) 126 years

6. Biotic potential refers to
   (A) an estimate of the maximum capacity of living things to survive and reproduce under optimal environmental conditions
   (B) the proportion of the population of each sex at each age category
   (C) the ratio of total live births to total population
   (D) a factor that influences population growth and that increases in magnitude with an increase in the size or density of the population
   (E) events and phenomena of nature that act to keep population sizes stable

7. The number of children an average woman would have, assuming that she livers her full reproductive lifetime, is known as the

   (A) birth rate
   (B) crude birth rate
   (C) TFR
   (D) RLF
   (E) zero population growth rate

8. The average American's ecological footprint is approximately

   (A) the size of a shoe
   (B) 0.5 acres
   (C) 3 acres
   (D) 6 acres
   (E) 12 acres

9. Which of the following statements is FALSE?

   (A) The United States, while having only 5% of the world's population, consumes 25% of the world's resources.
   (B) Up to 50% of all plants and animals could become extinct within the next 100 years.
   (C) In 1990, 20% of the world's population controlled 80% of the world's wealth.
   (D) The theoretical maximum number of people on Earth that the planet could support is 15 billion.
   (E) They are all true statements.

10. Pronatalists (people in favor of having many children) would include all of the following EXCEPT

   (A) many children die early due to health and environmental conditions
   (B) children are expensive and time intensive
   (C) children provide extra income for families
   (D) children provide security for parents when the parents reach old age
   (E) status of the family is often determined by the number of children

11. The most successful method of controlling a country's population has been

   (A) required sterilization
   (B) government quotas on children produced
   (C) birth control
   (D) financial incentives
   (E) all of the above

12. The following age-structure diagram would be typical of

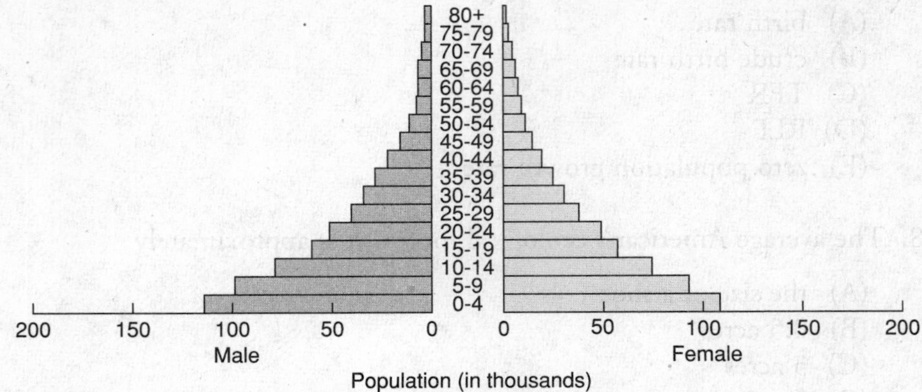

Source: U.S. Census Bureau, International Data Base.

(A) Russia
(B) China
(C) the United States
(D) Niger or Peru
(E) Canada

13. Examine the following age-structure diagram.

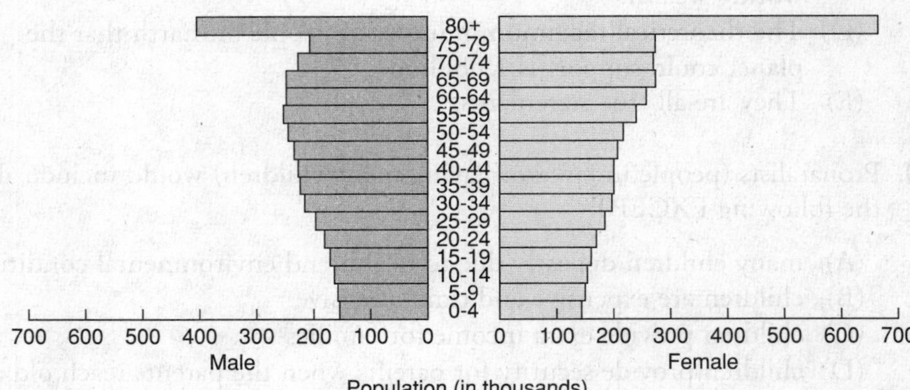

Source: U.S. Census Bureau, International Data Base.

This population will be

(A) declining rapidly in the future
(B) growing slowly in the future
(C) remaining stable
(D) growing rapidly in the future
(E) declining slowly in the future

14. The main reason that the population of Russia is declining is

(A) the government instituted taxation incentives for having smaller families
(B) the success of birth control
(C) mandatory sterilization
(D) the standard of living has declined
(E) massive emigration

15. A density-dependent factor would include all of the following EXCEPT

(A) drought
(B) fires
(C) predation
(D) flooding
(E) all of the above are density-dependent

16. Between 1963 and 2000, the rate of the world's annual population change ______ but the human population size ______ .

(A) dropped 40%; rose 90%
(B) rose 90%; dropped 40%
(C) dropped 90%; rose 40%
(D) remained stable; tripled
(E) rose 40%; dropped 2.9%

17. RLF for a couple is

(A) 1.0
(B) 2.0
(C) 2.1
(D) 3.0
(E) varies depending on country

18. Which of the following factors is associated with the highest potential for population growth?

(A) High percentage of people under age 18
(B) High percentage of people in their 30s
(C) High percentage of people in their 50s
(D) High percentage of people in high-income groups
(E) High percentage of people in low-income groups

19. Using the demographic transition model, what stage would be characteristic of death rates falling while birth rates remain high?

(A) Preindustrial
(B) Industrial
(C) Postindustrial
(D) Transitional
(E) None of the above

20. All of the following factors tend to cause women to have fewer children EXCEPT
    (A) higher education
    (B) high infant mortality
    (C) better prenatal care
    (D) birth control education
    (E) human rights are protected

FREE-RESPONSE QUESTION

By: William Aghassi
Brooklyn Technical High School, Brooklyn, NY
B.S. (Mechanical Engineering- Polytechnic University)
M.S. (Environmental Engineering- New Jersey Inst. of Technology)

The 1990 U.S. census revealed that for the first time, a majority of Americans live in suburbs of major cities or in suburban-like communities.

(a) List three factors why this trend may not be sustainable from a resource point of view.
(b) Explain why each of the three reasons you have listed is not sustainable and the possible consequence of each.
(c) What national public policy decision(s) made the United States into a suburban country as opposed to Europe, which is largely urbanized?
(d) Many environmentalists think of the 21st century as the "Century of the City." State whether or not you agree or disagree with this statement and include your reasons.

MULTIPLE-CHOICE ANSWERS AND EXPLANATIONS

1. **(C)** Dynamic equilibrium defines a population that is in balance with the carrying capacity of the environment.
2. **(D)** Arithmetic or linear growth is characterized by a constant increase per unit of time. In this case, the constant is an increase of 20.
3. **(A)** A 1% growth rate would cause a population to double in 70 years. Hint: divide 70 by the annual percentage growth rate to get the doubling time in years.
4. **(D)** Population size = original size (500) + births (60) – deaths (10) + immigration (0) – emigration (0) = 550. Net growth *rate* = 550/500 = 1.1.
5. **(D)** Doubling time = 70/1.1 = 63.6 years.
6. **(A)** The maximum reproductive rate is called the biotic potential.
7. **(C)** Global TFR is approximately 2.6.

8. (E) An ecological footprint is a metaphor used to depict the amount of land a person would hypothetically need to provide the resources required to support himself.
9. (E) No explanation needed.
10. (B) Pronatalists urge people to have many children. Choice (B) is the only argument provided that does not promote having children.
11. (E) All of the choices listed have had some measure of success.
12. (D) Age-structure diagrams with a wide base are populations that have a high proportion of young, which results in a powerful, built-in momentum to increase population size, assuming death rates do not unexpectedly increase.
13. (A) This graph is the reverse of the graph in Question 12. In this case, the majority of the population is beyond reproductive years and the death rate exceeds the birth rate. This projected age structure diagram is for Hong Kong in 2050.
14. (D) Russia's population is estimated to be approximately 143 million people. Russia is declining in population at a rate of about 700,000 per year. If the trend continues, Russia is predicted to have a population of the size of the Republic of the Congo in 50 years. In 2006, President Putin said Russia's population had witnessed an annual decline because of low birth rates, high mortality, and emigration. A 10-year national program has been instituted to try to reverse the situation with a key element being to increase childcare benefits to support young mothers, especially those who have a second child.
15. (C) Density-independent factors influence population growth and do not depend on the size or density of the population. Predation rates are affected by population size.
16. (A) Population change is an increase or decrease in the size of a population. It is equal to (births + immigration) – (deaths + emigration). Population size is the number of individuals in the population.
17. (C) A TFR of 2.1 is considered the replacement rate. Once the TFR of a population reaches 2.1, the population will remain stable, assuming no immigration or emigration takes place.
18. (A) Population momentum is the tendency for changes in population growth rates to lag behind changes in childbearing behavior and mortality conditions. Momentum operates through the population age distribution. A population that has been growing rapidly for a long time acquires a "young" age distribution that will result in positive population growth rates for many decades into the future.
19. (D) In the preindustrial stage, living conditions are harsh, birth and death rates are high, and there is little increase in population size. In the transitional stage, living conditions improve, the death rate drops, and birth rates remain high. In the industrial stage, growth slows. In the postindustrial stage, zero population growth is reached and the birth rate falls below the death rate.
20. (B) Countries with high infant mortality rates generally have high TFRs.

FREE-RESPONSE ANSWER

(a) Three factors may limit the trend to American suburbanization:

1. Limit to the amount of open space available near cities
2. Non-point-source pollution (runoff) that could jeopardize the integrity of the region's watershed
3. Traffic congestion
4. Diversity of plant and animal ecosystems could be jeopardized
5. Degradation of air quality
6. Loss of agricultural lands
7. Scarcity of available and inexpensive energy sources

(b)

1. The two features that distinguish land use in suburban areas are low-density housing and that over 60% of available land is devoted to the needs of the automobile. As suburbanization increases, these two factors tend to diminish the availability of open space.
2. Suburbanization near cities tends to encroach upon the watersheds of these regions. Suburban lawns, parking lots, and roads are responsible for a much greater nutrient and chemical loading than undeveloped land and greatly affect the water quality of the region's drinking water.
3. Suburban communities are low density and widely dispersed. Since suburban residents do not commute from or to a central location, they are usually ill served by mass transit and carpooling is sometimes not practical. Commuting times therefore lengthen to intolerable levels, resulting in high air pollution levels.
4. Suburban sprawl can destroy many unique and/or productive ecosystems such as wetlands and forests, thus limiting regional diversity.

5. Since suburbs are decentralized, residents must rely on the automobile for every errand, and all of their goods must be transported by truck. This often results in traffic congestion and air quality degradation that leads to respiratory distress in humans, damage to plants, and damage to buildings.
6. Land goes toward the highest value use. As suburbs encroach on agricultural areas, farmers may be forced to sell to developers due to rezoning and a resulting increase in real estate taxes or simply a higher rate of return than that of farming.
7. Low-density single-family housing and exclusive reliance on the automobile to get around are energy inefficient. As imported oil supplies become expensive or hard to obtain, this could limit suburban growth.

(c) After World War II, the decision to invest infrastructure dollars on roads and disinvest in mass transit made suburbanization possible.

The public policy decision to keep energy prices low through relatively low taxation compared with Europe, which has high energy taxation and high energy prices, made suburban sprawl possible.

(d) Agree: Cities are more energy efficient and, if well planned, are more convenient and offer a better sense of community than many suburbs. This will attract more Americans to move back to cities.

Disagree: Americans are reluctant to forgo the American dream of the single-family home and the automobile to ever return to cities in large numbers.

UNIT IV: LAND AND WATER USE (10–15%)

Areas on Which You Will Be Tested

A. **Agriculture**
   1. Feeding a growing population—human nutritional requirements, types of agriculture; green revolution, genetic engineering, crop production, deforestation, irrigation, and sustainable agriculture.
   2. Controlling pests—types of pesticides, costs and benefits of pesticide use, integrated pest management, and relevant laws.

B. **Forestry**—tree plantations, old-growth forests, forest fires, forest management, and national forests.

C. **Rangelands**—overgrazing, deforestation, desertification, rangeland management, and federal rangelands.

D. **Other Land Use**
   1. Urban land development—planned development, suburban sprawl, and urbanization.
   2. Transportation infrastructure—federal highway system, canals and channels, roadless areas, and ecosystem impacts.
   3. Public and federal lands—management, wilderness areas, national parks, wildlife refuges, forests, and wetlands.
   4. Land conservation options—preservation, remediation, mitigation, and restoration.
   5. Sustainable land use strategies.

E. **Mining**—mineral formation, extraction, global reserves, and relevant laws and treaties.

F. **Fishing**—fishing techniques, overfishing, aquaculture, and relevant laws and treaties.

G. **Global Economics**—globalization, World Bank, *Tragedy of the Commons*, and relevant laws and treaties.

Land and Water Use

CHAPTER 7

One does not sell the Earth upon which the people walk.
—Crazy Horse

FEEDING A GROWING POPULATION

In order to feed a population adequately, several factors must be taken into account. The following sections discuss these in detail.

Human Nutritional Requirements

A healthy diet generally requires 2,500 calories for the average male and 2,000 calories for the average female. Proper nutrition also requires a balanced intake of protein, carbohydrates, and fat. Protein produces 4 calories of energy per gram and should make up about 30% of all calories. Carbohydrates also produce 4 calories of energy per gram and should make up approximately 60% of the daily diet. Fats produce 9 calories of energy per gram and should not make up more than 10% of the total daily caloric intake.

Only about 100 species of plants of the 350,000 known are commercially grown to meet human nutritional needs. Of these, wheat and rice supply over half the human caloric intake. Just 8 species of animal protein supply over 90% of the world's needs. It takes about 16 pounds (7 kg) of grain to produce 1 pound (0.5 kg)of edible meat, and 20% of the world richest countries consume 80% of the world's meat production. About 90% of the grain grown in the United States is grown for animal feed. By consuming the grain directly instead of consuming the animals that feed upon it, there would be a 20-fold increase in the amount of calories available and an 8-fold increase in the amount of protein available. The benefit of consuming meat products is that they are concentrated sources of protein that are broken down through digestion into amino acids.

In terms of famine and malnutrition, about 11 million children die each year from starvation, with 850 million people considered malnourished. Chronic undernourishment and vitamin or mineral deficiencies result in stunted growth, weakness, and increased susceptibility to illnesses.

Types of Agriculture

AGROFORESTRY

A system of land use in which harvestable trees or shrubs are grown among or around crops or on pastureland as a means of preserving or enhancing the productivity of the land.

ALLEY CROPPING

A method of planting crops in strips with rows of trees or shrubs on each side. Alley cropping increases biodiversity, reduces surface water runoff and erosion, improves the utilization of nutrients, reduces wind erosion, modifies the microclimate for improved crop production, improves wildlife habitat, and enhances the aesthetics of the area.

CROP ROTATION

Planting a field with different crops from year to year to reduce soil nutrient depletion. Example: Rotating corn or cotton, which removes large amounts of nitrogen from the soil, with soybeans which then add nitrogen to the soil.

HIGH-INPUT AGRICULTURE

Includes the use of mechanized equipment, chemical fertilizers, and pesticides.

INDUSTRIAL AGRICULTURE OR CORPORATE FARMING

A system characterized by mechanization, monocultures, and the use of synthetic inputs such as chemical fertilizers and pesticides, with an emphasis on maximizing productivity and profitability.

INTERCROPPING

To grow more than one crop in the same field, especially in alternating rows or sections.

INTERPLANTING

Growing two different crops in an area at the same time. To interplant successfully, plants should have similar nutrient and moisture requirements.

LOW INPUT

Depends on hand tools and natural fertilizers; lacks large-scale irrigation.

LOW-TILL, NO-TILL, OR CONSERVATION-TILL AGRICULTURE

Soil is disturbed little or not at all to reduce soil erosion. Has lower labor costs, reduces the need for fertilizer, and saves energy.

MONOCULTURE

The cultivation of a single crop.

PLANTATION

A commercial tropical agriculture system that is essentially export oriented. The local government and foreign/international companies exploit the natural resources of the tropical rain forest for profit, usually short-term economic gain. It often involves the deliberate introduction and cultivation of economically desirable species of tropical plants at the expense of widespread replacement of the original native and natural flora. Plantation practices include modifications or disturbance of the natural landscape through such artificial practices as the permanent removal of natural vegetation, changes in drainage channels, application of chemicals to the soil, and so on.

POLYCULTURE

Polyculture uses *different* crops in the same space, in imitation of the diversity of natural ecosystems, and avoids large stands of a single crop (monoculture). It includes crop rotation, multicropping, intercropping, and alley cropping. Polyculture, though it often requires more labor, has several advantages over monoculture. The diversity of crops avoids the susceptibility of monocultures to disease. The greater variety of crops provides habitat for more species, increasing local biodiversity.

POLYVARIETAL CULTIVATION

Planting a plot of land with several varieties of the *same* crop.

SUBSISTENCE

Agriculture carried out for survival—with few or no crops available for sale. It is usually organic, simply for lack of money to buy industrial inputs such as fertilizer, pesticides, or genetically modified seeds.

TILLAGE

Conventional method in which the surface is plowed which then breaks up and exposes the soil. This is then followed by smoothing the surface and planting. This method exposes the land to water and wind erosion.

Green Revolution

The first green revolution occurred between 1950 and 1970. It involved planting monocultures, using high applications of inorganic fertilizers and pesticides, and the widespread use of artificial irrigation systems. Before the first green revolution, crop production was correlated with increases in acreage under cultivation. After the first green revolution, crop acreage increased about 25%, but crop yield increased 200%. Crop yield then reached a plateau since it was easier and more economical to increase crop production through various agricultural techniques than to buy and clear new land.

The second green revolution began during the 1970s and is continuing today. It involves growing genetically engineered crops that produce the most yields per acre. It is in contrast with past agricultural practices in which farmers planted a variety of locally adapted strains. For example, of all wheat grown in the United States today, 50% comes from 9 different genotypes.

Genetic Engineering and Crop Production

Genetic engineering involves moving genes from one species to another or designing gene sequences with desirable characteristics. These include pest, drought, mold, and saline resistance, higher protein yields, and higher vitamin content. About 75% of all crops grown derive from genetically engineered or transgenic crop species.

Genetically Engineered Crops

Pros	Cons
May require less water and fertilizer	Unknown ecological effects
Higher crop yields	Less biodiversity
Less spoilage	May harm beneficial insects
Faster growth which may mean greater productivity, resulting in lower operating costs	May pose allergen risk
More resistant to disease, drought, frost, and insects	May result in mutations with unknown consequences
May be able to grow in saltier soils	May cause pesticide-resistant strains

CASE STUDY

Golden rice is produced by splicing three foreign genes, two from the daffodil and one from a bacterium, into a variety of rice that supplies vitamin A to populations that frequently suffer from vitamin A deficiency.

Irrigation

Three-quarters of all freshwater used on Earth is used for agriculture. Worldwide, approximately 40% of all crop yields come from 16% of all cropland that is irrigated. The use of irrigation depends on the climate and the degree of industrialization. For example, Canada irrigates about 10% of its crops, whereas India requires 90% of its crops to be irrigated. With inefficiencies such as seepage, leakage, and evaporation, up to 70% of all irrigation water can be lost. A drip irrigation system, which solves many of these problems but is more expensive to install, is used on approximately 1% of crops worldwide.

Increases in human population growth are outpacing the rate of land that is being irrigated. Sustainable irrigation is limited as a result of increases in costs, depletion of current sources of water, competition for water by urban areas, restoration of wetlands and fisheries, waterlogging, and salinization. Future water capacity will increase through increases in efficiency.

FACTOIDS

- Texas has lost approximately 15% of its irrigable land due to aquifer depletion.
- That 122 fish species in the western United States are endangered or threatened is due to changes in water level, salinization, or silting caused by irrigation.
- The Aral Sea has decreased 75% in volume and its salinity has increased 30% due to demand for irrigation of cotton.

Sustainable Agriculture

Sustainable agriculture involves a variety of approaches. It integrates three main goals: environmental health, economic profitability, and social and economic equity. Specific strategies must take into account topography, soil characteristics, climate, pests, local availability of inputs, and the individual grower's goals. Despite the site-specific and individual nature of sustainable agriculture, several general principles can be applied to help growers select appropriate management practices. These are described below.

Changes in agriculture have had many positive effects and reduced many risks in farming. However, there have also been significant costs. Prominent among these are topsoil depletion, groundwater contamination, the decline of family farms, continued neglect of the living and working conditions for farm laborers, increasing costs of production, and the disintegration of economic and social conditions in rural communities.

EFFICIENT USE OF INPUTS

Sustainable farmers maximize reliance on natural, renewable farm inputs with the goal to develop efficient, biological systems that do not need high levels of material inputs. Sustainable approaches are those that are the least toxic and least energy intensive and yet maintain productivity and profitability. Preventive strategies and other alternatives (such as integrated pest management) should be employed before using chemical inputs from any source.

SELECTION OF SITE, SPECIES, AND VARIETY

Preventive strategies, when adopted early, can reduce inputs and help establish a sustainable production system. When possible, pest-resistant crops should be selected that are tolerant of existing soil or site conditions. When site selection is an option, factors such as soil type and depth, previous crop history, and location (climate and topography) should be taken into account before planting.

SOIL MANAGEMENT

Proper soil, water, and nutrient management can help prevent some pest problems brought on by crop stress or nutrient imbalance. Furthermore, crop management systems that impair soil quality often result in greater inputs of water, nutrients, pesticides, and/or energy for tillage to maintain yields. In sustainable systems, the soil is viewed as a fragile and living medium that must be protected and nurtured to ensure its long-term productivity and stability. Methods to protect and enhance the productivity of the soil include using cover crops, compost, and/or manures, reducing tillage, and maintaining soil cover with plants and/or mulches. Regular additions of organic matter or the use of cover crops can increase soil aggregate stability, soil tilth and the diversity of soil microbial life.

SPECIES DIVERSITY

By growing a variety of crops, farmers spread out the economic risk and are less susceptible to the radical price fluctuations associated with changes in supply and demand. For example, in annual cropping systems, crop rotation can be used to suppress weeds, pathogens, and insect pests. Also, cover crops can have stabilizing effects on the agroecosystem by holding soil and nutrients in place, conserving soil moisture with dead mulches, and increasing the water infiltration rate and soil water-holding capacity. Optimum diversity may be obtained by integrating both crops and livestock in the same farming operation. Growing row crops on more level land and pasture or forages on steeper slopes will reduce soil erosion. Planting pasture and forage crops in rotation enhances soil quality and reduces erosion. Livestock manure, in turn, contributes to soil fertility. Livestock can buffer the negative impacts of low rainfall periods by consuming crop residue that in plant-only systems would have been considered crop failures. Feeding and marketing are flexible in animal production systems. This can help cushion farmers against trade and price fluctuations and, in conjunction with cropping operations, make more efficient use of farm labor and resources.

CONTROLLING PESTS

Pesticides can be used to control pests, but their use has drawbacks. Integrated pest management is another strategy to control pests.

Types of Pesticides

Pesticides differ in several ways. Their chemistry, how long they remain effective in the environment (environmental persistence), and their effect on the food web (bioaccumulation and biomagnification) are just a few concerns. Others include what type of organisms are affected, how the pesticides work (nervous system, reproductive cycles, blood chemistry), how fast they work, and their application.

BIOLOGICAL

Living organisms are used to control pests. Examples include bacteria, ladybugs, parasitic wasps, and certain viruses.

CARBAMATES

Carbamates, also known as urethanes, affect the nervous system of pests. 100 grams of a carbamate has the same effect as 2,000 grams of a chlorinated hydrocarbon such as DDT. Carbamates are more water soluble than chlorinated hydrocarbons, which brings a greater risk of them being dissolved in surface water and percolating into groundwater.

CHLORINATED HYDROCARBONS

Chlorinated hydrocarbons are synthetic organic compounds that affect the nervous system of pests. They are highly resistant to decomposition and can remain in the ecosystem for up to 15 years or longer. During the 1950s, DDT was linked with the thinning of eggshells in certain species of birds, such as the bald eagle.

FUMIGANTS

Used to sterilize soil and prevent pest infestation of stored grain.

INORGANIC

Broad-based pesticides. Includes arsenic, copper, lead, and mercury. Highly toxic and accumulate in the environment.

ORGANIC OR NATURAL

Natural poisons derived from plants such as tobacco or chrysanthemum.

ORGANOPHOSPHATES

Extremely toxic but remain in the environment for only a brief time. Examples include malathion and parathion.

Costs and Benefits of Pesticide Use

Despite the use of pesticides, many pests worldwide have increased in numbers. This is due to genetic resistance, reduced crop rotation, increased mobility of pests due to increased world trade, and reduction in crop diversity. The following table shows the pros and cons of pesticide use.

The Pros and Cons of Pesticides

Pros	Cons
Kill unwanted pests that carry disease	Accumulate in food chains
Increase food supplies	Pests develop resistance and create a pesticide treadmill
More food means food is less expensive	Estimates range from $5 to $10 in damage done to the environment for every $1 spent on pesticides; pesticides are expensive to purchase and apply
Newer pesticides are safer and more specific	Pesticide runoff and its effect on aquatic environments through biomagnification
Reduces labor costs	Inefficiency—only 5% of a pesticide reaches a pest
Agriculture is more profitable	Threatens endangered species and pollinators; also affects human health

Integrated Pest Management

Integrated pest management (IPM) is an ecological pest control strategy that uses a variety of methods. When used in combination, these methods working together can reduce or eliminate the use of traditional pesticides. The aim of IPM is not to eradicate pests but to control their numbers to acceptable levels. In integrated pest management, chemical pesticides are the last resort. Methods employed in IPM include:

- Polyculture
- Intercropping
- Planting pest-repellent crops
- Using mulch to control weeds
- Using pyrethroids or naturally occurring microorganisms (such as Bt) instead of toxic pesticides
- Natural insect predators
- Rotating crops often to disrupt insect cycles
- Using pheromones or hormone interrupters
- Releasing sterilized insects
- Developing genetically modified crops that are more insect resistant
- Regular monitoring through visual inspection and traps followed by record keeping
- Construction of mechanical controls such as traps, tillage, insect barriers, or agricultural vacuums equipped with lights

FACTOIDS

- Each year 2.5 million tons of 600 different types of pesticides are used for agricultural purposes.
- U.S. agriculture uses 77% of all pesticides but feeds only 5% of the world's population.
- Between 1945 and 1990, pesticide usage increased 300% on American farms.
- Crop loses due to pests in 1990 were about the same (37%) as in 1945.
- Nonnative high-yield crops require more pesticides than native plants that have evolved natural adaptations to local pests.
- Approximately 25 million farm workers worldwide are poisoned at some level by pesticides each year with estimates of over 220,000 deaths.
- The Environmental Protection Agency has ranked pesticide residue in foods as the third most serious cancer risk.
- The U.N. Food and Agriculture Organization reports nearly 500,000 tons of old and unused toxic pesticides have been abandoned in landfills.

RELEVANT LAWS

Federal Insecticide, Fungicide and Rodenticide Control Act (FIFRA) (1947): Regulates the manufacture and use of pesticides. Pesticides must be registered and approved. Labels require directions for use and disposal.

Federal Environmental Pesticides Control Act (1972): Requires registration of all pesticides in U.S. commerce.

Food Quality Protection Act (FQPA) (1996): Emphasizes the protection of infants and children in reference to pesticide residue in food.

FORESTRY

Forestry involves the management of forests. Sometimes this involves planting new trees, and sometimes it involves fires. This section will delve into this topic.

Ecological Services

Forests are an important global reserve. The ecological services of forests include:

1. Providing wildlife habitats
2. Carbon sinks
3. Affecting local climate patterns
4. Purifying air and water
5. Reducing soil erosion as they serve as a watershed, absorbing and releasing controlled amounts water
6. Providing energy and nutrient cycling

Tree Plantations

Tree plantations are large, managed commercial or government-owned farms with uniformly aged trees of one species (monoculture). Trees may not be native to the area and may be hybrids (genetically modified). The primary use of plantation trees is for pulp and lumber. Pine, spruce, and eucalyptus are widely used due to their fast growth rates and can be used for paper and timber. Trees are harvested by clear-cutting. Short rotation cycles of 25–30 years or 6–10 years in the tropics are economically important factors. Just 5% of the world's forests are tree plantations, but they account for 20% of the current world wood production. In comparison, 63% of the world's forests are secondary-growth forests, and 22% are old-growth forests.

Annually, tropical tree plantations yield much more wood (25 m^3/hectare) than traditional forests (1–3 m^3/hectare). Some of the natural closed forests—7%—are being lost in the tropics due to land conversion to tree plantations. Tree plantations do not support food webs found in old-growth forests, and they contain is little biodiversity. Decaying wood is absent, which provides a vital link in an old-growth forest. Conversion to tree plantations may result in draining wetlands and replacing traditional hardwoods. Newer techniques allow leaving blocks of native species within the plantation or retaining corridors of natural forest. The Kyoto Protocol encourages use of tree plantations to reduce carbon dioxide levels although carbon dioxide may eventually re-enter the atmosphere after harvesting.

Old-Growth Forests

Old-growth forests are forests that have not been seriously impacted by human activities for hundreds of years. Old-growth forests are rich in biodiversity. Old-growth forests are characterized by:

- Older and mixed-aged trees
- Minimal signs of human activity
- Multilayered canopy openings due to tree falls
- Pit-and-mound topography due to trees falling and creating new microenvironments by recycling carbon-rich organic material directly to the soil and providing a substrate for mosses, fungi (necessary for *in-situ* recycling), and seedlings
- Decaying wood and ground layer that provides a rich carbon sink
- Dead trees (snags) that are necessary nesting sites for woodpeckers and spotted owls
- Healthy soil profiles
- Indicator species
- A fungal ecosystem.

Depletion of old-growth forests increases the risk of climatic change. Many old-growth forests contain species of trees that have high economic value but that require a long time to mature (mahogany, oak, and so on).

Forest Fires

Forests are unique in terms of their ecological significance and the number and frequency of forest fires. Current wildfire frequency in the United States is about 4 times the average of 1970–1986. The total area burned by current fires is about 7 times its previous level. The U.S. Forest Service has lengthened the wildfire season by 78 days. The change in wildfire frequency appears to be linked to annual spring and summer temperatures. Longer, warmer summers are documented with an increase in the number of forest fires. A correlation exists between early arrivals of the spring snowmelt in the mountainous regions and the incidence of large forest fires. An earlier snowmelt can lead to an earlier and longer dry season, which provides greater opportunities for larger fires. As forests burn, they release their stored carbon as carbon dioxide into the atmosphere, further compounding the problems of global warming. Another reason for the increase in forest fires is a change in fire management philosophy. Any naturally started fire on federal land that is not threatening resources (homes or commercial structures) is allowed to burn.

CROWN FIRES

Occur in forests that have not had surface fires for a long time. Extremely hot. Burn entire trees and leap from treetop to treetop. Kills wildlife, increases soil erosion, and destroys structures.

GROUND FIRES

Fires that occur underground and burn partially decayed leaves. Common in peat bogs. Difficult to detect and extinguish.

SURFACE FIRES

Burns undergrowth and leaf litter. Kills seedlings and small trees. Spares older trees, and allows many wild animals to escape. Advantages: burns away flammable ground litter, reducing larger fires later; releases minerals back into soil profile; stimulates germination for some species with serotinous cones (requires heat to open up and release the seeds, such as giant sequoia and jack pine); helps to keep pathogens and insects in check; and allows vegetation to grow in clearings that provides food for deer, moose, elk, muskrat, and quail.

METHODS TO CONTROL FIRES

Two methods are used to control forest fires: prevention and prescribed burning. Prevention involves burning permits, closing parts of the forest during times of the year when the number of visitors is high and during periods of drought, and educating the public. Prescribed burning involves purposely setting controlled surface fires and setting small, prescribed fires to thin out underbrush in high-risk areas. It requires careful planning and monitoring. Other strategies include allowing fires to burn themselves out and creating large, clear areas around structures.

RELEVANT LAW

Healthy Forest Initiative (2003): Allows timber companies to cut down economically valuable trees in most national forests for 10 years. Timber companies in return must clear out small, more fire-prone trees and underbrush. Law may have consequences of increasing fires by accumulation of slash and increasing the number of fire-prone, younger trees.

Deforestation

Deforestation is the conversion of forested areas to nonforested areas. They are then used as grasslands for livestock grazing, grain fields, mining, petroleum extraction, fuel wood cutting, commercial logging, tree plantations, and urban sprawl. Natural deforestation can be caused by tsunamis, forest fires, volcanic eruptions, glaciation, and desertification. Deforestation results in a degraded environment with reduced biodiversity and reduced ecological services. Deforestation threatens the extinction of species with specialized niches, reduces the available habits for migratory species of birds and butterflies, decreases soil fertility brought about by erosion, and allows runoff into aquatic ecosystems. It also causes changes in local climate patterns and increases the amount of carbon dioxide released into the air from burning and tree decay. In addition to the direct effects brought about by deforestation, indirect effects caused by edge effects and habitat fragmentation can also occur.

Methods that are currently employed to manage and harvest trees include:

- **Even-age management**—Essentially the practice of tree plantations.
- **Uneven-age management**—Maintain a stand with trees of all ages from seedling to mature.
- **Selective cutting**—Specific trees in an area are chosen and cut.
- **High grading**—Cutting and removing only the largest and best trees.
- **Shelterwood cutting**—Removes all mature trees in an area within a limited time.
- **Seed tree cutting**—Majority of trees are removed except for scattered, seed-producing trees used to regenerate a new stand.
- **Clear-cutting**—All of the trees in an area are cut at the same time. This technique is sometimes used to cultivate shade-intolerant tree species.
- **Strip cutting**—Clear-cutting a strip of trees that follows the land contour. The corridor is allowed to regenerate.

Clear-cutting has several pros and cons, as presented in the following table.

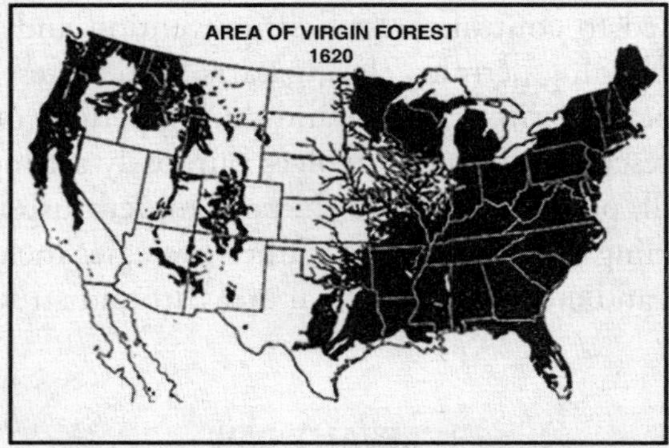

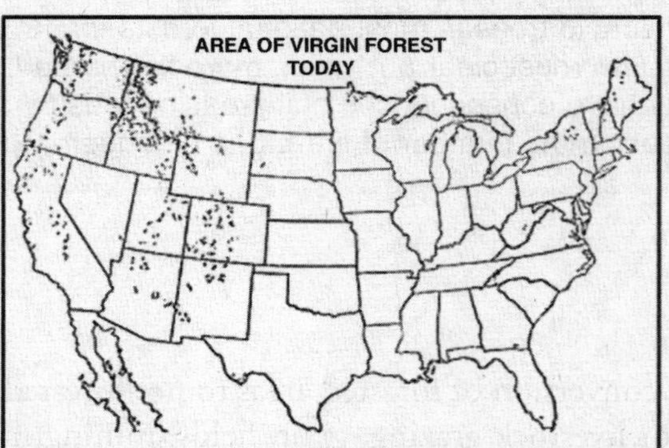

Figure 7.1 Deforestation in the United States (1620–present)

Deforestation alters the hydrologic cycle, potentially increasing or decreasing the amount of water in the soil and groundwater. This then affects the recharge of aquifers and the moisture in the atmosphere. Shrinking forest cover lessens the landscape's capacity to intercept, retain, and transport precipitation. Instead of trapping precipitation, which then percolates to groundwater systems, deforested areas become sources of surface water runoff, which moves much faster than subsurface flows. The faster transport of surface water can translate into flash flooding and more extreme floods than would occur with the forest cover.

Tree Plantations

Pros	Cons
Practical method for trees that require full or moderate sunlight in order to grow.	Reduces recreational value of land.
Practical method for tree plantations.	If done on steeply sloped areas, will often cause soil erosion, water pollution, and flooding.
Genetically improved species of trees that resist disease and grow faster can be grown.	Causes habitat fragmentation.
Increases economic returns on investments.	Reduces biodiversity.
Produces a high yield of timber at the lowest cost, and provides jobs.	Promotes monoculture and tree plantations that are prone to disease or infestation through the lack of diversity.

Deforestation also contributes to decreased evapotranspiration. This lessens atmospheric moisture and precipitation levels. It also affects precipitation levels downwind from the deforested area as water is not recycled to downwind forests but, instead, is lost in runoff and returns directly to the oceans. Forests are also important carbon sinks. Forests can extract carbon dioxide and pollutants from the air, thus contributing to biosphere stability and reducing the greenhouse effect. Forests are also valued for their aesthetic beauty and as a cultural resource and tourist attraction.

Three schools of thought exist with regards to the causes of deforestation—the impoverished school, the neoclassical school, and the political-ecology school. The impoverished school believes that the major cause of deforestation is the growing number of poor. The neoclassical school believes that the major cause is "open-access property rights." The political-ecology school believes that the major cause of deforestation is due to entrepreneurs.

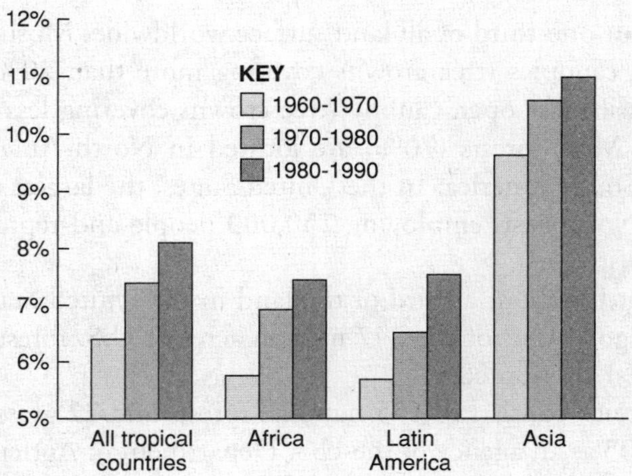

Figure 7.2 Rate of deforestation in tropical rain forests

CASE STUDY

The Hubbard Brook Experimental Forest centered on how deforestation affects nutrient cycles. The forest consisted of several watersheds each drained by a single creek. Impervious bedrock was close to the surface, which prevented seepage of water from one forested hillside, valley, and creek ecosystem to the other. Several conclusions were made. First, an undisturbed mature forest ecosystem is in dynamic equilibrium with respect to chemical nutrients. Nutrients leaving an ecosystem are balanced by nutrients entering the ecosystem. Second, inflow and outflow of nutrients was low compared with levels of nutrients being recycled within the ecosystem. Third, when deforestation occurred, water runoff increased. Consequently, soil erosion increased, which caused a large increase in the outflow of nutrients from the ecosystem. Increases in outflow of nutrients caused water pollution. Fourth, nutrient loss could be reduced by clearing trees and vegetation in horizontal strips. Remaining vegetation reduced soil erosion.

FACTOIDS

- Tropical forests hold an immense amount of carbon, which joins with oxygen to form carbon dioxide. The plants and soil of tropical forests hold approximately 550 billion metric tons of carbon worldwide. Each acre of tropical forest stores about 180 metric tons of carbon.
- From 1850 to 1990, deforestation worldwide (including the United States) released 122 billion metric tons of carbon into the atmosphere, with the current rate being about 1.6 billion metric tons per year. In comparison, all of the fossil fuels (coal, oil, and gas) burned during a year release about 6 billion tons.
- Tropical deforestation releases about 25% of the carbon dioxide amount released by fossil fuel burning.

Forest Management

Forests cover about one-third of all land surface worldwide. Most—80%—of these forests are closed canopies (tree crowns covering more than 20% of the ground), and 20% are classified as open canopy (tree crowns covering less than 20% of the ground surface). Most forests (70%) are located in North America, the Russian Federation, and South America. In the United States, the largest area of timbering is in the Pacific Northwest, employing 150,000 people and representing a $7 billion per year industry.

Forests account for about a third of the land in the United States, the largest of any land use category. Out of the 747 million acres of U.S. forest land, two-thirds (500 million acres) are nonfederal.

The Forest Service consists of 155 national forests and 22 grasslands, and it was established in 1905 as an agency of the U.S. Department of Agriculture. The Forest Service manages public lands in national forests and grasslands and encompasses

193 million acres (close to the size of Texas). These resources are used for logging, farming, recreation, hunting, fishing, oil and gas extraction, watersheds, mining, livestock grazing, farming, and conservation purposes.

The Forest Service protects and manages natural resources on National Forest System lands. It sponsors research on all aspects of forestry, rangeland management, and forest resource utilization. It provides community assistance and cooperation with state and local governments, forest industries and private landowners to help protect and manage nonfederal forests, associated ranges, and watersheds to improve conditions in rural areas. The Forest Service also provides international assistance in formulating policy and coordinating U.S. support for the protection and sound management of the world's forest resources.

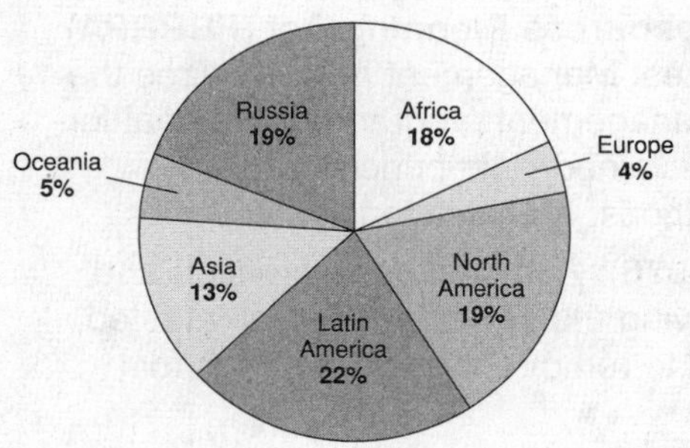

Figure 7.3 World forest distribution

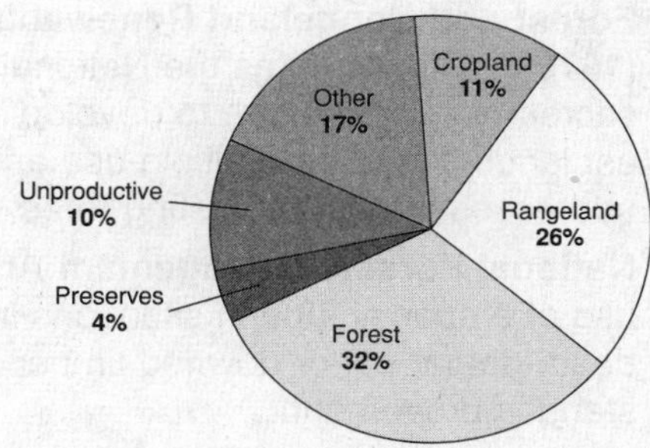

Figure 7.4 World land use

FACTOIDS

- Each American uses the equivalent of a 100-foot (30 m) tree each year.
- The average single-family home uses 5,824 board feet (1,775 board m) of timber.
- To grow a pound of wood, a tree uses 1.5 pounds (0.7 kg) of carbon dioxide and gives off 1 pound (0.5 kg) of oxygen.
- One acre (0.4 ha) of forest produces 4,000 pounds (1,800 kg) of wood per year, uses 6,000 pounds (2,700 kg) of carbon dioxide, and gives off 4,000 pounds (1,800 kg) of oxygen.
- Developed countries use 80% of all wood products but produce less than 50% of the wood required.
- More than 50% of the world's population relies on firewood or charcoal for heating and/or cooking. Inexpensive ceramic stoves would save 4 times the amount of wood as cooking over an open fire.
- Most of the total bird, amphibian, and fish species—90%—and 80% of mammal and reptile species are found in forested lands.

RELEVANT LAWS

Forest Reserve Act (1891): Gave the president authority to establish forest reservations from public domain lands.

Multiple Use and Sustained Yield Act (1960, 1968): Directs the U.S. Secretary of Agriculture to manage national forests for recreation, wildlife habitat, and timber production through principles of multiple use and sustained yield.

Federal Land Policy and Management Act (FLPMA) (1976): Along with the Taylor Grazing Act, outlines policy concerning the use and preservation of public lands in the United States. Grants federal government jurisdiction on consequences of mining on public lands. Grants Bureau of Land Management (Department of the Interior) responsibility to manage all public lands not within national forests or national parks—a multiple-use policy.

Forest and Rangeland Renewable Resources Planning Act (FRRRPA) (1976): Also known as the National Forest Management Act. Requires the secretary of agriculture to develop a management program for national forest lands based on multiple-use and sustained yield principles. Also addresses timber-harvesting rates, methods, and locations.

National Forests Management Act (1976): Authorized the creation and use of a special fund in situations involving the salvage of insect-infested, dead, damaged, or downed timber and to remove associated trees for stand improvement.

RANGELANDS

Rangelands are being compromised by overgrazing and desertification. The federal government is trying to manage and sustain the rangelanads.

Overgrazing

Overgrazing, which was the theme of the essay "The Tragedy of the Commons," occurs when plants are exposed to grazing for too long without sufficient recovery periods. When a plant is grazed severely, it uses energy stored in its roots to support regrowth. As this energy is used, the roots die back. The degree to which the roots die back depends on the severity of the grazing. Root dieback does add organic matter to the soil, which increases soil porosity, the infiltration rate of water, and the soil's moisture-holding capacity. If sufficient time has passed, enough leaves will regrow and the roots will regrow as well. A plant is considered overgrazed when it is regrazed before the roots recover. Overgrazing can reduce root growth by up to 90%.

These are consequences to overgrazing. Pastures are less productive, soils have less organic matter and become less fertile, and soil porosity decreases. The infiltration rate and moisture-holding capacity of the soil drops and susceptibility to soil compaction increases. Additionally, desirable plants become stressed, while weedier species thrive in these harsher conditions. Overgrazing causes biodiversity to decrease by reducing native vegetation, which leads to erosion. Riparian areas are affected by cattle destroying banks and streambeds, thereby increasing silting. Eutrophication increases due to cattle wastes. Therefore, aquatic environments are negatively impacted. Predator-prey relationships and the balance achieved through predator control programs are affected. Overgrazing increases the incidence of disease in native plant species. Finally, land is affected to the point where sustainability is threatened.

Desertification

Desertification is the conversion of marginal rangeland or cropland to a more desert-like land type. It is often caused by overgrazing, soil erosion, prolonged drought, or climate changes as well as the overuse of available resources such as nutrients and water. Desertification proceeds with the following steps: First, overgrazing results in animals eating all available plant life. Next, rain washes away the trampled soil, since nothing holds water anymore. Wells, springs, and other sources of water dry up. What vegetation is left dies from drought or is taken for firewood. Then weeds that are unsuitable for grazing may begin to take over. The ground becomes unsuitable for seed germination. Finally, wind and dry heat blow away the topsoil.

Federal Rangeland Management

Rangelands comprise about 40% of the landmass of the United States and are the dominant type of land in arid and semiarid regions. Nearly 80% of the lands of the western United States are classified as rangelands, whereas only 7% of some areas near the East Coast are classified as rangelands. Rangelands provide valuable grazing lands for livestock and wildlife. Rangelands serve as a source of high-quality water, clean air, and open spaces. They benefit people as a setting for recreation and as an economic means for agriculture, mining, and living communities. Rangelands serve multiple purposes:

- A habitat for a wide array of game and nongame animal species
- A habitat for a diverse and wide array of native plant species
- A source of high-quality water, clean air, and open spaces
- A setting for recreational hiking, camping, fishing, hunting, and nature experiences
- The foundation for low-input, fully renewable food production systems for the cattle industry

Jurisdiction of public grazing rangelands is coordinated through the Forest Service and the Bureau of Land Management (BLM). Before 1995, grazing policies were determined by rancher advisory boards composed of permit holders. After 1995, resource advisory councils were formed made up of diverse groups representing different viewpoints and interests. Forty percent of all federal grazing permits are owned by 3% (or approximately 2000) of all livestock operators. Federal grazing permits average about 5 cents per day per animal through federal subsidies. The true cost of doing business would make this fee closer to $10 to $20 per animal per day.

Methods of rangeland management include:

1. Controlling the number and distribution of livestock so that the carrying capacity is not exceeded
2. Restoring degraded rangeland
3. Moving livestock from one area to another to allow the rangeland to recover
4. Fencing off riparian (stream) areas to reduce damage to these sensitive areas

5. Suppressing the growth of invasive plant species
6. Replanting barren rangeland with native grass seed to reduce soil erosion
7. Providing supplemental feed at selected sites
8. Locating water holes, water tanks and salt blocks at strategic points that do not degrade the environment

Land administered by the BLM is inhabited by 219 endangered species of wildlife. Livestock grazing is the fifth-rated threat to endangered plant species, the fourth leading threat for all endangered wildlife, and the number one threat to all endangered species in arid regions of the United States.

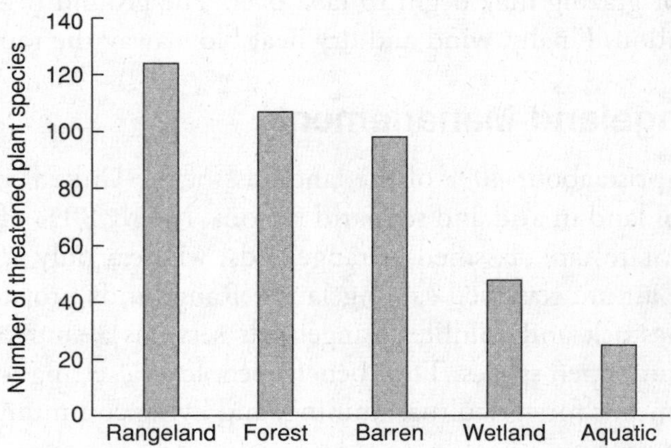

Figure 7.5 Endangered plant species as a function of biome

FACTOIDS

- Government subsidies to cattle ranching do not trickle down. Only 0.06% of all jobs concerning rangelands and grazing involve cattle ranching.

RELEVANT LAWS

Renewable Resources Planning Act (RPA) (1974): Mandates periodic assessments of forests and rangelands in the United States. Directs that the assessment be conducted by the U.S. Forest Service and consider a broad range of renewable resources, including outdoor recreation, fish, wildlife, water, range, timber, and minerals.

Public Rangelands Improvement Act (1978): Established and reaffirmed a commitment to manage, maintain, and improve rangelands so that they become as productive as feasible.

Taylor Grazing Act (1934): Requires grazing permits on federal land.

URBAN LAND DEVELOPMENT

The following sections discuss another form of land use—urban land development.

Planned Development

There are more than 76 million residential buildings and approximately 5 million commercial buildings in the United States alone. Together, these buildings use one-third of all the energy and two-thirds of all the electricity consumed in the United States. Energy needs of buildings account for almost half of the sulfur dioxide emissions, one-fourth of the nitrous oxide emissions, and one-third of the carbon dioxide emissions.

Green building and city characteristics focus on whole-system approaches. They include:

- Energy conservation through government and private industry rebates and tax incentives for solar and other less-polluting forms of energy
- Resource-efficient building techniques and materials
- Indoor air quality
- Water conservation through the use of xeriscaping
- Designs that minimize waste while utilizing recycled materials
- Placing buildings whenever possible near public transportation hubs that use a multitude of venues such as light rail, subways, and park and rides
- Creating environments that are pedestrian friendly by incorporating parks, greenbelts, and shopping areas in accessible areas
- Preserving historical and cultural aspects of the community while at the same time blending into the natural feeling and aesthetics of a community

FACTOIDS

- A standard wood-framed house consumes over 1 acre (0.4 ha) of forest to build with 3–7 tons of construction wastes.
- If only 10% of homes in the United States used solar water-heating systems, it would eliminate 8.5 million tons (8.5 million t) of carbon emissions per year.
- Proper landscaping can reduce heating needs by 30%, air-conditioning needs by 75%, and water consumption by 80%.
- More than 1,500 bacterial, fungal, and chemical air pollutants have been identified in buildings.

Suburban Sprawl and Urbanization

Urbanization refers to the movement of people from rural areas to cities and the changes that accompany it. Areas that are experiencing the greatest growth in urbanization are countries in Asia and Africa. Asia alone has close to half of the world's urban inhabitants even though 60% of its population still lives in rural areas. Africa, which is generally considered overwhelmingly rural, now has a larger urban population than North America. Reasons for this include access to jobs, higher standards of living, easier access to health care, mechanization of agriculture, and access to education. Nations with the most rapid increases in their urbanization rates are generally those with the most rapid economic growth. From 1950 to 1990, the world's economy increased fivefold. The following table lists the pros and cons of urbanization.

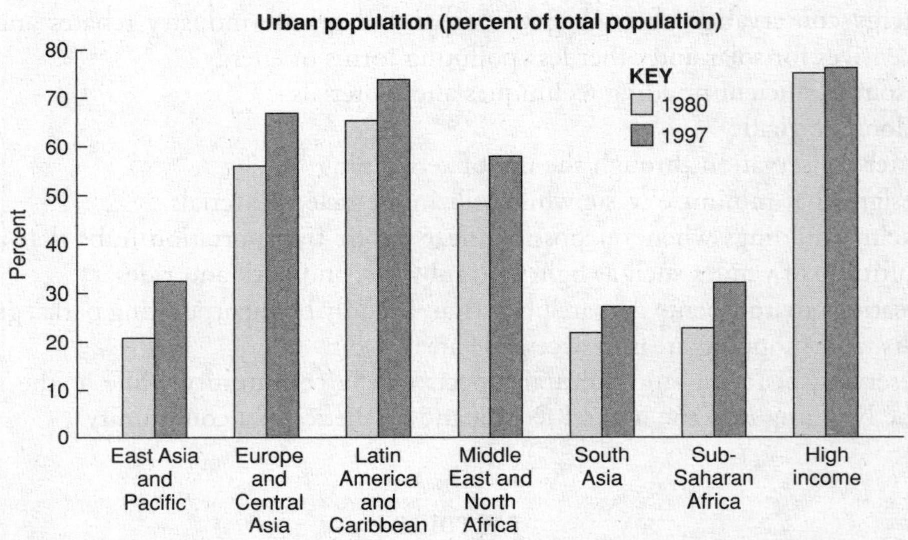

Figure 7.6 Urban population as a function of geography and income

FACTOIDS

- In 1850, 2% of the world's population (10,000 people) lived in urban areas. In 2000, this percentage had increased to 50%. By the year 2025, estimates are that 63% will live in urban areas.
- In the United States, urban population was about 5% in 1850 and had increased to 75% in 2000.
- In 1900, there were 19 cities worldwide that had populations greater than 1 million. In 2000, that number had increased to 400 cities, 19 of which had populations greater than 10 million.
- 17% of the people who live in urban areas worldwide live in slums or shantytowns.
- 90% of untreated urban sewage in less-developed countries is released into local surface waters.

Urbanization

Pros	Cons
Uses less land—less impact on the environment.	Impact on land is more concentrated and more pronounced. Examples include water runoff and flooding.
Better educational delivery system.	Overcrowded schools.
Mass transit systems decrease reliance on fossil fuels—commuting distances are shorter.	Commuting times are longer because the infrastructure cannot keep up with growth.
Better sanitation systems.	Sanitation systems have greater volumes of wastes to deal with.
Recycling systems are more efficient.	Solid-waste buildup is more pronounced. Landfill space becomes scarce and costly.
Large numbers of people generate high tax revenues.	Large numbers of poor people place strains on social services. This results in wealthier people moving away from urban areas into suburbs and decreasing the tax base.
Urban areas attract industry due to availability of raw materials, distribution networks, customers, and labor pool.	Higher population densities increase crime rates. Population increase may be higher than job growth.
Much of the pollution comes from point sources, enabling focused remediation techniques.	Since population densities are high, pollution levels are also high (urban heat islands, ozone levels, and water and soil pollution).

TRANSPORTATION INFRASTRUCTURE

Transportation can be via roadways or water channels. Areas without ransportation infrastructure suffers an ecosystem impact.

Federal Highway System

The federal highway system contains approximately 160,000 miles (256,000 km) of roadway important to the nation's economy, defense, and mobility. Although interstate highways usually receive substantial federal funding and comply with federal standards, they are owned, built, and operated by the states in which they are located. Current federal highway taxes include an 18-cents per gallon tax on gasoline, a 25-cent per gallon tax on diesel, and a tax on heavy vehicles.

The system serves all major U.S. cities. Unlike counterparts in most industrialized countries, interstates go through downtown areas and facilitate urban sprawl. The distribution of virtually all goods and services involve interstate highways at some point. Residents of American cities commonly use urban interstates to travel to their employment.

An efficient and well-maintained federal highway system can have the following impacts on the environment:

LESS POLLUTANTS

Vehicles caught in stop-and-go traffic emit far more pollutants, particularly carbon monoxide, nitrogen oxides, and smog-forming volatile organic compounds, than they do without frequent braking and acceleration.

REDUCE GREENHOUSE GASES

Improving traffic flow and reducing congestion will decrease atmospheric carbon dioxide.

IMPROVE FUEL ECONOMY AND REDUCE FOREIGN OIL DEPENDENCE

For vehicles stuck in traffic, not only do tailpipe emissions go up but fuel economy goes down. Modest improvements to the nation's worst traffic bottlenecks would save 1 billion gallons of fuel each year.

IMPROVE THE ECONOMY

Interstates return $6 in economic productivity for every $1 invested.

IMPROVE THE QUALITY OF LIFE

An interstate highway system allows products (both perishable and nonperishable) to be distributed throughout the country in a short period of time.

> **RELEVANT LAW**
>
> **Federal Aid Highway Act (1956):** Authorized construction of the Interstate Highway System. Under the new law, the federal government agreed to fund 90% of the construction costs for the interstates. States, in turn, would provide the remaining funds, administer the construction projects, and own and operate the completed interstate highways.

Canals and Channels

The term channel is another word for strait, which is defined as a relatively narrow body of water that connects two larger bodies of water. Channels can occur naturally or be constructed. Repeated dredging of canals and channels is often necessary because of silting.

In the United States, channels frequented by ships are generally maintained by the United States Department of the Interior and monitored and policed by the United States Coast Guard. Smaller channels are maintained by the various states or local governments.

The two largest canals in the world of major economic value are the Panama Canal and the Suez Canal. The 48-mile (77 km) Panama Canal connects the Pacific Ocean with the Atlantic. It allows water transport without having to circumnavigate South America. The 163-mile (190 km) Suez Canal connects the Red Sea with the Mediterranean. It allows water transport between Europe and Asia without traveling around Africa, and 8% of the world's shipping runs through the Suez Canal.

CASE STUDY

Gatún Lake is an artificial lake created for the operation of the locks in the Panama Canal. A significant environmental problem that occurs with Gatún Lake is the drainage of water (up to 52 million gallons [197 million L] of freshwater leaves the lake every time a ship passes through the canal). Although there is sufficient annual rainfall to replenish the water used by the canal, the seasonal nature of the rainfall requires storage of water from one season to the next. The rainforest surrounding the lake traditionally played a major role by absorbing rainwater and releasing it slowly. However, due to recent deforestation of the land around the lake, rain flows rapidly down through the deforested slopes, carrying silt into the lake. The excess water is spilled out into the ocean. This results in a shortfall during the dry season and silt buildup in the lake, which must be periodically dredged.

Roadless Areas and Ecosystem Impacts

Roadless areas are places where no roads have been built and where, as a result, no logging or other development can occur. Roadless areas are havens for fish and wildlife whose habitats in many other forest areas has been fragmented or entirely destroyed. They provide habitats for more than 1,600 threatened, endangered, or sensitive plant and animal species and include watersheds that supply clean drinking water. The roadless rule protects 60 million acres, or 31%, of National Forest System lands—about 2% of the total land base of the United States. National wilderness areas are roadless, such as the Bob Marshall Wilderness Area in the western United States.

RELEVANT LAW

Roadless Area Conservation Rule (2001): This rule places about one-third of the national forest system's total acreage off-limits to virtually all road building and logging. More than half of the national forest land is already open to such activity. The plan protects 59 million acres (23 million ha) of unspoiled national forest land in 39 states. It preserves all current opportunities for public access and recreation, including hiking, fishing, hunting, camping, and mountain biking, as well as the revenue and jobs that these activities generate in local areas.

PUBLIC AND FEDERAL LANDS

The federal government manages public lands. It sets aside areas as national parks, wildlife refuges, and wetlands.

Management

The Bureau of Land Management (BLM) is responsible for managing 262 million acres (105 million ha) of land, about one-eighth of the land in the United States. The BLM also manages about 300 million additional acres (120 million ha) of subsurface mineral resources. The bureau is also responsible for wildfire management and preservation on 400 million acres (162 million ha). Most of the lands the BLM manages are located in the western United States, including Alaska. They are dominated by extensive grasslands, forests, high mountains, arctic tundra, and deserts. The BLM manages a wide variety of resources and uses including energy and minerals, timber, forage, wild horse and burro populations, fish and wildlife habitats, wilderness areas, and archaeological, paleontological, and historical sites.

> **RELEVANT LAW**
>
> **Federal Land Policy and Management Act (1976):** Outlined policy concerning the use and preservation of public lands. Granted federal jurisdiction on consequences of mining on public lands.

National Parks

There are over 1,100 national parks in the world today. However, many of them do not receive proper protection from poachers, loggers, miners, or farmers due to the costs involved.

The United States National Park System encompasses approximately 84 million acres (34 million ha), of which more than 4 million acres (1.6 million ha) remain in private ownership. The largest area is in Alaska and is more than 16% of the entire system. U.S. National parks are threatened by high demand by large numbers of visitors, which leads to congestion, eroded trails, noise that disrupts wildlife, and pollution from autos and visitors. Other threats include off-road vehicles, introduction of nonnative species that impact biodiversity, and commercial activities such as mining, logging, livestock grazing, and urban development.

There are several solutions to these problems:

- Reducing the amount of private land within national parks through incentives to current owners
- Providing education programs to the public
- Setting quotas on attendance through advanced reservation systems
- Adopting a fee system that covers external costs
- Banning off-road vehicles
- Banning autos and instead provide shuttle buses to control traffic
- Providing tax incentives for property owners near national parks to use land grants
- Conducting periodic and detailed wildlife and plant inventories

RELEVANT LAWS

Yellowstone National Park Act (1872): Preserves the watershed of the Yellowstone River "for the benefit and enjoyment of the people." For the first time, public lands were preserved for public enjoyment and were to be administered by the federal government.

National Park Service Act (1916): Established that national parks are to be maintained in a manner that leaves them unimpaired for future generations and established the National Park Service to manage the parks.

Outdoor Recreation Act (1963): Laid out the Interior Department's role as coordinator of all federal agencies for programs affecting the conservation and development of recreation resources.

Wilderness Act (1964): Wilderness was defined by its lack of noticeable human modification or presence. Federal officials are required to manage wilderness areas in a manner conducive to retention of their wilderness character.

Land and Water Conservation Fund Act (1965): Established a fund, administered by the National Park Service, to assist the states and federal agencies in meeting present and future outdoor recreation demands and needs of the American people.

National Trails System Act (1968): Established a national system of recreational, scenic, and historic trails.

Wild and Scenic Rivers Act (1968): Established a system of areas distinct from the traditional park concept to ensure the protection of each river's unique environment. It also preserves certain selected rivers that possess outstanding scenic, recreational, geological, cultural, or historic values and maintains their free-flowing condition.

Wildlife Refuges

President Theodore Roosevelt designated 4-acre (1.6 ha) Pelican Island, off Florida, in 1903 as the first national wildlife refuge, designed to protect breeding birds. Roosevelt designated another 52 wildlife refuges before he left office in 1909. The early refuges were established primarily to protect wildlife such as the overhunted bison and birds killed by market hunters, such as egrets and waterfowl. During the drought years of the Great Depression, refuges were created to protect waterfowl. The system developed piecemeal largely in response to such wildlife crises. The National Wildlife Refuge System, consisting today of 547 refuges encompassing more than 93 million acres (37 million ha), is managed by the U.S. Fish and Wildlife Service.

Wetlands

Wetlands are areas that are covered by water and support plants that can grow in water-saturated soil. High plant productivity supports a rich diversity of animal life. Countries with the most wetlands are Canada (14% of land area), the Russian Federation (including Siberia), and Brazil. Wetlands were once about 10% of the land area in the United States but have been reduced to about 5%, with most of them in Louisiana and Florida. Most wetland habitat loss—90%—is due to conversion of the land to agriculture with the rest of the loss due to urbanization. Wetlands are home to a wide variety of species. One-third of all endangered species in the United States spend part of their life span in wetlands. Wetlands serve as natural water purification systems removing sediments, nutrients, and toxins from flowing water. Wetlands along lakes and oceans stabilize shorelines and reduce damage caused by storm surges, reduce the risks of flooding, and reduce saltwater intrusion.

Fens are wetlands characterized by continuous sources of groundwater rich in magnesium and calcium, which makes a fen very alkaline. This groundwater comes from glaciers that have melted, depositing their water in layers of gravel and sand. The water sits upon layers of soil (glacial drift) that are not permeable, thus keeping the water from sinking beneath the surface. The water is then forced to flow sideways along the surface, where it picks up minerals in its path.

A bog is a type of wetland that accumulates acidic peat, a deposit of dead plant material that can be dried and burned for fuel. Bogs are located in cold, temperate climates. They are usually in boreal biomes such as western Siberia, parts of Russia, Ireland, Canada, and the states of Minnesota and Michigan. Bogs are generally low in nutrients and highly acidic. Carnivorous plants have adapted to these conditions by using insects as their nutrient source.

LAND CONSERVATION OPTIONS

Preservation, remediation, mitigation, restoration and sustainable land use strategies are land conservation options. Several principles can be employed in land conservation using a land-use ethic model:

1. Protect biodiversity, wildlife habitats, and the ecological functioning of public land ecosystems through careful monitoring and enforcement.
2. Adopt a user pay approach for extracting resources from public lands. Eliminate government subsidies and tax breaks to corporations that extract publicly owned resources.
3. Institute fair compensation for resources extracted from public land. Instead of the government subsidizing the extraction of resources, the corporations should be paying the government fair market value for natural resources.
4. Require responsibility for any user who damages or alters public land.
5. Adopt uneven-aged management forestry practices that foster maintaining a variety of tree species at various ages and sizes. Uneven-aged management fosters biological diversity, long-term sustainable production of high-quality timber, selective cutting, and the principle of multiple use of the forests for recreation, watershed protection, wildlife, and timber.
6. Include ecological services of trees in estimating valuation.
7. Reduce road building into uncut forest areas. Require restoration plans for those roads that are currently in place, and require such plans for any future roads.
8. Coordinate with the Forest Service on leaving fallen timber and standing dead trees in place to promote nutrient cycling and providing wildlife habitats.

9. Grow timber on longer rotations.
10. Reduce or eliminate clear-cutting, shelter wood cutting, or seed tree cutting on sloped land.
11. Rely on more sustainable tree-cutting methods such as selective and strip cutting.
12. Reduce fragmentation of remaining large forests.
13. Require certification of lumber that is cut according to sustainable forest practices.
14. Use sustainable techniques for tropical forests. These include: educating settlers about sustainable forest practices and their advantages, monitoring and enforcing cutting based on sound ecological principles, and reducing subsidies that encourage tropical deforestation. Other techniques include instituting debt-for-nature and conservation easements, creating subsidies for sustainable practices, and rehabilitating areas that have already been degraded.
15. Solutions to urban land use problems include zoning. Various parcels of land are designated for certain uses but can be influenced by developers. Local governments can limit permits, require environmental impact analyses, require developer fees for services, tax farmland on the basis of its actual (not potential) use, promote compact development and preservation of open space, and establish green spaces, urban boundaries, and land trusts.

The following items describe some land conservation options:

PRESERVATION OR SUSTAINABLE

To keep or maintain intact.

REMEDIATION

The act or process of correcting a fault or deficiency.

MITIGATION

To moderate or alleviate in force or intensity.

RESTORATION

To restore to its former good condition. Ecosystem restoration involves management actions designed to facilitate the recovery or reestablishment of native ecosystems. A central premise of ecological restoration is that restoration of natural systems to conditions consistent with their evolutionary environments will prevent their further degradation while simultaneously conserving their native plants and animals.

MINING

The following table provides an overview of mining.

Overview of Mining

Steps	Descriptions	Environmental Effects and Issues
Mining	Removing mineral resource from the ground. Can involve underground mines, drilling, room-and-pillar mining, long-wall mining, open pit, dredging, contour strip mining, and mountaintop removal.	Mine wastes—acids and toxins. Displacement of native species. Reclamation of land and recycling.
Processing	Removing ore from gangue. Involves transportation, processing, purification, smelting, and manufacturing.	Pollution (air, water, soil, and noise). Human health concerns, risks, and hazards.
Use	Involves distribution to end user.	

Extraction

Before mining begins, economic decisions are made to determine whether a site will be profitable. Factors that enter into the decision include current and projected price, amount of ore at the site, concentration, type of mining required, cost of transporting the ore to a processing facility, and cost of reclamation. After all factors are analyzed, several steps are employed.

SITE DEVELOPMENT

Samples are taken from an area to determine the quality and quantity of minerals in a location. Roads and equipment are brought in.

EXTRACTION

Three methods can be employed. Surface mining is the first. Earth-moving equipment is employed. This is a safe form of mining and a cost-effective one. However, topsoil and vegetation are removed, air quality issues develop regarding dust, large amounts of waste materials are involved, and pits are dug below the water table so groundwater is pumped out. The water that is pumped out affects fish and wetlands, through erosion, sedimentation, dewatering of wetlands, and diverting and channelizing streams.

The second method is underground mining. Large shafts are dug into the Earth. There is less surface destruction and waste rock produced than in surface mining, but it is unsafe. Acid mine drainage containing pyrites and other sulfides are produced, affecting air and water quality.

The third method is *in-situ* leaching. Small holes are drilled into the site. Water-based chemical solvents are used to flush out desired minerals. Waste rock is minimal. Safer procedures are used by miners. It is less expensive since rocks do not have to be broken up or removed. There are shorter lead times to production, less surface ground disturbance, and less required remediation. On the negative side, fluids injected into Earth are toxic and enter the groundwater supply.

PROCESSING

Processing involves intensive chemical processing during smelting. This is the method by which a metal is obtained from its ore, either as an element or as a simple compound. It is usually accomplished by heating beyond the melting point, ordinarily in the presence of reducing agents such as coke or oxidizing agents such as air. A metal whose ore is an oxygen compound (iron, zinc, or lead oxides) is heated (reduction smelting) in a blast furnace to a high temperature. The oxide combines with the carbon in the coke, escaping as carbon monoxide or carbon dioxide. Other impurities are removed by adding flux, with which they combine to form slag. If the ore is a sulfide mineral (copper, nickel, lead, or cobalt sulfides) air or oxygen is introduced to oxidize the sulfide to sulfur dioxide and any iron to slag, leaving the metal.

In cyanide heap leaching, gold ore is heaped into a large pile. Cyanide solution is then sprayed on top of the pile. As the cyanide percolates downward, the gold leaches out of the ore and collects in pools at the bottom. The gold extracted may be only 0.01% of the total ore processed. Liquid wastes containing cyanide and other toxins are kept in tailing ponds, which eventually leak and enter groundwater supplies.

Global Reserves

Two billion tons (2 billion t) of minerals are extracted and used each year in the United States (about 10 tons [10 t] for every American). At the same time, the United States imports more than 50% of its most needed minerals. As mineral reserves become depleted, lower grades of ore are mined, which causes more processing and consequently more pollution.

The United States, Germany, and Russia represent 8% of the world's population, yet they consume 75% of the most widely used metals, with the United States consuming 20%.

OIL

A large portion of Earth's global crude oil reserve—45% to 70%—has already been depleted. It is estimated that there is a 50-year supply left on Earth, with countries in the Middle East owning about half of what's left. The United States owns 3% of the world's oil reserves but uses 30% of the oil extracted in the world each year. Increased competition for foreign oil by China and India increase the world's cost of oil. Two-thirds of the oil used in the United States is used for transportation (gasoline, diesel, jet fuel, and so on). About one-fourth of it is used in industry (such as plastics, medicines, and cosmetics). Oil imports in the United States increased from 52% in 1996 to a projected 70% by the year 2010.

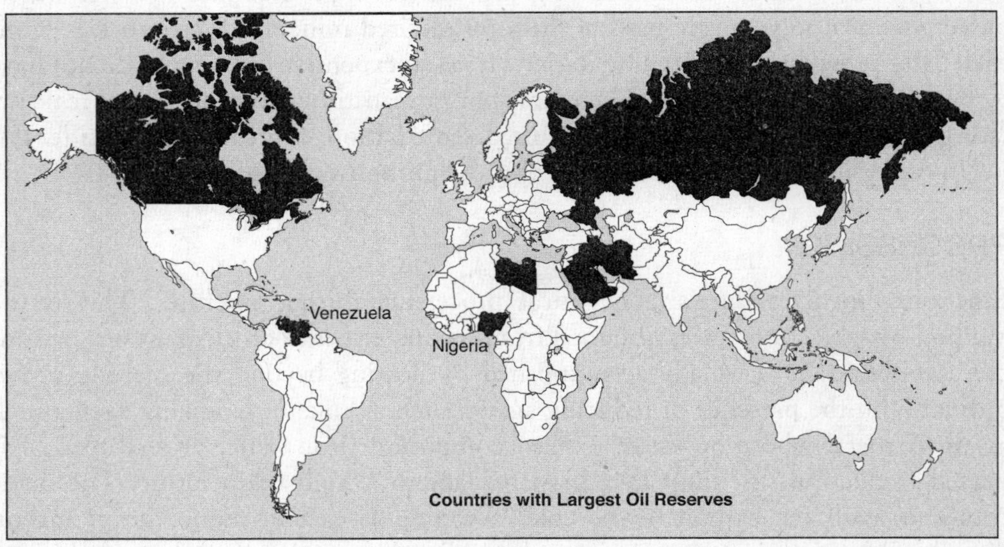

Figure 7.7 Countries with the largest oil reserves

COAL

Coal is currently the world's single largest source of fuel used to produce electricity. China is the world's largest producer. Global coal reserves are estimated to last about 300 years at current levels of extraction.

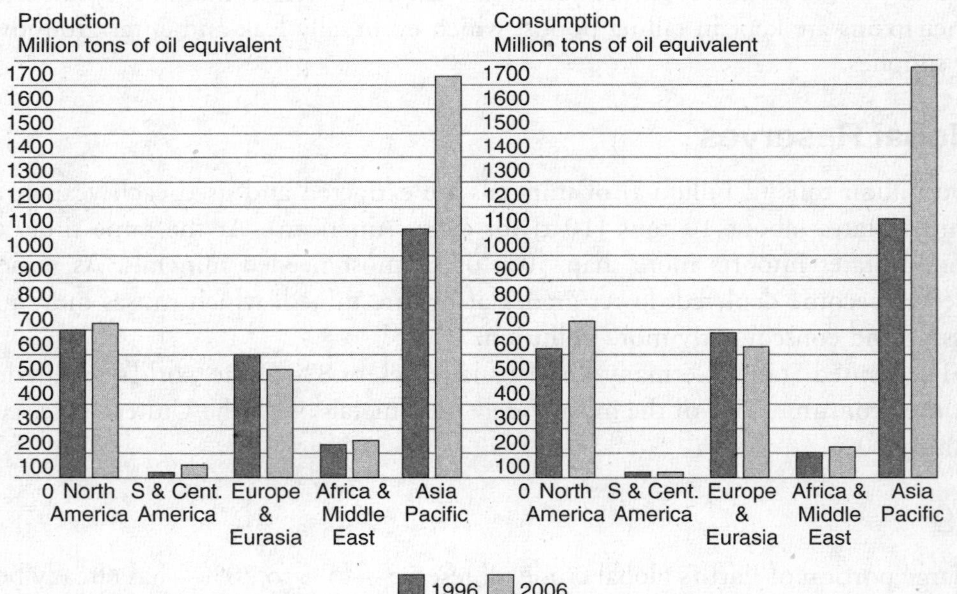

Figure 7.8 Coal production and consumption, 2005
Source: British Petroleum.

NATURAL GAS

Most of the world's natural gas reserves are located in the Middle East (34% of the world total). Europe, the Russian Federation, and the former countries of the U.S.S.R. own 42% of total world reserves. The United States possesses 3% of the world's total natural gas reserves. Given U.S. production levels, there is enough natural gas in the United States to meet approximately 75 years of domestic production. This estimate does not taking into account expected increasing levels of domestic production or the potential opening up of access to currently restricted land such as the Arctic National Wildlife Refuge.

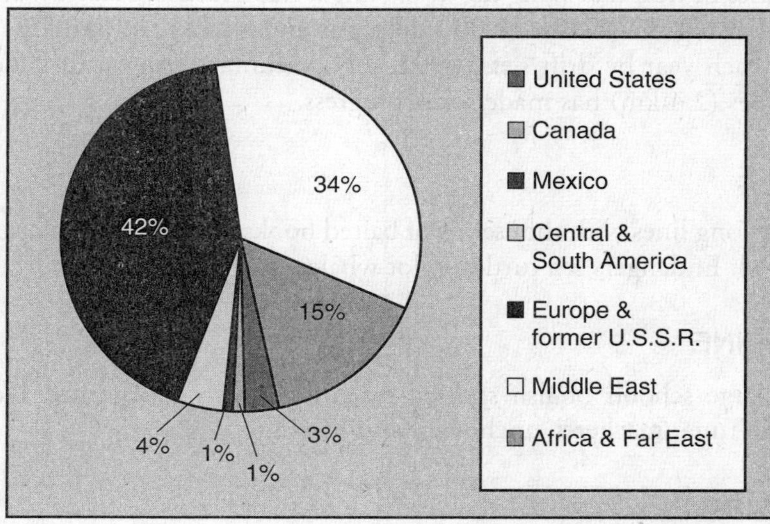

Figure 7.9 Natural gas reserves

> **RELEVANT LAWS**
>
> **General Mining Law (1872):** Grants free access to individuals and corporations to prospect for minerals in public domain lands and allows them, upon making a discovery, to stake a claim on that deposit.
>
> **Surface Mining Control and Reclamation Act (1977):** Established a program for regulating surface coal mining and reclamation activities.

FISHING

Different techniques are used to fish the planet's waters:

BOTTOM TRAWLING

Uses a funnel-shaped net to drag the ocean bottom. Shrimp, cod, flounder, and scallops. Analogous to clear-cutting forests. Species not wanted is called bycatch.

DRIFT NET

Long expanses of nets that hang down in water. Traps turtles, seabirds, and marine mammals. During the 1980s, 10,000 dolphins and whales and millions of sharks were killed each year by drift nets. 1992 U.N. voluntary ban on drift nets longer than 1.5 miles (2.4 km) has made some progress.

LONG LINE

Placing very long lines with thousands of baited hooks. Swordfish, tuna, sharks, halibut, and cod. Endangers sea turtles, pilot whales, and dolphins.

PURSE SEINE

Surrounds large schools of fish spotted by aircraft with a large net. Net is then drawn tight. Tuna, mackerel, anchovies, and herring.

Overfishing

The oceans have been looked on as unlimited resources. Ocean productivity is generally low and results from spatial separation of required plant nutrients. Light is restricted to surface waters.

The oceans supply 1% of all human food and represent 10% of the world's protein source. China is responsible for about one-third of all fish harvesting from the oceans. About one-third of the total catch of fish is used for purposes other than human consumption, such as fish oil, fish meal, and animal feed. Another one-third of global catches consists of bycatch. These are marine mammals, sea turtles, birds, noncommercial fish, and shellfish that are ensnared in fishing nets or dredged up by trawling and discarded.

Maximum sustained yield is the largest amount of marine organisms that can be continually harvested without causing the population to crash. This yield generally occurs when a population is maintained at half the carrying capacity.

Methods to manage fisheries in a sustainable manner include:

1. Regulate locations and numbers of fish farms and monitor their pollution output
2. Encourage the production of herbivorous fish species
3. Require and enforce labeling of fish products that were raised or caught according to sustainable methods
4. Set catch limits far below maximum sustainable yields
5. Eliminate government subsidies for commercial fishing

6. Prevent importation of fish from foreign countries that do not adhere to sustainable-harvesting methods
7. Place trading sanctions on foreign countries that do not respect the marine habitat, including countries that hunt whales
8. Assess fees for harvesting fish and shellfish from public waters
9. Increase the number of marine sanctuaries and no-fishing areas
10. Increase penalties for fishing techniques that do not allow escape of bycatch, including unwanted fish species, marine mammals, sea birds, and sea turtles
11. Ban the throwing back of bycatch
12. Monitor and destroy invasive species transported through ship ballast

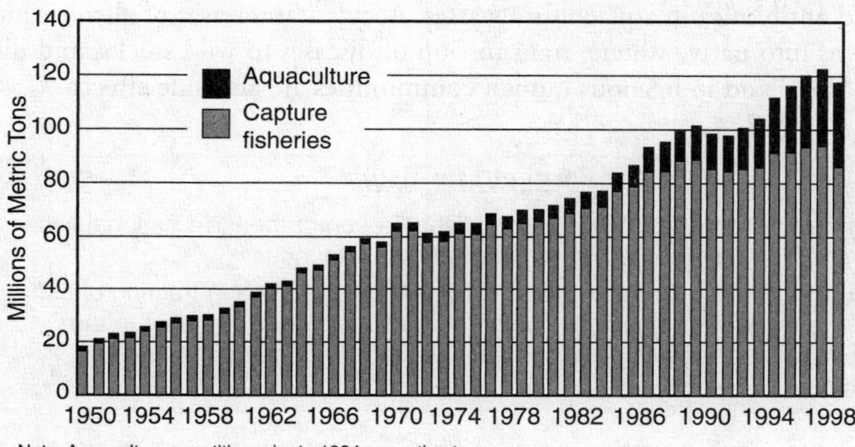

Figure 7.10 World fishing and aquaculture production

Several methods can restore habitats suitable for freshwater fish. These include: planting native vegetation on stream banks, rehabilitating in-stream habitats, controlling erosion, and controlling invasive species. Other restsorative methods include restoring fish passages around human-made impediments, monitoring, regulating, and enforcing recreational and commercial fishing; and protecting coastal estuaries and wetlands.

Aquaculture

Aquaculture, common known as mariculture or fish farming includes the commercial growing of aquatic organisms for food. It involves stocking, feeding, protection from predators, and harvesting. Aquaculture is growing about 6% annually and provides 5% of the total food production worldwide, most of it coming from less-developed countries. Currently, the most popular products being produced through aquaculture include seaweeds, mussels, oysters, shrimp, and certain species of fish (primarily salmon, trout, and catfish). Kelp makes up about 17% of aquaculture output and is used as a food product and as a source of various products used in the food industry. Aquaculture is used to raise 80% of all mollusks, 40% of all shrimp, and 75% of all kelp.

Aquaculture offers several advantages over raising livestock in that cold-blooded organisms convert more feed to usable protein. For example, for every 1 million calories of feed required, a trout raised on a farm produces about 35 grams of protein whereas a chicken produces 15 grams of protein and cattle produce 2 grams of protein. For every hectare of ocean, intense oyster farming can produce 58,000 kg of protein while natural harvesting of oysters produces 10 kg of protein.

For aquaculture to be profitable, the species must be marketable, inexpensive to raise, trophically efficient, at marketable size within 1 to 2 years, and disease resistant. Aquaculture creates dense monocultures that reduce biodiversity within habitats and requires large levels of nutrients in the water.

Aquaculture offers possibilities for sustainable protein-rich food production and for economic development to local communities. However, aquaculture on an industrial scale may pose several threats to marine and coastal biological diversity. It creates wide-scale destruction and degradation of natural habitats and leaves nutrients and antibiotics in aquaculture wastes. Accidental releases of alien or modified organisms into native waters, transmission of diseases to wild stocks, and displacement of local and indigenous human communities are also side effects.

RELEVANT LAWS

Fish and Wildlife Act (1956): Established a comprehensive national fish and shellfish resource policy directed primarily to industry.

Fish and Wildlife Coordination Act (1980): Assistance in training of state fish and wildlife enforcement personnel and assistance to states in the development and revision of conservation plans for nongame fish and wildlife.

GLOBAL ECONOMICS

The economy and the environment are intrinsically linked such that both are simultaneously causes and effects and are inputs and outputs of each other. The environment contains all the resources that can be used in the economy. The use of resources for economic purposes continuously creates new environmental situations. For example, while some resources are depleted and transformed from usable to unusable states, economic resources are used to expand additional resources. This occurs through increasing the available supply of materials, opening land to agricultural production, transporting resources from locations where they are in surplus to areas of shortage, and so on.

Increased levels of economic activity and improvements in living standards have occurred since the end of World War II. People in the wealthiest countries constitute 15% of the global population and enjoy average levels of income that are about 20 times greater than the 85% of the global population that live in poorer countries.

If the income of the poorest 85% were raised to only one-third that of the richest countries, the level of total world production and consumption would have to double, with a similar increase in the use of Earth's resources. The conclusion is that continued increases in living standards in poorer countries will increase pressure on the carrying capacity of the planet.

Until recently, developments in local economies and local environments were dispersed and, to a great extent, isolated. They did not typically result in cumulative processes that had widespread or global impact. However, as with the economy, over the past century or so, the growth of human populations throughout the world (and the greater vigor with which these populations have assaulted the environment in their pursuit of higher production and consumption levels) has led to a significant increase in global environmental disruption. The effects of these disruptions have become increasingly interlinked. A global environment—a set of interrelated processes, causes and effects, not simply a group of unrelated ecological events—has come into being.

World Bank

The World Bank is a source of financial and technical assistance to developing countries around the world. The World Bank (owned by 184 member countries) provides low-interest loans, interest-free credit, and grants to developing countries for education, health, infrastructure, communications, and environmental issues. In 2001, the World Bank Board of Directors endorsed an environment strategy to guide the bank's actions in environmental areas. The strategy emphasizes three objectives: improving the quality of life, improving the quality of growth, and protecting the quality of the regional and global commons through the "greening" of investments in agriculture, water sanitation, and other environmental projects. The World Bank in 2005 has distributed $13.8 billion in public and private funds in the areas of biodiversity, conservation, climate change, and international waters; funded $740 million in projects to phase out ozone-depleting substances; and funded $1.6 billion into projects that reduce greenhouse gas emissions.

The World Bank is also the greatest single source of funds for large dam projects, having provided more than $50 billion for construction of more than 500 large dams in 92 countries. Since 1948, theses large dam projects have forcibly displaced approximately 10 million people from their land.

The following diagram shows how the World Bank distributes economic resources to less-developed countries:

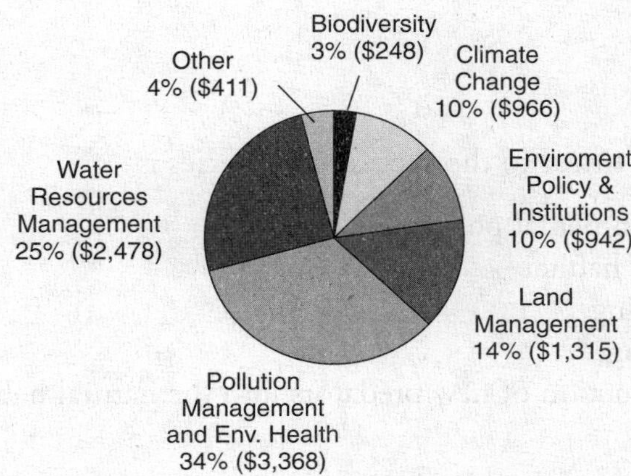

Figure 7.11 World Bank environmental investments (in millions)

"Tragedy of the Commons"

Garrett Hardin wrote the "Tragedy of the Commons" in 1968, and it appeared in the journal *Science*. The story parallels what is happening worldwide in regard to resource depletion and pollution. The seas, air, water, animals, and minerals are all the commons. They are there for humans to use. Those who exploit them become rich. The price of depleting the resources of the commons is an external cost paid by all people on Earth. Limits to the "Tragedy of the Commons" include the following:

- Economic decisions are generally short term, based on reactions in the world market. Environmental decisions are long term.
- Land that is privately owned is subject to market pressure. For example, if privately owned timberland is increasing in value at an annual rate of 3% but interest rates on loans to purchase the land are 7%, this could result in the land being sold or the timber being harvested for short-term profits.
- Some commons are easier to control than others. Land, lakes, rangeland, deserts, and forests are geographically defined and easier to control than air or the open oceans that do not belong to any one group. This is the problem with the United States passing the Kyoto Protocol.
- Incorporating discount rates into the valuation of resources would be an incentive for investors to bear a short-term cost for a long-term gain.
- Breaking a commons into smaller, privately owned parcels fragments the policies of governing the entire commons. Different standards and practices used on one parcel may or may not affect all other parcels.

MULTIPLE-CHOICE QUESTIONS

1. Chronically undernourished people are those who receive approximately ____ calories or less per day.

   (A) 500
   (B) 1,000
   (C) 1,500
   (D) 2,000
   (E) 3,000

2. The greatest threat to the success of a species is

   (A) environmental pollution
   (B) loss of habitat
   (C) poaching
   (D) hunting
   (E) introduction of new predators into the natural habitat

3. The second law of thermodynamics would tend to support

   (A) people eating more meat than grain
   (B) people eating more grain than meat
   (C) people eating about the same amount of grain as meat
   (D) people being undernourished
   (E) people eating a balanced diet from all food groups

4. The majority of nutrients and calories in the average human diet come from

   (A) potatoes, corn, and rice
   (B) wheat, rice, and soybeans
   (C) wheat, corn, and rice
   (D) wheat, corn, and oats
   (E) wheat, corn, and soybeans

5. Planting trees and/or shrubs between rows of crops is known as

   (A) contour farming
   (B) planting windbreaks
   (C) strip cropping
   (D) alley cropping
   (E) tillage farming

6. The one area of the world that is NOT expected to increase food production soon is

   (A) Asia
   (B) Latin America
   (C) sub-Saharan Africa
   (D) China
   (E) India

7. What percentage of the crops grown in the world today are transgenic or genetically modified?

   (A) 0
   (B) 1%–15%
   (C) 20%–35%
   (D) 40%–60%
   (E) >75%

8. During the last 150 years, urban population has increased

   (A) 10%
   (B) 20%
   (C) 50%
   (D) 100%
   (E) 250%

9. Chronically undernourished people represent

(A) 1 in every 10
(B) 1 in every 8
(C) 1 in every 5
(D) 1 in every 3
(E) 1 in every 2

10. Adding more fertilizer does NOT necessarily increase crop production is an example of

(A) law of supply and demand
(B) Leibig's law of minimum
(C) limiting factors
(D) second law of thermodynamics
(E) Gaia hypothesis

11. Soil that is transported by the wind is

(A) alluvial
(B) feral
(C) gangue
(D) rill
(E) aeolian

12. Most of Earth's land area is

(A) urban
(B) forest
(C) desert
(D) rangeland
(E) agricultural

13. Cities experiencing the greatest urban growth are found in

(A) Asia
(B) Africa
(C) Europe
(D) North America
(E) both choices (A) and (B)

14. Which of the following is NOT a concept designed to create a sustainable city?

(A) Conserve natural habitats
(B) Focus on energy and resource conservation
(C) Design affordable and fuel-efficient automobiles
(D) Provide ample green space
(E) All are correct

15. Approximately what percentages of the world depends upon wood or charcoal for heating and/or cooking?

    (A) Less than 10%
    (B) Between 20% and 30%
    (C) Between 50% and 60%
    (D) More than 75%
    (E) Has not been determined

16. Of all the jobs in the U.S. Forest Service, the majority are concerned with

    (A) lumber management
    (B) fire management
    (C) mining management
    (D) recreation
    (E) administrative functions

17. In which of the following states or areas would you NOT find significant amounts of old-growth forests?

    (A) New England
    (B) Appalachia
    (C) British Columbia
    (D) Washington State
    (E) California

18. Most grain that is grown in the United States is used

    (A) for export to countries that need grain
    (B) for cereals and baked goods
    (C) to feed cattle
    (D) for trade with other countries
    (E) to create fuel and liquor

19. What is the number one source of soil erosion?

    (A) Physical degradation
    (B) Chemical degradation
    (C) Water erosion
    (D) Wind erosion
    (E) All forms contribute equally

20. A process in which small holes are drilled into Earth and water-based chemical solvents are used to flush out desired minerals is known as

    (A) chemical leaching
    (B) *ex-situ* leaching
    (C) beneficiation
    (D) heap leaching
    (E) *in-situ* leaching

TIP

The Free-Response section of the APES exam consists of four required questions: one data-set question, one document-based question, and two synthesis and evaluation questions. Be sure you are comfortable with and know what is required for each type of question.

FREE-RESPONSE QUESTION

by: Sarah Utley

Ten points necessary for full-credit.
Section 1: worth 2 points. Describe either surface or subsurface mining.
Section 2: worth 4 points. Explain at least two remediation techniques.
Section 3: worth 4 points. Two points awarded for each completely described environmental impact.

Deposits of coal were recently discovered in a small valley in the eastern foothills of the Rocky Mountains. These deposits were found both near the surface and underneath the ground.

(a) Describe a method that could be used to mine the coal.

(b) Using coal as an energy source can have serious environmental consequences. Discuss current technological methods employed to reduce the impact of using coal as an energy source.

(c) The residents of this small town have mixed opinions about the impact of the mine on their community. Some residents support the mine because of the increased employment opportunities and increased tax base that supports projects such as schools and roads. Others are concerned about current or possible future environmental impacts on this region of the state. Discuss two of these possible environmental impacts.

MULTIPLE-CHOICE ANSWERS AND EXPLANATIONS

1. (D) Chronic malnutrition is defined as receiving fewer than 2,000 calories per day. People who live in developed nations receive an average of 3,340 calories per day.
2. (B) There are many reasons why certain species decline and become endangered; however, most environmentalists agree that the most significant factor is habitat loss and degradation.
3. (B) The second law of thermodynamics states that there is about a 10% loss in each successive higher trophic level in the energy pyramid.
4. (C) Wheat and rice supply approximately 60% of worldwide human calories.
5. (D) See the table in the section "Types of Agriculture" for a more detailed explanation about alley cropping.
6. (C) Droughts, pestilence, AIDS, and civil strife in this area have had serious impacts on food production in this part of the world.
7. (E) Transgenic plants are designed to incorporate desirable qualities into commercially grown crops. The majority of agricultural plants grown today are genetically modified to some extent.
8. (D) No explanation needed.

9. (C) In 1970, 1 in every 3 people in the world was classified as undernourished. The current estimates are that 1 in every 5 is undernourished. Higher crop yields combined with higher standards of living and more efficient distribution networks have had a positive impact.
10. (C) There is a limit to crop yield versus fertilizer application. Additional fertilizer past a certain point can actually harm plants.
11. (E) Aeolian soils are sand-sized particles transported by wind action.
12. (B) About one-third of Earth's surface is covered by forests.
13. (E) In 2015, the largest cities in the world are projected to be Tokyo (Japan), Bombay (India), Lagos (Nigeria), and Dhaka (Bangladesh).
14. (C) Designing affordable and fuel-efficient cars would increase the number of automobiles already in cities since more people could afford them. More private vehicles results in congestion, pollution, and so on.
15. (C) As late as the 1850s, wood supplied over 90% of the energy requirements in the United States. Half of the energy currently used in Africa is fuel wood.
16. (D) The number of recreational visitors in all National Forests in 1993 was 730 million. By the year 2045, the number is projected to rise by approximately 63 percent.
17. (A) Only 20% of the worlds' original forests remain intact. In New England, the first area to be colonized by Europeans, 99% of the frontier forests have been destroyed.
18. (C) There are about 1.5 billion cattle on Earth. They graze on 25% of the landmass and consume enough grain to feed hundreds of millions of people. In the United States, cattle consume 70% of all grain produced.
19. (C) Water erosion is about half of the problem.
20. (E) See the section "Extraction" for a more detailed explanation.

FREE-RESPONSE ANSWER

Notable points:

Description of mining type.

Specific description of coal removal.

(a) 1. Coal that is located close to the surface is frequently extracted by strip-mining. In surface mining, the rock, soil, and vegetation layers that are on top of the coal are scraped away and removed. The removed rock and soil are called overburden. Specifically, in strip mining, a long trench is dug, removing the overburden and the coal. Then a second trench is dug parallel to the first, and the overburden from the second is used to fill the first trench. The hill of loose overburden that fills the previous trench is called a spoil bank. The process continues until the coal is depleted or the cost of future extraction is prohibitive.

or

Description of mining type.

Description of coal removal.

2. Coal that is located deep underneath the surface of Earth is removed through underground or subsurface mining techniques. A deep shaft must be dug or blasted into the area. An extensive network of tunnels (which are supported with wooden or metal beams) is then dug, and these extend out from the main shaft. Sometimes part of the coal vein is left intact to support the shafts and tunnels. If the mine is large enough, electricity is wired, an elevator is installed to take workers to the active part of the mine, and tracks are laid on which the loaded bins of coal travel. The coal itself can be removed with special mining machines equipped with large drills. The mined coal is then loaded into the mining bins and taken to the surface.

Though an answer may contain more than two remediation techniques, at a minimum, a full-credit answer should contain two techniques.

Introduction of techniques.

Purpose of washing.

Description of washing (performed after excavation and prior to combustion).

Purpose of scrubbers.

Description of scrubbers.

(b) Coal can contain contaminants such as rocks, ash, mineral dust, and sulfur. These contaminants decrease the efficiency of the coal and can increase the environmental impact of burning the fossil fuel. When coal with high sulfur content is burned, the sulfur can enter the atmosphere as sulfur dioxide—a component of acid rain. Several techniques can decrease the environmental impact of coal use. When coal is initially mined, it is usually washed. Washing involves separating the impurities from the coal based on density. The contaminants are generally denser than the coal and sink in the separation fluid. The process involves first grinding the coal into small pieces. Then the pieces are placed onto a separation table with a fluid. The coal is skimmed off the top. The contaminants, such as pyrite (a sulfur derivative), sink to the bottom. In addition to the initial washing of coal, remediation techniques can decrease the amount of pollutants that result from the burning or use of coal.

The Clean Air Act (revised in 1990) requires that all new coal-burning plants built after 1978 must be equipped with air quality structures that remove the sulfur from the combustion gases before they are released through the smokestack. These structures are called scrubbers. In a scrubber, a mixture of water and crushed limestone is sprayed into the gases. The sulfur from the emissions and the limestone chemically combine to form a solid (calcium sulfate). This solid is then disposed in a solid waste disposal facility or can be used in the production of some building products.

Description of fluidized bed combustion.

Purpose of fluidized bed combustion.

In addition to scrubbers, there is additional technology that can decrease the environmental impact of coal burning. In a fluidized bed combustion plant, the coal particles are suspended within the boiler on jets of air. This tumbling of the burning coal allows limestone to be introduced into the plant prior to the smokestack. The limestone combines with the sulfur and chemically bonds with it to capture a greater amount of sulfur pollutants. In addition to removing the sulfur prior to entering the smokestack, a fluidized bed boiler burns at a cooler temperature than a conventional boiler. This cooler temperature prevents the smog-forming NO_x pollutants from becoming volatile and being released into the air.

(c) (Give two implications.)

(1) Water pollution is frequently a concern with the mining of coal. Water is used throughout the mining process, from cooling machinery and smelters to washing coal. If this contaminated water is not contained and purified, it can seep into the local waterways or water table. In addition, precipitation that percolates through the area can leach soil contaminants from the mining

The question specifically asks for two environmental implications.

Name of each concern.

Cause of each impact.

Environmental result of each.

process into the local water table or waterways. The water table of the areas located close to the mines must be carefully monitored for quality.

(2) Air pollution is also a concern in the mining industry. The mining machinery, which is frequently powered by diesel engines, contributes to air pollution in the mine vicinity. In addition, the burning of coal often releases air toxins such as sulfur dioxide, nitrogen oxides, and particulate matter into the atmosphere. New combustion units are now mandated by the Clean Air Act to be fitted with scrubbers and other clean air apparatuses. Many states are now requiring that even older equipment must meet new clean air requirements.

(3) There is also the concern of habitat destruction and wildlife–native plant displacement. The process of surface mining, which requires the removal of many tons of soil, disturbs the natural ecosystem and displaces native animals. Subsurface mining is often less disruptive to soil and the soil surface but has its own concerns, such as subsidence and land instability. Though mining corporations are required by law to return the land to the original topography, since the soil structure has been altered, native plant revegetation techniques often fail. When the native plants fail to thrive, reintroducing native wildlife is difficult.

(4) Mining can also be harmful to human health. With subsurface mining, workers must be protected against unstable mine shafts, generally dangerous working conditions, and health concerns such as poor air quality. Safely ventilating the mines below the ground can be difficult, and care must be taken to protect the workers from the respiratory diseases such as emphysema and lung cancer. With the large amount of mining equipment (large cranes, bulldozers, and so on) required in surface mining, human safety is always a concern.

UNIT V: ENERGY RESOURCES AND CONSUMPTION (10–15%)

Areas on Which You Will Be Tested

A. **Energy Concepts**—energy forms, power units, conversions, and laws of thermodynamics.

B. **Energy Consumption**
   1. History—Industrial Revolution, exponential growth, and energy crisis.
   2. Present global energy use.
   3. Future energy needs.

C. **Fossil Fuel Resources and Use**—formation of coal, oil, and natural gas, extraction/purification methods, world reserves and global demand, synfuels, and environmental advantages/disadvantages of sources.

D. **Nuclear Energy**—nuclear fission process, nuclear fuel, electricity production, nuclear reactor types, environmental advantages/disadvantages, safety issues, radiation and human health, radioactive wastes, and nuclear fusion.

D. **Hydroelectric Power**—dams, flood control, salmon, silting, and other impacts.

F. **Energy Conservation**—energy efficiency, CAFE standards, hybrid electric vehicles, and mass transit.

G. **Renewable Energy**—solar energy, solar electricity, hydrogen fuel cells, biomass, wind energy, small-scale hydroelectric, ocean waves and tidal energy, geothermal, and environmental advantages/disadvantages.

Energy

Climb the mountains and get their good tidings. Nature's peace will flow into you as sunshine flows into trees. The winds will blow their own freshness into you and the storms their energy, while cares will drop away from you like the leaves of autumn.

—John Muir

ENERGY CONCEPTS

The following table describes the different energy forms.

Forms of Energy

Form	Description
Mechanical	There are two types of mechanical energy: potential energy (a book sitting on a table) and kinetic energy (a baseball flying through the air).
Thermal	Heat is the internal energy in substances—the vibration and movement of the atoms and molecules within substances.
Chemical	Chemical energy is stored in bonds between atoms in a molecule.
Electrical	Electrical energy results from the motion of electrons.
Nuclear	Nuclear energy is stored in the nuclei of atoms. It is released by either splitting or joining of atoms.
Electromagnetic	Electromagnetic energy travels by waves.

Power and Units

Power is the amount of work done per time. The most common unit of power is the kilowatt-hour (kWh).

Units of Energy/Power

Unit or Prefix	Description
Btu (British Thermal Unit)	Btu is a unit of energy used in the United States. In most countries it has been replaced with the joule. A Btu is the amount of heat required to raise the temperature of 1 pound of water by 1°F. 1 watt is approximately 3.4 Btu/hr. 1 horsepower is approximately 2,540 Btu/hr. 12,000 Btu/hr. is referred to as a "ton" in many air-conditioning applications.
Horsepower	Primarily used in the automobile industry. 1 horsepower (HP) is equivalent to 746 watts.
Kilo-	Means 1,000 or 10^3. 1 kW = 10^3 watts.
Mega- (M)	Means 1,000,000 or 10^6. 1 MW = 10^6 watts.
Watt (electrical)	A kilowatt-hour (kWh) is the amount of energy expended by a 1 kilowatt (1000 watts) device over the course of one hour. Often measured in the context of power plants and home energy bills.
Watt (thermal)	Nuclear power plants produce heat measured in thermal watts.

Conversions

Let's do a sample conversion problem that you might experience on the APES exam. We will break it down into steps and show you how it might be graded.

TIP

Be sure to practice these types of conversion problems. They appear very frequently in the Free-Response Question on the APES test. Also, be sure you are comfortable with scientific notation and the factor-label method.

Thorpeville is a rural community with a population of 8,000 homes. It gets its electricity from a small, municipal coal-burning power plant just outside of town. The power plant's capacity is rated at 20 megawatts with the average home consuming 10,000 kilowatt hours (kWh) of electricity per year. Residents of Thorpeville pay the utility $0.12 per kWh. A group of entrepreneurs is suggesting that the residents support a measure to install 10 wind turbines on existing farmland. Each wind turbine is capable of producing 1.5 MW of electricity. The cost per wind turbine is $2.5 million dollars to purchase and operate for 20 years.

(a) The existing power plant runs 8,000 hours per year. How many kWh of electricity is the current plan capable of producing?

2 points. 1 point for correct setup. 1 point for correctly calculating the amount of electricity generated. You must correctly convert MW to kW. Points may be earned if you write the answer as a word problem. No points will be awarded without showing your work. Alternative setups are acceptable.

$$\frac{20\ \cancel{MW}}{1} \times \frac{(1\times10^{6}\ \cancel{watts})}{1\ \cancel{MW}} \times \frac{1\ kW}{10^{3}\ \cancel{watts}} = 2 \times 10^{4}\ kW$$

$$\frac{(2\times10^{4}\ kW)}{1} \times \frac{8{,}000\ hours}{1\ yr.} = 16{,}000 \times 10^{4}\ kWh/yr$$

$$= 1.6 \times 10^{8}\ kWh/yr.$$

(b) How many kWh of electricity do the residents of Thorpeville consume in one year?

2 points. 1 point for correct setup. 1 point for correctly calculating the amount of electricity generated. You must correctly convert MW *to* kW*. Points may be earned if you write the answer as a word problem. No points will be awarded without showing your work. Alternative setups are acceptable.*

$$\frac{8\times10^{3}\ \cancel{homes}}{1} \times \frac{1\times10^{4}\ kWh/\cancel{home}}{1\ yr.} = 8 \times 10^{7}\ kWh/yr.$$

(c) Compare answers (a) and (b). What conclusions can you make?

2 points, plus 1 possible elaboration point. 1 point for comparing answers (a) with (b) with an explanation of why the numbers in parts (a) and (b) would be the same or different (must be a viable reason).

OR

1 point for a solid or accurate explanation of why (a) and (b) are different even if the calculations were not attempted.

1 possible elaboration point for explanations that go into great detail about why the numbers differ.

Note: If you say that (a) and (b) are the same, you must state that this can only occur if the households have backup systems that will produce energy for them if they exceed the power generated by the plant.

The power plant produces 1.6×10^{8} kWh per year. The residents, however, only use 8×10^{7} kWh per year. This leaves a surplus of $1.6 \times 10^{8} - 8 \times 10^{7} = 8 \times 10^{7}$ kWh in one year which can be sold to other towns. At a rate of \$0.12 per kWh, this provides a surplus of 8×10^{7} kWh × \$0.12/kWh = \$0.96 × 10^{7} = \$9,600,000.

Additional revenues could be generated by running the power plant 24 hours a day, 365 days a week

Differences between Thorpeville's consumption and the power plant's output could be attributed to the following:

- The power plant needs to produce higher amounts of power to compensate for line loss.
- The power plant needs to produce higher amounts of power to supply energy to the town during peak hours, not just the average usage.
- The power plant needs to plan for possible future growth of the town.
- The power plant was built over capacity to provide a source of income to the town.

(d) Assuming that the population of Thorpeville remains the same for the next 20 years, and that electricity consumption remains stable per household, what would be the cost (expressed in $ per kWh) of electricity to the residents over the next 20 years if they decided to go with wind turbines?

2 points. 1 point for correct setup. 1 point for correct answer with calculations. Alternative setups are acceptable. If your answer in part (b) is incorrect but you appropriately use it as the basis for the calculations for answering the question in part (d), you will receive full credit for answering part (d) if the setup and calculations are correct, even if the answer is not correct.

Based on current community consumption of 8 × 10⁷ kWh/yr. from part (b).

$$\text{kWh for 20 years} = \frac{8\times10^7\ \text{kWh}}{\cancel{\text{year}}} \times 20\ \cancel{\text{years}} = 1.6\times10^9\ \text{kWh}$$

$$\text{Direct cost for 20 years} = 10\ \cancel{\text{turbines}} \times \frac{\$2.5\times10^6}{\cancel{\text{turbine}}} = \$2.5\times10^7$$

$$\text{cost/kWh} = \frac{\$2.5\times10^7}{1.6\times10^9\ \text{kWh}} = \$1.6\times10^{-2}/\text{kWh} = \$0.016/\text{kWh}$$

(e) What are the pros and cons of the existing coal-burning plant compared with the proposed wind farm?

Pros	**Cons**
WIND: The electricity produced from the wind turbines costs $0.016 per kWh, but each homeowner would also have to pay $25,000,000/8,000 homes = $3,125.00 over 20 years ($156.25/yr.) to pay for the wind turbines. 10,000 kWh at $0.016 per kWh for electricity produced from wind turbines = $160 plus $156.25 per year to pay for the wind turbines = $316.25 per year.	COAL: Electricity costs $0.12 per kWh produced from the coal-burning plant. 10,000 kWh of electricity per year from the coal-burning plant at $0.12 per kWh = $1,200 per year. Clearly, electricity produced from wind turbines is much cheaper.
WIND: Zero emissions. Acid rain would be reduced as well as photochemical smog.	COAL: Produces air pollution, specifically SO_2 and NO_x.
WIND: The wind is free.	COAL: As labor prices increase, the price of coal would be expected to increase over the next 20 years.
WIND: No heavy-metal or radioactive emissions.	COAL: Coal-burning plants produce heavy metals such as mercury, lead, and cadmium pollution along with radioactive contaminants.
WIND: No thermal pollution.	COAL: Can produce thermal pollution to local streams. However, cooling towers can be installed to reduce this form of pollution.
WIND: Multiple use of the land.	COAL: Cannot utilize the concept of multiple use of land.

Laws of Thermodynamics

FIRST LAW

Energy cannot be created or destroyed.

SECOND LAW

When energy is converted from one form to another, a less useful form results (energy quality). Energy cannot be recycled to a higher quality. Only 20% of the energy in gasoline is converted to mechanical energy. The rest is lost as heat and is known as low-quality energy.

ENERGY CONSUMPTION

Wood (a renewable energy source) served as the predominant form of energy up until the Industrial Revolution. During the Industrial Revolution, coal (a nonrenewable energy source) surpassed wood's usage. Coal was overtaken by petroleum during the middle of the 20th century, with petroleum continuing to be the primary source of energy worldwide. Natural gas and coal experienced rapid development in the second half of the 20th century.

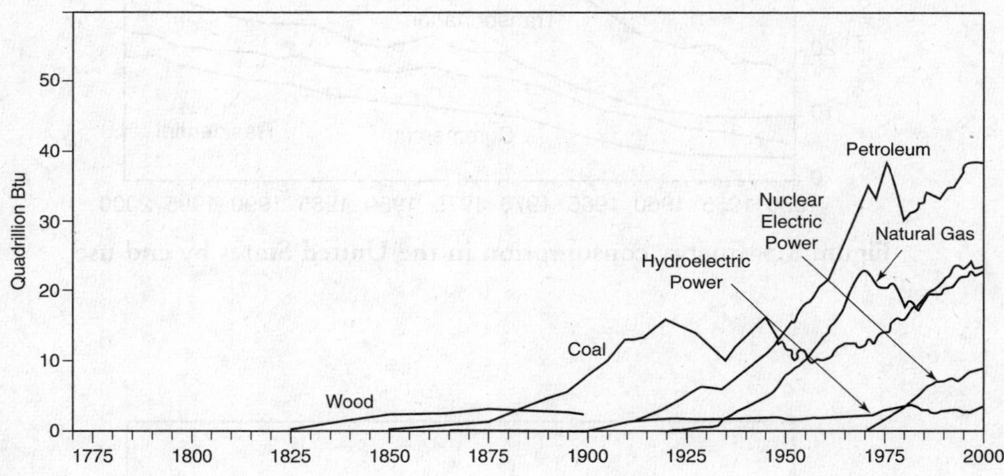

Figure 8.1 Energy consumption by source
Source: Energy Information Administration

The United States was self-sufficient in energy until the late 1950s. At that time, energy consumption began to outpace domestic production, which then led to oil imports.

The industrial sector in the United States has traditionally used the largest share of energy, followed by transportation and then residential and commercial uses. While coal was once the prominent form of energy in the industrial sector, it gave way to natural gas and petroleum in the late 1950s, with rapid increases occurring through the 1970s.

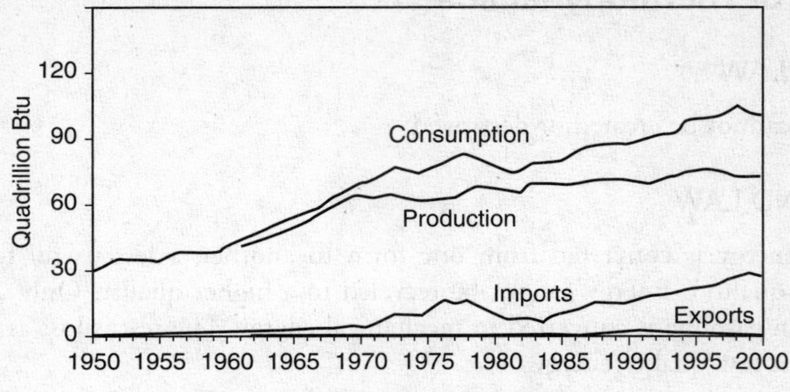

Figure 8.2 U.S. energy overview

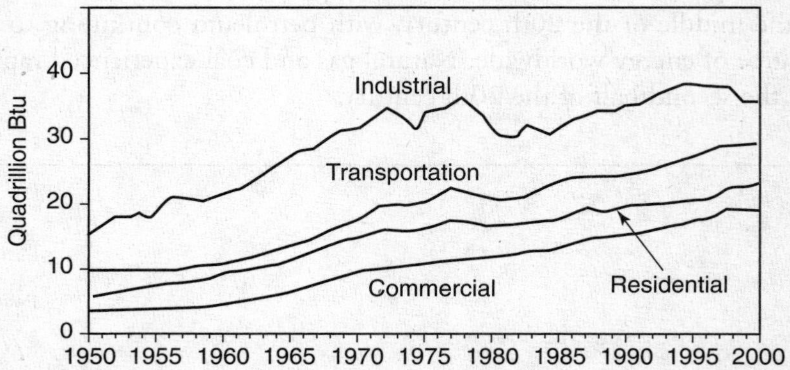

Figure 8.3 Energy consumption in the United States by end use

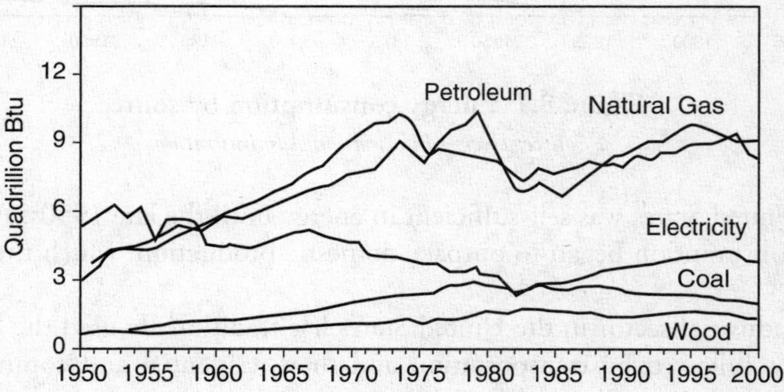

Figure 8.4 Industrial energy consumption

Beginning in 1998, net imports of oil surpassed the domestic oil supply in the United States. The United States accounts for 25% of the world consumption of petroleum.

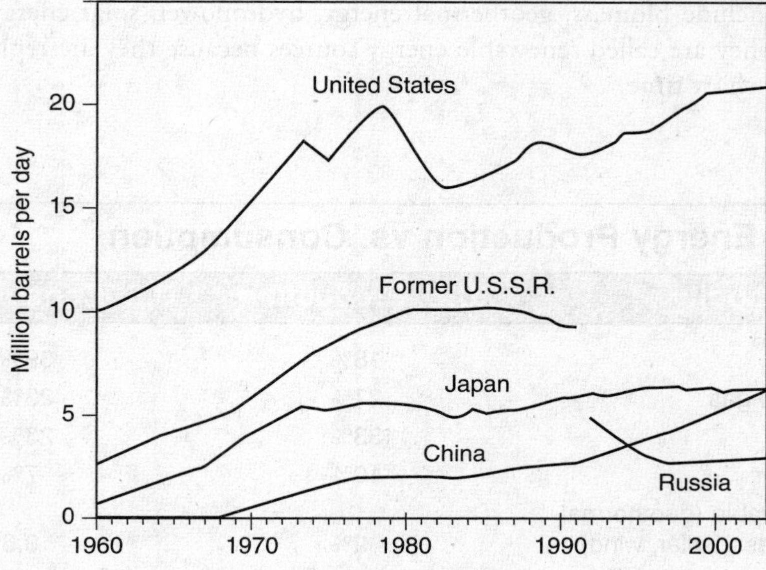

Figure 8.5 Leading petroleum consumers

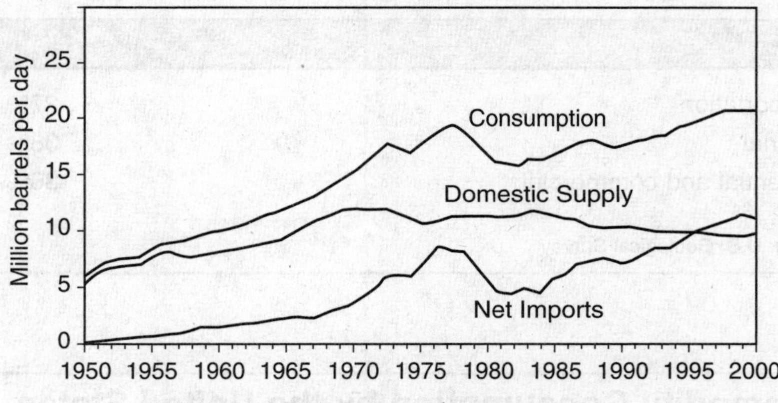

Figure 8.6 U.S. petroleum overview

Present Global Energy Use

In the United States, most of the energy comes from nonrenewable energy sources such as coal, petroleum, natural gas, propane, and uranium. These energy sources are called nonrenewable because their supplies are limited. Renewable energy sources include biomass, geothermal energy, hydropower, solar energy, and wind energy. They are called renewable energy sources because they are replenished in a relatively short time.

U.S. Energy Production vs. Consumption

Commodity	U.S. Production	U.S. Consumption
Oil	18%	39%
Natural gas	27%	23%
Coal	33%	23%
Nuclear	10%	7%
Renewable (geothermal, biomass, solar, wind)	9%	3.6%
Hydroelectric	5%	4%

U.S. Energy Production by Sector

Sector	%
Transportation	27%
Industrial	38%
Residential and commercial	36%

*Source: U.S. Geological Survey

Commodity Consumption by the United States

Commodity	%
(% of Total World Usage)	
Oil	40%
Natural gas	23%
Coal	23%

Future Energy Needs

Although the United States's energy history is one of large-scale change as new forms of energy were developed, the outlook for the next few decades is for continued growth and reliance on the three major fossil fuels: petroleum, natural gas, and coal. There will be modest expansion in renewable resources and relatively flat generation from nuclear power. The most realistic, economical and viable resources of future energy needs for the immediate future are clean coal, methane hydrates, oil shale, and tar sands.

CLEAN COAL

The world's supply of coal is substantial and can be expected to meet the world's energy needs for many years to come. Clean-coal technology refers to processes that reduce the negative environmental effects of burning coal. The processes include washing the coal to remove minerals and impurities and capturing the sulfur dioxide and carbon dioxide from the flue gases. FutureGen is a project designed to build a prototype zero-emission, coal-fired 275 MW power plant that produces hydrogen and electricity using carbon capture and storage technology. Other clean-coal technology is focusing on natural gas or microbial fuel cells charged from biomass or sewage.

METHANE HYDRATES

Methane hydrates (methane locked in ice) are a recently discovered source of methane that form at low temperature and high pressure. They are found in two types of geologic settings: on land in permafrost regions where cold temperatures persist in shallow sediments and beneath the ocean floor at water depths greater than 1,640 feet (500 m) where high pressures dominate. The hydrate deposits themselves may be several hundred meters thick. Methane bound in hydrates amount to approximately 3,000 times the volume of methane in the atmosphere. Some believe there is enough methane in the form of hydrates to supply energy for hundreds or thousands of years. Conventional natural gas reserves in the United States are estimated to be 1,400 trillion cubic feet (39 trillion m^3). Estimates of methane hydrates in waters belonging to the United States are estimated at 200,000 trillion cubic feet (5,600 trillion m^3). Worldwide, estimates of the natural gas potential of methane hydrates approach 400 million trillion cubic feet (11 million trillion m^3) compared with the 5,000 trillion cubic feet (140 trillion m^3) that make up the world's current known gas reserves.

Natural gas is expected to take on a greater role in power generation. This is largely because of increasing pressure for clean fuels and the relatively low capital costs of building new natural gas–fired power equipment. The United States will consume increasing volumes of natural gas well into the 21st century. U.S. natural gas consumption is expected to increase from almost 23 trillion cubic feet (6 billion m^3) in 1996 to more than 32 trillion cubic feet (9 billion m^3) in 2020—a projected increase of 40%. Also, natural gas demand is expected to grow because of its expanded use as a transportation fuel and potentially, in the long-term, as a source of alternative liquid fuels and a source of hydrogen for fuel cells. The primary waste product of burning natural gas is carbon dioxide—a greenhouse gas that contributes to global warming.

OIL SHALE

Oil shales contain an organic material called kerogen. If the oil shale is heated in the absence of air, the kerogen converts to oil. There are approximately 3 trillion barrels of recoverable oil from oil shale in the world, with 750 billion of it being located in the United States. Oil shale can be extracted through either surface mining or *in-situ* methods that consist of heating the oil shale underneath the ground and extracting the oil and gases through pumping. Most of the oil shale in the United States is found in Wyoming, Utah, and Colorado. The largest world reserves are found in Estonia, Australia, Germany, Israel, and Jordan.

Surface mining of oil shale negatively impacts the environment. The net energy yield of producing oil through oil shale is moderate since energy is required for blasting, drilling, crushing, heating the material, disposing of waste material, and environmental restoration. *In-situ* methods have the potential of affecting aquifers. Even though the world has large oil shale reserves, the problem remains that once the oil is obtained from shale, traditional issues of environmental pollution, acid rain, and global warming will continue.

TAR SANDS

Tar sands contain bitumen—a semisolid form of oil that does not flow. Specialized refineries are capable of converting bitumen to oil. Tar sand deposits are mined using strip-mining techniques. *In-situ* methods using steam can also be used to extract bitumen from tar sands. The sulfur content of oil obtained from tar sands is about 5%. Most of the tar sand deposits are located in Canada and Venezuela, with those in Canada being the most concentrated and therefore the most economical to mine. The oil in tar sands represents about two-thirds of the world's total oil reserves. The net-energy yield of producing oil through tar sands is moderate since energy is required for blasting, drilling, crushing, heating the material, disposing of waste material, and environmental restoration. As with oil shale, once the oil has been extracted from tar sands, the problems of environmental pollution, acid rain, and global warming continue.

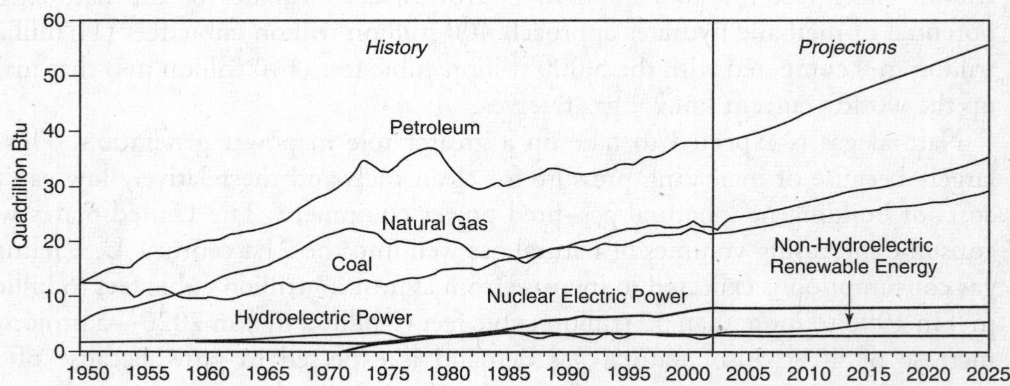

Figure 8.7 Energy consumption history and outlook, 1949–2025

Energy Crisis

In a free-market economy, the price of energy is driven by the principle of supply and demand. Sudden changes in the price of energy can occur if either supply or demand changes. In some cases, an energy crisis is brought on by a failure of world markets to adjust prices in response to shortages. Oil supply is largely controlled by nations with significant reserves of easily extractable oil, such as Saudi Arabia and Venezuela, who belong to an association of oil-producing countries known as OPEC (Organization of Petroleum Exporting Countries). When OPEC reduces the output quotas of its member countries, the price of oil increases as the supply diminishes. Similarly, OPEC can boost oil production in order to increase supplies, which drives down the price. When OPEC raises the price of oil too high, demand decreases and the production of oil from alternative sources becomes profitable. Historically, there have been several energy crises.

Most of the world's energy is supplied by burning oil. At current rates of consumption, world oil reserves are predicted to last 50 years with oil reserves in the United States predicted to last 25 years. As supply decreases, prices will increase. Higher prices for oil may make other sources (shale oil and tar sands) more economical.

U.S. Commodity Consumption

Year	Description
1973	Oil crisis. Export embargo by OPEC in response to western support of Israel.
1979	Oil crisis caused by the Iranian Revolution.
1990	First Gulf War.
2000–2001	California electricity crisis. Deregulation of the industry and corporate corruption.
2006	Oil price increases due to increased demand by India and China, political instability in Iran, Iraq, and Venezuela.

FOSSIL FUEL RESOURCES AND USE

Coal is produced by decomposition of ancient (286 million-year-old) organic matter under high temperature and pressure. Sulfur from the decomposition of hydrogen sulfide (H_2S) by anaerobic bacteria became trapped in coal. There are three types of coal: lignite, bituminous, and anthracite. Lignite or brown is the softest and has the lowest heat content. Bituminous is soft, has a high sulfur content, and constitutes 50% of the U.S. reserve. Anthracite is hard, has a high heat content and low sulfur content, and makes up 2% of the U.S. reserve. Peat is precoal and is used in some countries for heat but it has low heat content. Coal supplies 25% of the world's energy, with China and the United States consuming the most. In the United States, 87% of the coal is used for power plants to produce electricity. The Clean Air Act requires up to a 90% reduction in the release of sulfur-containing gases. This can be achieved by cleaning (washing) coal prior to burning, redesigning boilers (fluidized bed combustion), and scrubbing or adding limestone or lime into the effluent.

Oil is a fossil fuel produced by the decomposition of deeply buried organic material (plants) under high temperatures and pressures for millions of years. Compounds derived from oil are known as petrochemicals. They are used in the manufacture of paints, drugs, plastics, and so on.

Natural gas (known as methane or CH_4) is produced by the decomposition of ancient organic matter under high temperatures and pressure. Conventional sources of methane are found associated with oil deposits. Unconventional sources include coal beds, shale, gas hydrates, and tight sands. Methane can be liquefied (LNG), which allows for worldwide distribution.

Extraction-Purification Methods

There are three extraction-purification methods:

COAL

There are two primary methods of mining coal: surface mining and underground mining. Coal that is going to be burned in solid form may go through a variety of preparation processes. These include removing foreign material, screening for size, crushing, and washing to remove contaminants. It is also possible to turn solid coal into a gas or liquid fuel through clean-coal technologies.

OIL

Oil occurs in certain geologic formations at varying depths in Earth's crust. In many cases, elaborate, expensive equipment is required to extract it. Oil is usually found trapped in a layer of porous sandstone, which lies just beneath a dome-shaped or folded layer of some nonporous rock such as limestone. In other formations, the oil is trapped at a fault, or break in the layers of the crust. Natural gas is usually present just below the nonporous layer and immediately above the oil. Below the oil layer, the sandstone is usually saturated with salt water. The oil is released from this formation by drilling a well and puncturing the limestone layer. The oil is usually under such great pressure that it flows naturally, and sometimes with great force, from the well. However, in some cases, this pressure later diminishes so that the oil must be pumped. Once the oil has been collected, it is sent to a refinery, where it is cracked. Cracking involves separating the components of oil by their boiling points. Refining crude oil produces gasoline, heating oil, diesel oil, asphalt, etc.

NATURAL GAS

Natural gas typically flows from wells under its own pressure. It is collected by small pipelines that feed into the large gas transmission pipelines. In the United States, about 20 trillion cubic feet (560 billion m^3) of gas is produced each year.

World Reserves and Global Demand

Coal, oil, and natural gas are nonrenewable energy resources. Following is a brief description of the known world reserves of these energy sources and their expected demands in the future.

COAL

Best estimates show that coal reserves are expected to last for about 300 years at current rates of consumption. The largest reserve of coal is in China.

OIL

Of all known oil reserves, 65% is found in 1% of all fields—primarily in the Middle East.

NATURAL GAS

Russia and Kazakhstan have approximately 40% of the world reserves, the Middle East has about 25%, and the United States has about 3%.

Synfuels

A synfuel is a liquid fuel synthesized from a nonpetroleum source such as coal, natural gas, oil shale, or waste plastics. Shale oil is an example of a synfuel as it is derived from shale oil that is heated and the vapor condensed. Synthetic natural gas (SNG) is produced from coal liquefaction.

SOLID COAL TO SYNTHETIC NATURAL GAS (SNG), METHANOL, OR SYNTHETIC GASOLINE

Pros

- Easily transported through pipelines.
- Produces less air pollution.
- Large supply of raw materials available worldwide to meet current demands for hundreds of years.
- Can produce gasoline, diesel, or kerosene directly without reforming or cracking.

Cons

- Low net energy yield and requires energy to produce SNG.
- Plants are expensive to build.
- Would increase depletion of coal due to inherent inefficiencies.
- Product is more expensive than petroleum products.

Environmental Advantages/Disadvantages of Sources

COAL

Pros

- Abundant, known world reserves will last approximately 300 years at current rate of consumption.
- Unidentified world reserves are estimated to last 1,000 years at current rate of consumption.
- United States reserves are estimated to last about 300 years at current rate of consumption.
- Relatively high-net energy yield.
- U.S. government subsidies keep prices low.
- Stable; nonexplosive; not harmful if spilled.

Cons

- Most extraction in the United States is done through either strip mining or underground mining. These methods cause disruption to the land through erosion, runoff, and decrease in biodiversity.
- Up to 20% of coal ends up as fly ash, boiler slag, or sludge. Burning coal releases mercury, sulfur, and radioactive particles into the air. Thirty-five percent of all CO_2 releases are due to the burning of coal, with 30% of all pollution due to NO_x.
- Underground mining is dangerous and unhealthy.
- Expensive to process and transport. Cannot be used effectively for transportation needs.
- Pollution causes global warming. Scrubbers and other antipollution control devices are expensive.

OIL

Pros

- Inexpensive; however, prices are increasing, making alternatives more attractive.
- Easily transported through established pipelines and distribution networks.
- High net-energy yield.
- Ample supply for immediate future.
- Large U.S. government subsidies in place.
- Versatile—used to manufacture many products (paints, medicines, plastics, etc.).

Cons

- World oil reserves are limited and declining.
- Produces pollution (SO_2, NO_x, and CO_2). Production releases contaminated wastewater and brine.
- Causes land disturbances in drilling process, which accelerates erosion.
- Oil spills both on land and in ocean from platforms and tankers.
- Disruption to wildlife habitats (e.g., Arctic Wildlife Refuge).
- Supplies are politically volatile.

NATURAL GAS

Pros

- Pipelines and distribution networks are in place. Easily processed and transported as LNG over rail or ship.
- Relatively inexpensive, but prices are increasing. Viewed by many as a transitionary fossil fuel as the world switches to alternative sources.
- World reserves are estimated to be 125 years at current rate of consumption.
- High net energy yield.
- Produces less pollution than any other fossil fuel.
- Extraction is not as damaging to the environment as either coal or oil.

Cons

- H_2S and SO_2 are released during processing.
- LNG processing is expensive and dangerous, and it results in lower net energy.
- Leakage of CH_4 has a greater impact on global warming than does CO_2.
- Disruption to areas where it is collected.
- Extraction releases contaminated wastewater and brine.
- Land subsidence.

NUCLEAR ENERGY

During nuclear fission, an atom splits into two or more smaller nuclei along with by-product particles (neutrons, photons, gamma rays, and beta and alpha particles). The reaction is exothermic. If controlled, the heat that is produced is used to produce steam that turns generators that then produce electricity. If the reaction is not controlled, a nuclear explosion can result.

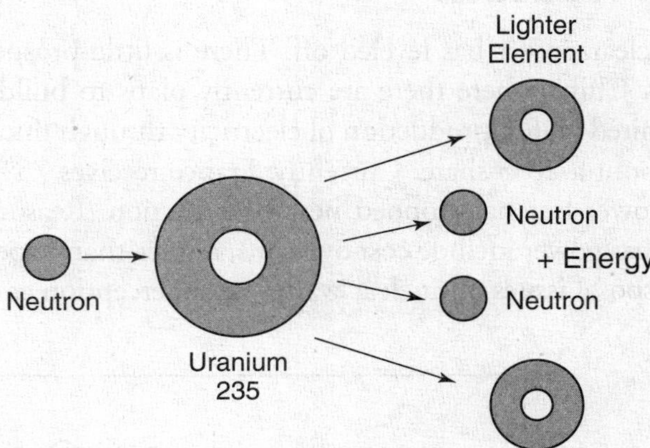

Figure 8.8 Nuclear fission

The amount of potential energy contained in nuclear fuel is 10 million times more than that of more traditional fuel sources such as coal and petroleum. The downside is that nuclear wastes remain highly radioactive for thousands of years and are difficult to dispose. The most common nuclear fuels are U-235, U-238, and Pu-239.

Nuclear Fuel

U-235

U-235 differs from U-238 in its ability to produce a fission chain reaction. The minimum amount of U-235 required for a chain reaction is called the critical mass. Low concentrations of U-235 can be used if the speed of the neutrons is slowed down through the use of a moderator. The fission of just one U-235 atom generates 200 MeV or 3.2×10^{-11} joules of energy. Less than 1% of all natural uranium on Earth is U-235. Uranium that has been processed to separate out U-235 is known as enriched uranium and is a subject of current controversy with Iran and North Korea. Nuclear weapons contain 85% or more U-235; nuclear power plants contain about 3% U-235. The half-life of U-235 is 700 million years.

U-238

U-238 is the most common (99.3%) isotope of uranium and has a half-life of 4.5 billion years. When hit by a neutron, it eventually decays into Pu-239, which is used as a fuel in fission reactors. Most depleted uranium is U-238.

PU-239

Pu-239 has a life-life of 24,000 years. It is produced in breeder reactors from U-238. Plutonium fission provides about one-third of the total energy produced in a typical commercial nuclear power plant. Control rods in nuclear power plants need to be changed frequently due to the buildup of Pu-239 that can be used for nuclear weapons and due to the buildup of Pu-240, a contaminant. International inspections of nuclear power plants regulate the amount of Pu-239 produced by power plants.

Electricity Production

Worldwide, nuclear power has leveled off. There is little prospect for building new plants except in China where there are currently plans to build 50 new reactors by 2020. In the United States, production of electricity through nuclear power plants has leveled off to about a 20% share. Currently, France receives 75% of its energy needs from nuclear power but has stopped new construction. Reasons for the decline in worldwide nuclear power include cost overruns, higher-than-expected operating costs, safety issues, disposal issues of nuclear wastes, and perception as a risky investment.

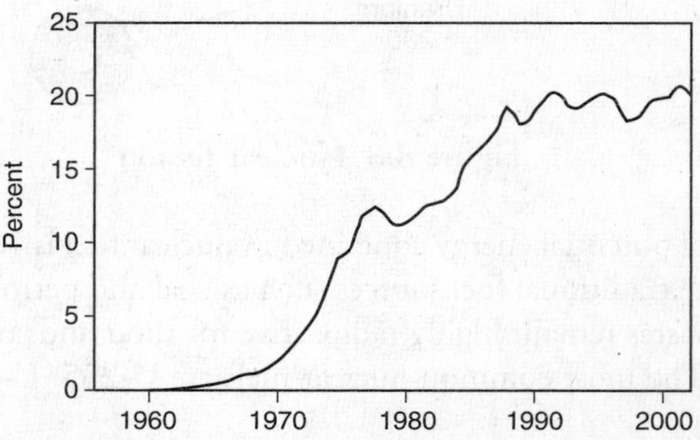

Figure 8.9 Nuclear share of electricity in the United States

Nuclear Reactor Types

Several different nuclear reactor types are in use: light-water reactors, heavy-water reactors, graphite-moderated reactors, and exotic reactors. However, they all have several features in common (see Figure 8.10).

A. The *core* contains up to 50,000 fuel rods. Each pellet has the energy equivalent of 1 ton (1 t) of coal.
B. Uranium oxide is the *fuel*: 97% use U-283 and 3% use U-235.
C. *Control rods* (usually made of boron) move in and out of the core to absorb neutrons and slow down the reaction.
D. A neutron *moderator* is a medium that reduces the velocity of fast neutrons, thereby turning them into thermal neutrons capable of sustaining a nuclear chain reaction. Moderators can be water, graphite (can produce plutonium for weapons), or deuterium oxide (heavy water).
E. *Coolant* removes heat and produces steam to generate electricity.

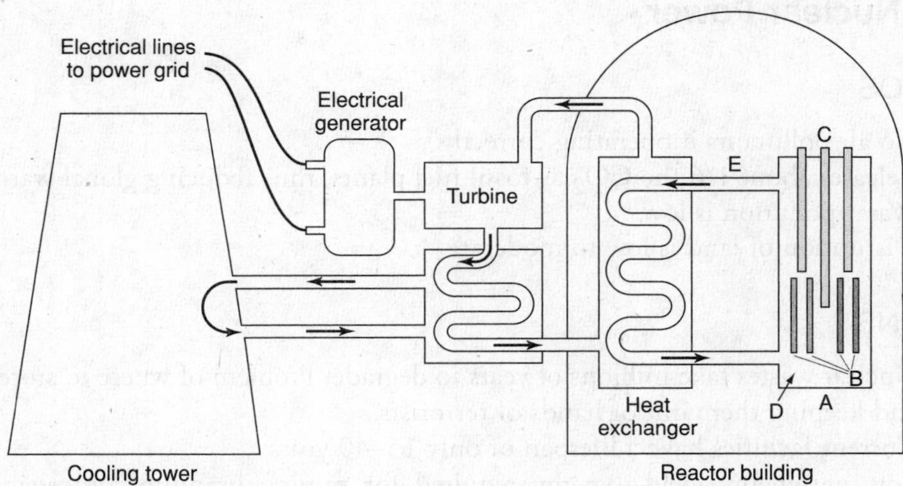

Figure 8.10 Diagram of nuclear power plant

LIGHT-WATER REACTORS

Both the moderator and coolant are light water (H_2O). To this category belong the pressurized-water reactors (PWR) and boiling-water reactors (BWR).

In PWR, the water coolant operates at a high pressure and is then pumped through the reactor core, where it is heated to about 620°F (325°C). The superheated water is pumped through a steam generator. Through heat exchangers, a secondary loop of water is heated and converted to steam. This steam drives one or more turbine generators, is condensed, and is pumped back to the steam generator. The secondary loop is isolated from the water in the reactor core and is not radioactive. A third stream of water from a lake, river, or cooling tower is used to condense the steam.

In BWR, the water coolant is permitted to boil within the core by operating at a lower pressure. The steam produced in the reactor pressure vessel is piped directly to the turbine generator, is condensed, and is then pumped back to the reactor. Although the steam is radioactive, there is no intermediate heat exchanger between the reactor and turbine to decrease efficiency. As in the PWR, the condenser cooling water has a separate source such as a lake or river.

HEAVY-WATER REACTORS

Both the coolant and moderator are heavy water (D_2O).

GRAPHITE-MODERATED REACTORS

This category uses light water for cooling, graphite for moderation, and uranium for fuel. These reactors require no separated isotopes such as enriched uranium or heavy water. This type of reactor, built by the Russians, was very unstable and is no longer being produced (see the Chernobyl case study).

EXOTIC REACTORS

Fast-breeder reactors and other experimental installations are in this group. Breeder reactors produce more fissionable material than they consume.

Environmental Advantages/Disadvantages of Nuclear Power

PROS

- No air pollutants if operating correctly.
- Releases about 1/6 the CO_2 as fossil fuel plants, thus reducing global warming.
- Water pollution is low.
- Disruption of land is low to moderate.

CONS

- Nuclear wastes take millions of years to degrade. Problem of where to store them and keeping them out of hands of terrorists.
- Current facilities have a lifespan of only 15–40 years.
- Low net-energy yield—energy required for mining uranium, processing ore, building and operating plant, dismantling plant, and storing wastes.
- Safety and malfunction issues.

RELEVANT LAW

Price-Anderson Nuclear Indemnity Act (1957): Covers all nonmilitary nuclear facilities constructed before 2026. It indemnifies the nuclear industry against all liability claims arising from nuclear accidents while ensuring compensation coverage for the general public through no-fault insurance—the first $10 billion coming from the nuclear industry and anything above $10 billion coming from the U.S. government.

Safety Issues (Radiation and Human Health)

The U.S. Department of Energy (DOE) estimates that up to 50,000 radioactive contaminated sites within the United States require cleanup with a projected cost of $1 trillion dollars. The situation is many times worse in the former Soviet Union.

Estimated Health Risks per Year in the United States

	Nuclear	Coal
Premature death	6,000	65,000
Genetic defects/damage	4,000	200,000

CASE STUDY

Chernobyl, Ukraine (1986): Explosion in a nuclear power plant sent highly radioactive debris throughout northern Europe. Estimates run as high as 32,000 deaths, and 62,000 square miles (161,000 sq km) remain contaminated. About 500,000 people were exposed to dangerous levels of radiation. Cost estimates run as high as $400 billion. The cause was determined to be both design and human error.

Nuclear Fusion

Nuclear fusion can occur when extremely high temperatures are used to force nuclei of isotopes of lightweight atoms to fuse together, which causes large amounts of energy to be released. A coal-fed electrical generating plant producing 1,000 megawatts of electricity in one day produces 30,000 tons of CO_2 gases, 600 tons of SO_2 gas, and 80 tons of NO_2 gas. In contrast, a fusion plant producing the same amount of electricity would produce 4 pounds of harmless helium as a waste product.

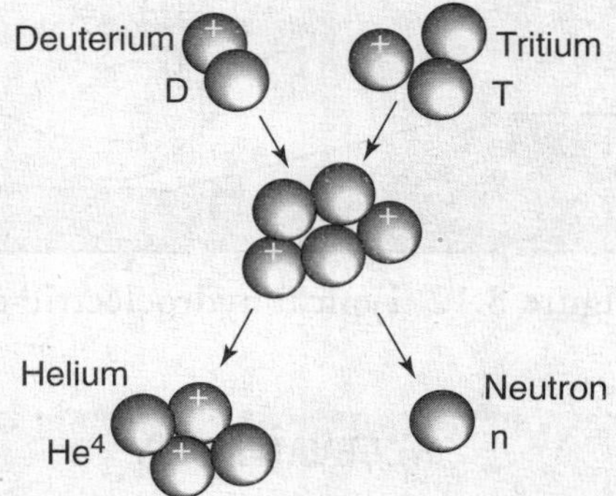

Figure 8.11 Deuterium-tritium fusion reaction

HYDROELECTRIC POWER

Dams are built to trap water, which in turn is then released and channeled through turbines that generate electricity. Hydroelectric power supplies about 10% of the electricity in the United States and approximately 3% worldwide.

Pros

- Dams control flooding.
- Low operating and maintenance costs.
- No polluting waste products.
- Long life spans.
- Moderate to high net-useful energy.
- Areas of water recreation.

Cons

- Dams create large flooded areas behind the dam from which people are displaced. Water is slow moving and can breed pathogens.
- Dams destroy wildlife habits and keep fish from migrating.
- Sedimentation requires dredging. Prevents sedimentation from reaching downstream and enriching farmland.
- Expensive to build.
- Destroys wild rivers.
- Large-scale projects are subject to earthquakes.

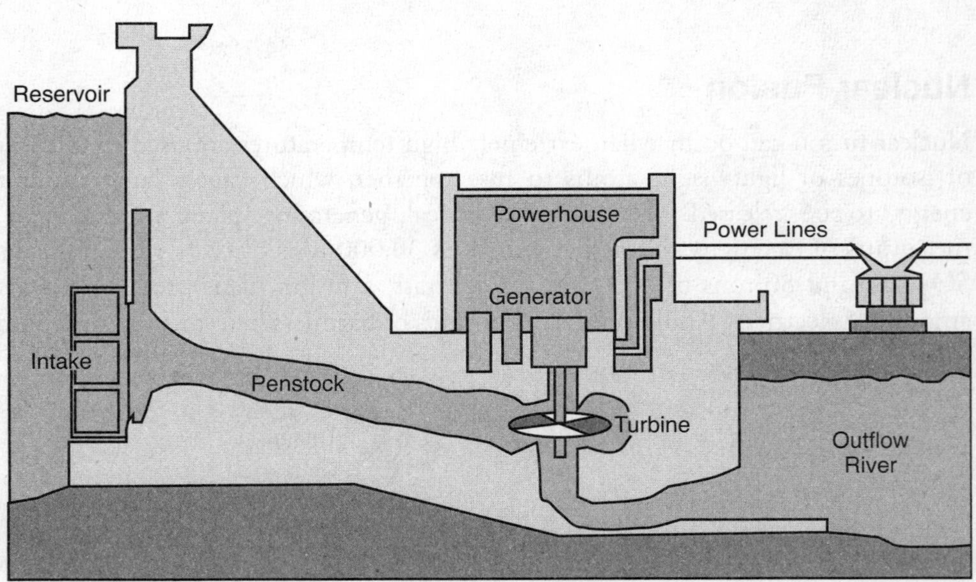

Figure 8.12 Typical hydroelectric dam

RELEVANT LAW

Water Resources Development Act (1986): Established dam safety programs and standards.

Flood Control

Methods to control floods include:

CHANNELIZATION

Straighten and deepen streams. Cons: removes bank vegetation and increases stream velocity, which causes erosion; may increase downstream flooding and sedimentation, which negatively impacts aquatic habitats.

DAMS

Dams store water in reservoirs. During periods of excessive rainfall, dams can be overwhelmed and excess water needs to be released.

IDENTIFY AND MANAGE FLOOD-PRONE AREAS

By identifying flood-prone areas, precautionary building practices such as floodways, building elevation, and pumping stations can be adopted.

LEVEES OR FLOODWALLS

Levees are raised embankments to prevent a river from overflowing. Levees contain river and stream flows but increase water velocity. Levees can break as they did in New Orleans during Hurricane Katrina in 2005.

PRESERVE WETLANDS

This technique preserves natural flood plains and maintains biodiversity.

Salmon

There are an estimated 74,993 dams in America, blocking 600,000 miles (965,000 km) of what had once been free-flowing rivers. Salmon are migratory fish that hatch in streams and rivers and then swim downstream to the ocean to live most of their lives. They return to the rivers and streams from which they hatched to spawn. Dams now block almost every major river system in the West. Many of those dams have destroyed important spawning and rearing habitats for salmon. In the Sacramento Valley in California, less than 5% of the salmon's original habitat is still available. In the Columbia River Basin, once the most productive salmon river system in the world, less than 70 miles (110 km) still remains free-flowing. As a result of habitat destruction, at least 106 major U.S. west coast salmon runs are extinct, and 25 more are now endangered. Dams also change the character of rivers, creating slow-moving, warm-water pools that are ideal for predators of salmon. Low water velocities in large reservoirs can also delay salmon migration and expose fish to higher water temperatures and disease. Cutting trees in forests near the streams and rivers clouds the water with silt and reduces water quality.

Things that have been done to reduce the impacts of dams on fish include fish passage facilities and fish ladders that help juvenile and adult fish migrate over or around many dams. Spilling water at dams over the spillway can help pass juvenile fish downstream because it avoids sending the fish through turbines. Water releases from upstream storage reservoirs have been used to increase water velocities and to reduce water temperatures in order to improve migration conditions through reservoirs. Juvenile fish also are collected and transported downstream in barges and trucks.

Silting and Other Impacts

DISEASE

Dam reservoirs in tropical areas, due to their slow movement, are literally breeding grounds for mosquitoes, snails and flies—the vectors that carry malaria, schistosomiasis, and river blindness.

DISPLACEMENT

Flooded areas behind dams destroy rich croplands and displace people.

EFFECTS ON WATERSHED

Downstream areas are deprived of the nutrient-rich silt that would revitalize depleted soil profiles.

IMPACT ON WILDLIFE

Migration and spawning cycles are disrupted.

SILTING

Silting occurs when silt (very fine particles intermediate in size between sand and clay) that is dissolved in river water settles out behind dams. Over time, the silt builds up and must be removed (dredged).

WATER LOSS

Large losses of freshwater occur through evaporation and seepage through porous rock beds.

ENERGY CONSERVATION

Energy Star is a joint program of the U.S. Environmental Protection Agency and the U.S. Department of Energy. It is designed to protect the environment through energy-efficient products and practices. Programs coordinated through Energy Star saved enough energy in 2005 to avoid greenhouse gas emissions equivalent to 23 million cars and $12 billion in utility bills. The symbol shown below appears on products that meet Energy Star standards.

CAFE Standards

Transportation needs consume two-thirds of the petroleum consumption in the United States. This sector of energy consumption is increasing faster than any other sector (1.8% growth per year). Imports of crude oil and other petroleum products are expected to increase 66% by 2020. CAFE (Corporate Average Fuel Economy) standards are the average fuel economies of a manufacturer's fleet of passenger cars or light trucks. The testing follows the guidelines established by the Environmental Protection Agency. It is estimated that CAFE standards result in savings of over 55 billion gallons (210 billion L) of fuel annually with a substantial reduction in carbon dioxide emissions of approximately 10%. CAFE standards are achieved through better engine design, efficiency, and weight reduction. The average CAFE standard of 27.5 miles per gallon (11.7 km/L) for automobiles has not increased in the United States since 1996 with SUV standards less than those for light trucks. Significant improvements in fuel mileage could be achieved by expanding CAFE standards to include:

1. Streamlining
2. Reduced tire-rolling resistance
3. Engine improvements, especially transitioning to a hybrid technology
4. Optimized transmission improvements
5. Transition to higher voltage automotive electrical systems
6. Performance-based tax credits

Hybrid Electric Vehicles

Cars should be able to drive at least 300 miles (482 km) between refuelings, be refueled quickly and easily, and keep up with the other traffic on the road. A gasoline-powered car meets these requirements but produces a relatively large amount of pollution and generally gets poor gas mileage. For example, 1 gallon (4 L) of gasoline weighs 6 pounds (2.7 kg). When burned, the carbon in it combines with oxygen from the air to produce nearly 20 pounds (9 kg) of carbon dioxide. An electric car, however, produces almost no pollution but has limited range between charges. A hybrid vehicle attempts to increase the mileage and reduce the emissions of a gas-powered car significantly while overcoming the shortcomings of an electric car.

Gasoline-electric hybrid cars contain five important parts. First, it has an engine. The gasoline engine on a hybrid is smaller than on a gas-only car and uses advanced technologies to reduce emissions and increase efficiency. Second, the fuel tank in a hybrid is the energy storage device for the gasoline engine. Gasoline has a much higher energy density than batteries do. For example, about 1,000 pounds (455 kg) of batteries are needed to store as much energy as 1 gallon/7 pounds (4 L/3 kg) of gasoline. Third, advanced electronics allow the electric motor to act as a generator. For example, when it needs to, the motor can draw energy from the batteries to accelerate the car. When acting as a generator, it can slow down the car and return energy to the batteries. Fourth, the generator is similar to an electric motor, but it acts only to produce electrical power. It is used mostly on series hybrids. Fifth, the batteries in a hybrid car are the energy storage device for the electric motor. Unlike the gasoline in the fuel tank, which can power only the gasoline engine, the electric motor on a hybrid car can put energy into the batteries as well as draw energy from them.

A parallel hybrid has a fuel tank that supplies gasoline to the engine and a set of batteries that supplies power to the electric motor. Both the engine and the electric motor power the car at the same time. In a series hybrid, the gasoline engine turns a generator, which charges the batteries and/or powers an electric motor. The gasoline engine never directly powers the vehicle.

Plug-in hybrid electric vehicles are hybrid cars with an added battery. Plug-in hybrids can be plugged in to a 120-volt outlet and charged. Plug-ins run on the stored energy for much of a typical day's driving—up to 60 miles per charge. When the charge is used up, plug-in hybrids automatically keep running on the fuel in the fuel tank. A plug-in does not entail any sacrifice of vehicle performance or driver amenities. A midsize plug-in can accelerate from 0 to 60 miles per hour (0–96 kph) in less than 9 seconds and sustain a top speed of 97 miles per hour (156 kph). Higher initial costs are partly offset by lower operating costs.

Mass Transit

Mass transit includes rail, bus services, subways, airlines, ferries, and so on. Mass transit often determines where people live, where they work, and how much air pollution they are subjected to. In the United States, private cars are the primary mode of transportation. In the rest of the world, mass transport is the primary form. For example, in the United States, only 3% of the population utilizes mass transit on a regular basis. In Japan, that figure expands to 47%. Land availability and whether cities expand vertically (land is not available) or expand horizontally (land is available) often determines the preferred mode of transportation. Use of mass transit use rises sharply with population density. Mass transit can be faster than private cars when either a separate infrastructure is reserved for mass transit or special lanes on shared highways are designated for mass transit vehicles, which results in higher speeds and less delay. In some areas, public transport systems are poorly developed and take significantly longer than an equivalent trip in a private vehicle. Perhaps the most efficient method to promote mass transit is to adopt a user-pay approach, where all external costs of operating a private vehicle are factored into license fees and/or vehicle taxes. However, this approach would be met with fierce private and political opposition.

LIGHT RAIL

Consists of trains that share space with road traffic and trains that have their own right-of-way and are separated from road traffic.

GROUP OR PERSONAL RAPID TRANSIT

Private vehicles similar to automobiles or buses able to travel under a driver's control but then be able to enter an automated guideway or track for extended distances. Could be powered by fuels when out of system and electricity when entering the automated guideway.

AUTOMATED HIGHWAY SYSTEMS

Sensors in the roadbed monitor and control traffic flow by adjusting vehicle speed and spacing to reduce congestion.

BUS RAPID TRANSIT

Includes bus-dedicated and grade-separated right-of-ways, bus lanes, bus signal preference and preemption, bus turnouts, bus-boarding islands, curb realignment, off-bus fare collection, and level boarding.

MAGLEV

Magnetically levitated trains that "float" above the rails to reduce friction.

TUBULAR RAIL

Trains that do not sit on tracks but rather travel through distantly spaced support structures.

RENEWABLE ENERGY

Several different forms of renewable enrgy can be used.

Solar Energy

Solar energy consists of collecting and harnessing radiant energy from the sun to provide heat and/or electricity. Electrical power and/or heat can be generated at home and industrial sites through photovoltaic cells or solar collectors or at a central solar-thermal plant.

Active solar collectors use the sun's energy to heat water or air inside a home or business. It requires an electrical input (for pumps and fans). Passive solar requires no moving parts. The structure is built to maximize solar capture, such as large, south-facing windows. Photovoltaic cells are used to generate electricity.

Pros

- Supply of solar energy is limitless.
- Reduces reliance on foreign imports.
- Only pollution is in manufacture of collectors. Little environmental impact.
- Can store energy during the day and release it at night—good for remote locations.

Cons

- Inefficient where sunlight is limited or seasonal.
- Maintenance costs are high.
- Systems deteriorate and must be periodically replaced.
- Current efficiency is between 10%–25% and not expected to increase soon.

Hydrogen Fuel Cells

Nine million tons of hydrogen is produced in the United States each day—enough to power 20 to 30 million cars or 5 to 8 million homes. Most of this hydrogen is used by industry in refining, treating metals, and processing foods.

The hydrogen fuel cell operates similar to a battery. It has two electrodes, an anode and a cathode, that are separated by a membrane. Oxygen passes over one electrode and hydrogen over the other. The hydrogen reacts with a catalyst on the anode that converts the hydrogen gas into negatively charged electrons and positively charged hydrogen ions. The electrons flow out of the cell to be used as electrical energy. The hydrogen ions move through the electrolyte membrane to the cathode, where they combine with oxygen and the electrons to produce water. Unlike batteries, fuel cells never run out.

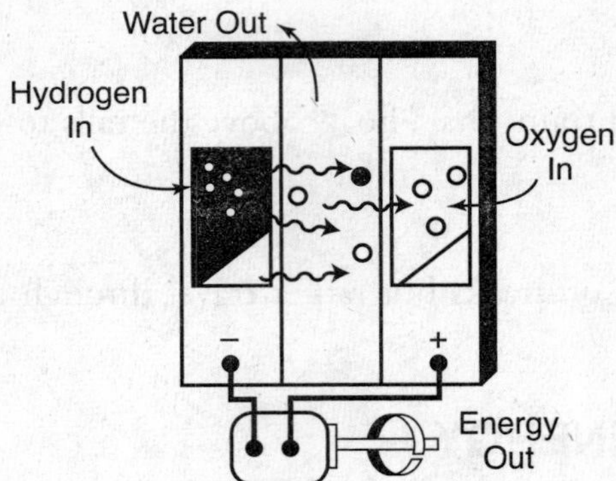

Figure 8.13 Hydrogen fuel cell

Pros

- Waste product is pure water.
- Ordinary water (either ocean or freshwater) can be used to obtain hydrogen.
- Does not destroy wildlife habitats and has minimal environmental impact.
- Energy to produce hydrogen could come from fusion reactor, solar, or other less-polluting source.
- Hydrogen is easily transported through pipelines.
- Hydrogen can be stored in compounds to make it safe to handle. Hydrogen is explosive, but so are methane, propane, butane, and gasoline.

Cons

- Takes energy to produce the hydrogen from either water or methane.
- Changing from a current fossil fuel system to a hydrogen-based system would be very expensive.
- Hydrogen gas is explosive.
- At the current time, it is difficult to store hydrogen gas for personal cars.

Biomass

Biomass is any carbon-based, biologically derived fuel source such as wood, manure, charcoal, bagasse grown for use as a biofuel. Examples include biodiesel, methanol, and ethanol. Plants that are suitable for biofuel include switch grass, hemp, corn, and sugarcane. Biomass can also be used for building materials and biodegradable plastics and paper. Approximately 15% of the world's energy supply is derived from biomass and is most commonly used in developing nations.

Pros

- Renewable energy source as long as used sustainably.
- Rate of use balanced with rate of renewal does not disrupt atmospheric CO_2 levels.
- Less SO_2 and NO_x produced than by burning fossil fuels.

- Can be sustainable if issue of deforestation and soil erosion are controlled.
- Could supply half of the world's demand for electricity.
- Biomass plantations (cottonwoods, poplars, sycamores, switch grass, and corn) can be located in less desirable locations and can reduce soil erosion and restore degraded land. In the U.S. alone, 200 million acres (80 million ha) are suitable for biomass plantations.

Cons

- Requires adequate water and fertilizer, of which sources are declining.
- Use of inorganic fertilizers, herbicides, and pesticides would harm environment.
- Would cause massive deforestation and loss of habitat, resulting in a decrease in biodiversity.
- Inefficient methods of burning biomass would lead to large levels of air pollution, especially particulate matter.
- Expensive to transport because it is heavy.
- Not efficient. About 70% of the energy derived from burning biomass is lost as heat.

Biomass can also be burned in large incinerators as an energy source.

Pros

- Crop residues are available (e.g., sugarcane in Hawaii).
- Ash can be collected and recycled.
- Reduces impact on landfills.

Cons

- Net-energy yield is low to moderate. Energy required drying and transporting material to a centralized facility is prohibitive.
- Severe air pollution if not burned in a centralized facility.
- CO_2 production would have a major impact on global warming.

CASE STUDY

About 90% of the cars in Brazil run on either alcohol or gasohol (a mixture of gasoline and ethanol). Flex fuel engines can run on either. The alcohol is produced from sugarcane, which grows in abundance in Brazil.

Wind Energy

Wind turns giant turbine blades that then power generators. Turbines can be grouped in clusters called wind farms.

Pros

- All electrical needs of the United States could be met by wind in North Dakota, South Dakota, and Texas.
- Wind farms can be quickly built and can also be built out on sea platforms.
- Maintenance is low and the farms are automated.

- Moderate-to-high net-energy yield.
- No pollution. Wind farms are in remote areas so noise pollution is minimal to humans.
- Land underneath wind turbines can be used for agriculture (multiple-use).

Cons

- Steady wind is required to make investment in wind farms economical. Few places are suitable.
- Backup systems need to be in place when the wind is not blowing.
- Visual pollution.
- May interferes with flight patterns of birds.
- May interfere with communication, such as microwaves, TV, and cell phones
- Noise pollution.

Small-Scale Hydroelectric

Small-scale hydropower utilizes small turbines connected to generators submerged in streams to generate power. Generally, the capacity of small-scale hydropower is 100 kW or less. This technology does not impede stream navigation or fish movement. This technology is especially attractive in remote areas where power lines are not available. Several factors should be considered when installing small-scale hydropower:

- The amount of water flow available on a consistent basis
- The amount of drop (head) the water has between the intake and output of the system
- Regulatory issues such as water rights and easements.

In many cases, there are economic incentives for installing small-scale hydropower systems through grants, loans, and tax incentives.

Ocean Waves and Tidal Energy

The natural movement of tides and waves spin turbines that generate electricity. Only a few plants are currently operating worldwide. They are on the north coast of France and in the Bay of Fundy between the United States and Canada).

Pros

- No pollution.
- Minimal environmental impact.
- Net-energy yield is moderate.

Cons

- Construction is expensive.
- Few suitable sites.
- Equipment can be damaged by storms and corrosion.

Geothermal

Heat contained in underground rocks and fluids from molten rock (magma), hot dry-rock zones, and warm-rock reservoirs produce pockets of underground dry steam, wet steam, and hot water. This steam can be used to drive turbines, which can then generate electricity. Geothermal energy supplies less than 1% of the energy needs in the United States. Geothermal energy is currently being used in Hawaii, Iceland, Japan, Mexico, New Zealand, Russia, and California. Areas of known geothermal resources tend to follow tectonic plate boundaries.

Pros

- Moderate net-energy yield.
- Limitless and reliable source if managed properly.
- Little air pollution.
- Competitive cost.

Cons

- Reservoir sites are scarce.
- Source can be depleted if not managed properly.
- Noise, odor, land subsidence.
- Can degrade ecosystem due to corrosive, thermal, or saline wastes.

RELEVANT LAW

Renewable Energy Law, China (2007): Rapid economic development throughout China has resulted in a significant increase in energy consumption, leading to a rise in harmful emissions and power shortages. China's 2007 Renewable Energy Law requires power grid operators to purchase resources from registered renewable energy producers. The law also offers financial incentives, such as a national fund to foster renewable energy development, and discounted lending and tax preferences for renewable energy projects. The Renewable Energy Law is designed to help protect the environment, prevent energy shortages, and reduce dependence on imported energy. The law includes the use of solar photovoltaics and active solar water heating. Finally, the law includes specific penalties for noncompliance. In 2003, China's renewable energy consumption accounted for only 3% of the country's total energy consumption. The government plans to increase this figure to 10% in 2020. The United States now stands at about 2% usage of renewable energy sources.

MULTIPLE-CHOICE QUESTIONS

1. Which of the following forms of energy is a renewable resource?

   (A) Synthetic oil
   (B) Breeder fission
   (C) Biomass
   (D) Oil shale
   (E) Synthetic natural gas

2. Which of the following forms of energy has low, short-term availability?

   (A) Solar energy
   (B) Synthetic oil derived from coal
   (C) Nuclear energy
   (D) Coal
   (E) Petroleum

3. Which of the following is a source of high net energy?

   (A) Tar sands
   (B) Wind
   (C) Fission
   (D) Synthetic natural gas
   (E) Geothermal

4. Which of the following alternatives would NOT lead to a sustainable energy future?

   (A) Phase out nuclear power subsidies.
   (B) Create policies to encourage governments to purchase renewable energy devices.
   (C) Assess penalties or taxes on continues use of coal and oil.
   (D) Decrease fuel-efficiency standards for cars, appliances, and HVAC systems.
   (E) Create tax incentives for independent power producers.

5. At today's rate of consumption, known U.S. oil reserves will be depleted in about

   (A) 100 years
   (B) 50 years
   (C) 25 years
   (D) 10 years
   (E) 3 years

6. Which country currently ranks number one in both coal reserves and use of coal as an energy source?

   (A) Russian Federation
   (B) United States
   (C) China
   (D) India
   (E) Brazil

7. The lowest average generating cost (cents per kWh) comes from what energy source?

   (A) Large hydroelectric facilities
   (B) Geothermal
   (C) Nuclear
   (D) Solar photovoltaic
   (E) Coal

8. The fastest-growing renewable energy resource today is

   (A) nuclear
   (B) coal
   (C) wind
   (D) large-scale hydroelectric
   (E) geothermal

9. The least-efficient energy conversion device listed is

   (A) steam turbine
   (B) fuel cell
   (C) fluorescent light
   (D) incandescent light
   (E) internal combustion engine

10. Which is NOT an advantage of using nuclear fusion?

   (A) Abundant fuel supply
   (B) No generation of weapons-grade material
   (C) No air pollution
   (D) No high-level nuclear waste
   (E) All are advantages

11. Only about 10% of the potential energy of gasoline is used in powering an automobile. The remaining energy is lost as low-quality heat. This is an example of the

    (A) first law of thermodynamics
    (B) second law of thermodynamics
    (C) law of conservation of energy
    (D) first law of efficiency
    (E) law of supply and demand

12. The law of conservation of mass and energy states that matter can neither be created nor destroyed and that the total energy of an isolated system is constant despite internal changes. Which society offers the best long-term solution to the constraints of this law?

    (A) Low-throughput society
    (B) High-throughput society
    (C) Matter-recycling society
    (D) Free-market economy
    (E) Global market economy

13. Which of the following methods CANNOT be used to produce hydrogen gas?

    (A) Reforming
    (B) Thermolysis
    (C) Producing it from plants
    (D) Coal gasification
    (E) All are methods of producing hydrogen gas

14. Energy derived from fossil fuels supplies approximately what percentage of the world's energy needs?

    (A) 10%
    (B) 33%
    (C) 50%
    (D) 85%
    (E) 99%

15. Which of the following terms is NOT a unit of power?

    (A) Btu
    (B) Horsepower
    (C) Kilowatt
    (D) Joule
    (E) All are units of power

16. Automobile manufacturers make money by selling cars. Cars pollute and use fossil fuels. Traditionally, American auto manufacturers have encouraged customers to buy bigger and more powerful cars. This lack of incentive to improve energy efficiency is known as a

    (A) harmful, positive-feedback loop
    (B) beneficial, positive-feedback loop
    (C) harmful, negative-feedback loop
    (D) beneficial, negative-feedback loop
    (E) mutualistic, positive-feedback loop

17. If you were designing a house in the United States, you lived in a cold climate, and you wanted it to be as energy efficient as possible, you would place large windows to capture solar energy on which side of the house?

    (A) North
    (B) South
    (C) East
    (D) West
    (E) It makes no difference which side

18. Which of the following is NOT an advantage of building a hydroelectric power plant?

    (A) Low pollution
    (B) High construction cost
    (C) Relatively low operating cost
    (D) Control flooding
    (E) Moderate-to-high net-useful energy

19. The country that has made the largest commitment to increasing its share of renewable energy resources is

    (A) the United States
    (B) the Russian Federation
    (C) China
    (D) Saudi Arabia
    (E) Iran

20. Which of the following nonrenewable energy sources has the least environmental impact?

    (A) Gasoline
    (B) Coal
    (C) Oil shale
    (D) Tar sands
    (E) Natural gas

FREE-RESPONSE QUESTION

A large, natural gas–fired electrical power facility produces 15 million kilowatt-hours of electricity each day it operates. The power plant requires an input of 13,000 Btus of heat to produce 1 kilowatt-hour of electricity. One cubic foot of natural gas supplies 1,000 Btus of heat energy.

(a) Showing all steps in your calculations, determine the:
- (i) Btus of heat needed to generate the electricity produced by the power plant in 2 hours.
- (ii) Cubic feet of natural gas consumed by the power plant each year.
- (iii) Cubic feet of carbon dioxide gas released by the power plant each day. Assume that methane combusts with oxygen to produce only carbon dioxide and water vapor and that the pressures and temperatures are kept constant.
- (iv) Gross profit per year for the power company. The power company is able to sell electricity at $50 per 500 kWh. They pay a wholesale price of $5.00 per 1,000 cubic feet of natural gas.

(b) What environmental effect might the production of electricity through the burning of natural gas pose?

(c) Describe two other methods of producing electricity, and provide technological, economic, and environmental pros and cons for each method discussed.

MULTIPLE-CHOICE ANSWERS AND EXPLANATIONS

1. **(C)** Renewable resources are those resources that theoretically will last indefinitely either because they are replaced naturally at a higher rate than they are consumed or because their source is essentially inexhaustible.
2. **(E)** At the current rate of consumption, global oil reserves are expected to last another 45–50 years. With projections of increased consumption in the near future, this figure will be even lower.
3. **(B)** Net-useful energy is defined as the total amount of useful energy available from an energy resource over its lifetime minus the amount of energy used in extracting it and delivering it to the end user.
4. **(D)** The key word in this question is "NOT." To foster a sustainable energy future, fuel efficiency standards would have to increase.
5. **(C)** World oil demand is increasing at a rate of about 2% per year. Known U.S. oil reserves are projected to last another 25 years.
6. **(C)** China gets approximately 75% of its energy from coal.
7. **(A)** Nonrenewable resources of energy (natural gas, oil) have had recent and dramatic price increases that have resulted in major increases in the cost of electricity. Renewable resources of energy, especially hydroelectric and wind, provide the least-expensive method of producing electricity.

8. **(C)** During the 1990s, wind power experienced an annual growth of 22%. Wind power supplies less than 2% of the energy used in the United States. The country with the largest sector of its energy needs met through wind power is Denmark at 8%.
9. **(D)** An incandescent lightbulb is only 5% efficient as compared with a fluorescent light at 22%. The typical internal combustion engine is 10% efficient, and a hydrogen fuel cell is 60% efficient. The United States wastes as much energy each day as two-thirds of the world consumes.
10. **(E)** The major fuel used in fusion reactors, deuterium, could be readily extracted from ordinary water. The tritium required would be produced from lithium, which can be extracted from seawater and land deposits.
11. **(B)** The second law of thermodynamics states that when energy changes from one form to another, some of the useful energy is always degraded into a lower-quality, more dispersed (higher entropy), and less-useful form.
12. **(A)** A low-throughput society, also known as a low-waste society, focuses on matter and energy efficiency through reusing and recycling, using renewable resources at a rate no faster than can be replenished, reducing unnecessary consumption, emphasizing pollution prevention rather than waste reduction, and controlling population growth.
13. **(E)** Reforming is a chemical process of splitting water molecules. Thermolysis is breaking water molecules apart at high temperatures. Hydrogen gas can be produced from algae by depriving the algae of oxygen and sulfur. Coal gasification is the conversion of coal into synthetic natural gas (SNG), which can then be converted into hydrogen.
14. **(D)** Oil supplies about 36%, coal about 26%, and natural gas about 23%.
15. **(D)** A joule is a unit of energy.
16. **(A)** Pollution is harmful, which is the first part of the answer. A negative-feedback loop tends to slow down a process, while a positive-feedback loop tends to speed it up. Speeding up or increasing sales would be a positive feedback, which is the second half of the answer.
17. **(B)** In cold climates, large south-facing windows allow significant solar energy into the house and also provide daylighting. Properly sized overhangs can prevent overheating in the summer. In hot climates, north-facing windows can provide daylighting without heating the house. East- and west-facing windows generally cause excessive heat gains in the summer and heat losses in the winter, so they are usually small. Although overhangs are impractical for east- and west-facing windows, vertical shading can be used or trees and shrubs can be strategically located to shade the windows.
18. **(B)** The construction of hydroelectric plants is initially expensive but the operating costs are low since there are no fuel costs.
19. **(C)** China has passed legislation to increase its percentage of renewable energy to 10%. The United States' percentage is about 2%.
20. **(E)** The combustion of natural gas releases very small amounts of sulfur dioxide and nitrogen oxides, virtually no ash or particulate matter, and lower levels of carbon dioxide, carbon monoxide, and other reactive hydrocarbons when compared with coal or oil. The following chart compares emission levels of various pollutants for three forms of energy: natural gas, oil, and coal.

Fossil Fuel Emission Levels

Pollutant	Natural Gas	Oil	Coal
(pounds of pollutant per billion Btu of energy input)			
Carbon dioxide	117,000	164,000	208,000
Carbon monoxide	40	33	208
Nitrogen oxides	92	448	457
Sulfur dioxide	1	1,122	2,591
Particulates	7	84	2,744
Mercury	0.000	0.007	0.016

FREE-RESPONSE ANSWER

(a)

(i)

$$\frac{24 \text{ hours}}{1 \text{ day}} \times \frac{1.5\times10^7 \text{ kWh}}{24 \text{ hours}} \times \frac{13{,}000 \text{ Btu}}{1 \text{ kWh}} = 2.0\times10^{11} \text{ Btu per day}$$

(ii)

$$\frac{2\times10^{11} \text{ Btu}}{\text{day}} \times \frac{1 \text{ ft.}^3}{1{,}000} \times \frac{1 \text{ day}}{24 \text{ hours}} = 8.3\times10^6 \text{ ft.}^3 \text{ natural gas/hr}$$

(iii) $CH_{4(g)} + 2O_{2(g)} \rightarrow CO_{2(g)} + 2H_2O_{(g)}$. If everything is a gas, and if temperature and pressure are kept constant, then the coefficients can represent volume. Therefore, we could say that 1 ft.^{3} of methane gas combines with 2 ft.^{3} of oxygen gas to produce 1 ft.^{3} of carbon dioxide gas and 2 ft.^{3} of water vapor.

$$\frac{9.3\times10^6 \text{ ft.}^3 \text{ CH}_4}{\text{hr}} \times \frac{1 \text{ ft.}^3 \text{ CO}_2}{1 \text{ ft.}^3 \text{ CH}_4} \times \frac{24 \text{ hours}}{1 \text{ day}} = 2.2\times10^8 \text{ ft.}^3 \text{ CO}_2 \text{ per day}$$

Income:

$$\frac{1.57\ \text{kWh}}{\text{day}} \times \frac{365\ \text{days}}{\text{year}} \times \frac{\$50}{500\ \text{kWh}} = \$5.48 \times 10^8\ (\$548\ \text{million})\ \text{per year}$$

Costs that the power company spends on natural gas:

$$\frac{8.36 \times 10^6\ \text{ft.}^3\ \text{natural gas}}{\text{hour}} \times \frac{24\ \text{hours}}{1\ \text{day}} \times \frac{365\ \text{days}}{\text{year}} \times \frac{\$5.00}{1{,}000\ \text{ft.}^3\ \text{natural gas}} =$$

$$\$3.66 \times 10^8\ (\$366\ \text{million})\ \text{per year}$$

Gross profit: $(\$5.48 \times 10^8) - (\$3.66 \times 10^8) = \$1.82 \times 10^8$ ($182 million)

(b) An environmental effect that results from the burning of natural gas to produce electricity is the production of carbon dioxide gas—a greenhouse gas. As a greenhouse gas, the carbon dioxide gas molecules absorb radiant energy reflected from Earth's surface and re-radiate this energy as long-wave infrared radiation back to Earth. This trapping of Earth's heat has serious environmental consequences in terms of its effect on global weather patterns, local climate conditions, and the glaciers and polar ice caps.

> **TIP**
>
> When answering essay and short answer questions, use the analyze, organize, and respond model.
>
> **Analyze**—Write down key phrases. Even if a key phrase pops into your mind for another question, write it down immediately or you risk forgetting it.
>
> **Organize**—Sequence yout ideas. This ensures that your answer is logically developed and will be clear to the grader.
>
> **Respond**—Try to provide support for claims that you make in your answer using Case Studies when possible. Star or underline important points. Write legibly. If your handwriting is poor, then print. Use proper grammar and be neat.

Other environmental effects produced by burning natural gas include the production of hydrogen sulfide gas (H_2S) and sulfur dioxide gas (SO_2) as production by-products. Sulfur dioxide gas when combined with atmospheric water vapor produces what is known as acid rain. Acid rain has serious environmental effects primarily on aquatic organisms in lower trophic levels and on the reproduction of developing aquatic organisms. Hydrogen sulfide has serious environmental effects as a toxic pollutant and respiratory irritant. Leakage of CH_4 during processing adds methane to the atmosphere, which has a more deleterious effect as a greenhouse gas than does carbon dioxide.

Drilling platforms to extract natural gas, the production of contaminated wastewater and brine, the possibility of land subsidence, and the disruption of wildlife and habitat refuges are other deleterious side effects of using natural gas. However, these negative side effects have less environmental impacts than does using other forms of fossil fuels.

(c) Other methods of producing electricity include burning coal and using dams. The advantages of using coal are that it is plentiful and relatively inexpensive. Currently, the world used 4 billion metric tons of coal per year. At this rate, the world has enough coal to last for the next 200 years. Drawbacks to using coal include that current extraction methods are environmentally damaging to the land and natural habitat and are potentially dangerous to miners. Burning coal also releases radioactive particles and mercury into the air. Carbon dioxide produced from burning coal adds to the problem of global warming. The nitrogen oxides released add to the problems of acid precipitation.

Dams create large flooded areas that destroy natural habitats and disrupt the lives of the people who inhabit the area. The collection of sediment by the dam, the loss of this valuable sediment downstream and thereby making the land less productive, the obstructions that dams cause to migrating fish populations, and the aesthetic loss of a wild river are detractors of hydroelectric power. On the positive side, hydroelectric power does not significantly add to global pollution.

UNIT VI: POLLUTION (25–30%)

Areas on Which You Will Be Tested

A. Pollution Types
   1. **Air pollution**—primary and secondary sources, major air pollutants, measurement units, smog, acid deposition—causes and effects, heat islands and temperature inversions, indoor air pollution, remediation and reduction strategies, Clean Air Act, and other relevant laws.
   2. **Noise pollution**—sources, effects, and control measures.
   3. **Water pollution**—types, sources, causes and effects, cultural eutrophication, groundwater pollution, maintaining water quality, water purification, sewage treatment/septic systems, Clean Water Act, and other relevant laws.
   4. **Solid waste**—types, disposal, and reduction.

B. Impacts on the Environment and Human Health
   1. **Hazards to human health**—environmental risk analysis, acute and chronic effects, dose-response relationships, air pollutants, smoking, and other risks
   2. **Hazardous chemicals in the environment**—types of hazardous waste, treatment/disposal of hazardous waste, cleanup of contaminated sites, biomagnification, and relevant laws.

C. **Economic Impacts**—cost-benefit analysis, externalities, marginal costs, and sustainability.

Pollution

CHAPTER 9

It isn't pollution that's harming the environment. It's the impurities in our air and water that are doing it.

—Former U.S. Vice President Dan Quayle

AIR POLLUTION

Primary pollutants are emitted directly into the air from natural sources such as volcanoes, mobile sources such as cars, or stationary sources such as industrial smokestacks. Examples include: particulate matter or soot (PM_{10}), nitric oxide (NO), nitrogen dioxide (NO_2), sulfur dioxide (SO_2), carbon dioxide (CO_2), and carbon monoxide (CO).

Secondary pollutants result from the reaction of primary pollutants in the atmosphere to form a new pollutant. Examples include sulfur trioxide (SO_3), sulfuric acid (H_2SO_4), ozone (O_3), and chemicals found in photochemical smog such as PANS and peroxyacyl nitrates.

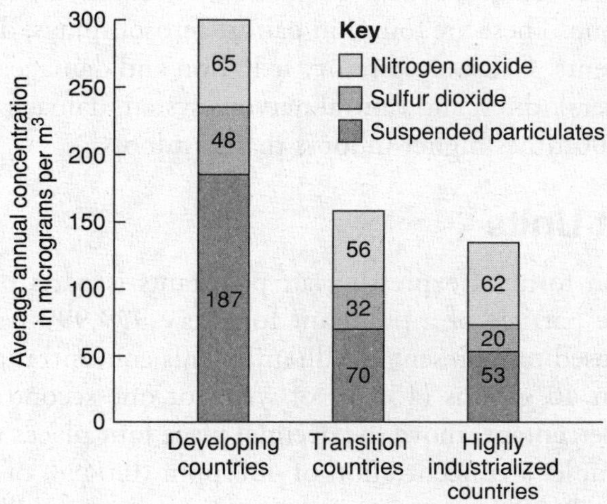

Figure 9.1 Particulate matter pollution in developing countries.

Source: World Resources Institute.

Major Air Pollutants

NITROGEN DIOXIDE (NO_2)

Forms when fuels are burned at high temperatures. Also results from forest fires, volcanoes, lightning, and bacterial action in soil. Forms nitric acid (HNO_3) in the air and contributes to acid deposition and cultural eutrophication. Results in lung irritation and damage, suppresses plant growth, and may be a carcinogen.

OZONE (O_3)

Major component of photochemical smog. Formed by sunlight reacting with NO_x and VOCs in the air. Causes lung irritation and damage, bronchial constriction, coughing, wheezing, and eye irritation. Damages plants, rubber, and plastics.

SULFUR DIOXIDE (SO_2)

Produced by burning high-sulfur oil or coal, smelting of metals, and paper manufacturing. Combines with water vapor in the air to produce acid precipitation, which reduces the productivity of plants. Causes breathing difficulties. Significant decreases in concentrations and emissions in SO_2 concentration in the United States reflect the success of the Acid Rain Program and the Clean Air Act.

SUSPENDED PARTICULATE MATTER (PM_{10})

PM_{10}s are particles with a diameter 1/7 the width of a human hair or less (< 10 μm) and include smoke, dust, diesel soot, lead, and asbestos. PM_{10}s cause lung irritation and damage. Many are known mutagens, teratogens, and carcinogens. Reduction in PM_{10}s would produce health benefits 10 times greater than similar reductions in all other air pollutants combined.

VOLATILE ORGANIC COMPOUNDS (VOCs)

Include organic compounds that have a high vapor pressure. Over 600 compounds have been identified. Examples of VOCs include toluene, xylene, formaldehyde, benzene, and acetone. These are found in paints, aerosol sprays, dry-cleaning fluids, and industrial solvents. Causes respiratory irritation and damage. Most are carcinogenic and cause liver, kidney, and central nervous system damage. Concentration of VOCs may be 1,000 times higher indoors than outdoors.

Measurement Units

The most common form of expressing air pollutants is parts per million (ppm), which denotes one particle of a pollutant for every 999,999 particles of air. The symbol μ is often used to represent a millionth. This concentration is equivalent to one drop of ink in 40 gallons (150 L) of water or one second in 280 hours. To change ppm to a percentage, move the decimal place four places to the left and add a % sign. For example, a concentration of 400 ppm (0.04%) of carbon monoxide may be fatal. Two other common measurements are parts per billion (ppb or nano) and parts per trillion (ppt or pico).

Smog

There are two forms of smog: industrial smog, and photochemical smog. Industrial smog tends to be sulfur-based and is also called gray-air smog. Photochemical smog is catalyzed by UV radiation and tends to be nitrogen-based. Photochemical smog is also called brown-air smog.

Formation of Industrial Smog

Step	Chemical Reaction
1. Carbon in coal or oil is burned in oxygen gas to produce carbon dioxide and carbon monoxide gas.	$C + O_2 \rightarrow CO_2$ $C + O_2 \rightarrow CO$
2. Unburned carbon ends up as soot or particulate matter (PM).	C
3. Sulfur in oil and coal reacts with oxygen gas to produce sulfur dioxide.	$S + O_2 \rightarrow SO_2$
4. Sulfur dioxide reacts with oxygen gas to produce sulfur trioxide.	$SO_2 + O_2 \rightarrow SO_3$
5. Sulfur trioxide reacts with water vapor in the air to form sulfuric acid.	$SO_3 + H_2O \rightarrow H_2SO_4$
6. Sulfuric acid reacts with atmospheric ammonia to form brown, solid ammonium sulfate.	$H_2SO_4 + NH_3 \rightarrow (NH_4)_2SO_4$

FORMATION OF PHOTOCHEMICAL SMOG

(Net result: $NO + VOCs + O_2 + UV \rightarrow O_3 + PANs$)

- 6 A.M.–9 A.M.
  As people drive to work, concentrations of nitrogen oxides and VOCs increase:

$$N_2 + O_2 \rightarrow 2NO$$

$$NO + VOCs \rightarrow NO_2$$

$$NO_2 \xrightarrow{uv} NO + O$$

- 9 A.M.–11 A.M.
  As traffic begins to decrease, nitrogen oxides and VOCs begin to react, forming nitrogen dioxide:

$$2NO + O_2 \rightarrow 2NO_2$$

- **11 A.M.–4 P.M.**
  As the sunlight becomes more intense, nitrogen dioxide is broken down and the concentration of ozone increases:

$$NO_2 \xrightarrow{uv} NO + O$$

$$O_2 + O \rightarrow O_3$$

  Nitrogen dioxide also reacts with water vapor to produce nitric acid and nitric oxide:

$$3NO_2 + H_2O \rightarrow 2HNO_3 + NO$$

  Nitrogen dioxide can also react with VOCs released by vehicles, refineries, gas stations, and so on to produce toxic PANs (peroxacyl nitrates):

$$NO_2 + VOCs \rightarrow PANs$$

- **4 P.M.–Sunset**
  As the sun goes down, the production of ozone is halted.

Acid Deposition—Causes and Effects

Acid rain is a broad term used to describe several ways that acids fall out of the atmosphere. A more precise term is acid deposition, which has two parts: wet and dry. Wet deposition refers to acidic rain, fog, and snow. As this acidic water flows over and through the ground, it affects a variety of plants and animals. The strength of the effects depends on many factors, including how acidic the water is, the chemistry and buffering capacity of the soils involved, and the types of organisms, such as fish, macroinvertebrates, trees, and other living things that rely on water and soil.

Dry deposition refers to acidic gases and particles. About half of the acidity in the atmosphere falls back to Earth through dry deposition. Wind blows these acidic particles and gases onto buildings, cars, homes, and trees. Dry-deposited gases and particles can also be washed from trees and other surfaces by rainstorms. When that happens, the runoff water adds those acids to the acid rain, making the combination more acidic than the falling rain alone.

Acid deposition due to sulfur dioxide (SO_2) begins with sulfur dioxide being introduced into the atmosphere by burning coal and oil, smelting metals, organic decay, and ocean spray. It then combines with water vapor to form sulfurous acid ($SO_2 + H_2O \rightarrow H_2SO_3$). Finally, the sulfurous acid reacts with oxygen to form sulfuric acid ($H_2SO_3 + \frac{1}{2}O_2 \rightarrow H_2SO_4$).

Acid deposition due to nitrogen oxides (NO_x) begins with nitrogen oxides formed by burning oil, coal, or natural gas. They also are found in volcanic vent gases and formed by forest fires, bacterial action in soil, and lightning-induced atmospheric reactions. Nitrogen monoxide, also known as nitric oxide (NO), reacts with oxygen gas to produce nitrogen dioxide gas ($NO + \frac{1}{2}O_2 \rightarrow NO_2$). Finally, nitrogen dioxide reacts with water vapor in the atmosphere to produce nitrous and nitric acids ($2NO_2 + H_2O \rightarrow HNO_2 + HNO_3$).

Acid rain causes acidification of lakes and streams. It contributes to the damage of trees at high elevations and many sensitive forest soils through nitrogen saturation and creating acidic conditions that are unhealthy for decomposers and mycorrhizal fungi. Acid shock, which is caused by the rapid melting of snow pack that contains dry acidic particles, results in acid concentrations in lakes and streams 5 to 10 times higher than acidic rainfall. In addition, acid rain accelerates the decay of building materials and paints, including irreplaceable statues and sculptures. Acid rain also leaches essential plant nutrients from the soil such as Ca^{2+}, K^+ and Mg^{2+}. Heavy metals such as Pb^{2+}, Cd^{2+}, and Hg^{2+} that are contained within rock structures may be leached out of the rocks and into the soil structure. Prior to falling to Earth, SO_2 and NO_x gases and their particulate matter derivatives (sulfates and nitrates) contribute to breathing difficulties and other health matters.

Heat Islands and Temperature Inversions

Urban heat islands occur in metropolitan areas that are significantly warmer than their surroundings. Urban air can be 2°F to 10°F (2°C to 6°C) warmer than the surrounding area. Since warmer air can hold more water vapor, rainfall can be as much as 30% greater downwind of cities when compared with areas upwind. One of the main reasons for higher-than-normal nighttime air temperatures in urban areas are buildings that reduce the radiation of urban heat to the night sky. Thermal properties of surface materials (bricks, concrete, and asphalt) store heat longer. The lack of vegetation and standing water in many urban areas also increase urban temperatures. The canyon effect results from buildings reflecting and absorbing heat and blocking winds that reduce heat through convection. Human activities that increase the heat island effect include the operation of automobiles, air conditioners, and industry. Since demand for air-conditioning rises during summer months, problems associated with energy availability and pricing become compounded.

High levels of pollution in urban areas can also create a localized greenhouse effect. Urban heat islands can directly influence the health and welfare of urban residents who cannot afford air-conditioning. As many as 1,000 people per year die in the United States due to excessive temperatures. Urban heat islands can produce secondary effects on local meteorology, including the altering of local wind patterns, the development of clouds and fog, the number of lightning strikes, and the rate of precipitation. The heat island effect can be slightly reduced by using white or reflective building materials and increasing the amount of landscaping and parks.

Temperature inversions occur when air temperature increases with height above the ground, as opposed to the normal decrease in temperature with height. This effect can lead to pollution such as smog being trapped close to the ground, with possible adverse effects on human health (asthma and increases in lung cancer). Temperature inversions commonly occur at night when solar heating ceases and the surface cools, which then cools the atmosphere immediately above it. A warm air mass then moving over a colder one keeps the cooler air mass trapped below, and the air becomes still. This results in dust and pollutants being trapped and their concentrations increasing. A nearly permanent temperature inversion occurs over Antarctica.

CASE STUDY

In 1948, 20 people were asphyxiated and over 7,000 were hospitalized or became ill as the result of severe air pollution over Donora, PA, a town of 14,000. Smog from the local zinc and steel smelting plants settled in the valley where the town was located. Four days later, winds finally cleared the toxins from the town. The investigation of this incident by state and federal health officials resulted in the first meaningful federal and state laws to control air pollution and marked the beginning of modern efforts to assess and deal with the health threats from air pollution.

Indoor Air Pollution

Many people spend the majority of their lives indoors sleeping, working, eating, and relaxing, where air circulation may be restricted. The California Air Resources Board estimates that indoor air pollutant levels may be 25% to 62% greater than outdoor levels. The most common pollutants found indoors include molds, bacteria, carbon monoxide, radon, allergens, formaldehyde (from carpeting, adhesives, and particleboard), asbestos, and tobacco smoke. Symptoms of indoor air pollution can range from headaches, breathing difficulties, and allergies to asthma, cancer, emphysema, and various nerve disorders.

Remediation and Reduction Strategies

Strategies designed to improve air quality in general include:

1. Emphasizing tax incentives for pollution control rather than fines and penalties
2. Setting legislative standards for energy efficiency
3. Increasing funding for research into renewable energy sources
4. Incorporating incentives for reducing air pollution into trade policies
5. Distributing solar cookstoves to developing countries to replace coal and firewood
6. Phasing out two-cycle gasoline engines
7. Modifying building codes to control materials used in construction
8. Providing incentives to use mass transit

Several strategies have been designed to reduce the effects of acid rain. They include designing more efficient engines to reduce NO_x emissions and increasing the efficiency of coal-burning plants to reduce SO_2, NO_x, and particulates through washing coal and using scrubbers, advanced filtration on smokestacks, electrostatic precipitators, and staged and catalytic burners with afterburners. Other strategies are increase penalties on stationary sources that do not reduce emissions, providing tax incentives to companies that do reduce pollutants, providing incentives to consumers to purchase Energy Star appliances and products that conserve energy, and increasing CAFE standards.

The EPA's Acid Rain Program is designed to achieve significant environmental and public health benefits through reductions in emissions of sulfur dioxide (SO_2) and nitrogen oxides (NO_x), the primary causes of acid rain. To achieve this goal at the lowest cost to society, the program employs both traditional and innovative, market-based approaches for controlling air pollution. In addition, the program encourages energy efficiency and pollution prevention. Specific strategies employed include an allowance trading system, an opt-in program that allows nonaffected industrial and small utility units to participate in allowance trading, setting new NO_x emissions standards for existing coal-fired utility boilers, and allowing emissions averaging to reduce costs. Another is a permit process that affords sources maximum flexibility in selecting the most cost-effective approach to reducing emissions. Continuous emission monitoring (CEM) requirements provide credible accounting of emissions to ensure the integrity of the market-based allowance system and to verify the achievement of the reduction goals. The excess emissions provision provides incentives to ensure self-enforcement. Another strategy is an appeals procedure that allows the regulated community to appeal decisions with which it may disagree.

The Clean Air Act was originally signed into law in 1963. Its goal was to protect public health from air pollution and limit the effects of air pollution on the environment. Early versions allowed individual states to set their own standards. Later versions of the act switched responsibility of setting uniform standards to the federal government. Primary standards protect human health. Secondary standards protect materials, crops, climate, visibility, and personal comfort. The 1990 version addressed acid rain, urban smog, air pollutants, ozone protection, marketing pollution rights, and VOCs. Estimates are that the Clean Air Act is responsible each year for saving 15,000 lives, reducing bronchitis cases by 60,000, and reducing 9,000 hospital admissions due to respiratory illnesses. The following chart shows some of the progress that the Clean Air Act has been responsible for since it was passed into law. Notice however, the increase in PM_{10} and NO_x levels.

Pollutant	% Change
Pb	–98%
VOC	–42%
SO_2	–37%
CO	–31%
NO_x	+17%
PM_{10}	+266 %

The 1997 Kyoto Protocol would have required the United States to reduce greenhouse emissions by 7% when compared with 1990 levels over a five-year period. Under this agreement, the United States would have faced penalties if it did not meet its emission cuts. The United States saw this as an unattainable target since carbon dioxide and greenhouse gases continue to increase and are projected to increase for the next 20 years. The United States felt that the protocol held developed nations responsible for meeting the cuts but did not apply the same standards to developing nations. Reasons given for not agreeing with the rest of the signing members of the Kyoto Protocol were the cost of meeting the emission targets would be too high, the time frame was too short for implementation, and there was no evidence of a correlation between greenhouse gases and global warming.

RELEVANT LAWS AND PROTOCOLS

Air Pollution Control Act (1955): The nation's first piece of federal legislation regarding air pollution. Identified air pollution as a national problem and announced that research and additional steps to improve the situation needed to be taken. It was an act to make the nation more aware of this environmental hazard.

Clean Air Act (1963): Dealt with reducing air pollution by setting emissions standards for stationary sources such as power plants and steel mills. It did not take into account mobile sources of air pollution, which had become the largest source of many dangerous pollutants. Amendments to the Clean Air Act were passed in 1965, 1966, 1967, and 1969. They set standards for auto emissions, expanded local air pollution control programs, established air quality control regions (AQCR), set air quality standards and compliance deadlines for stationary source emissions, and authorized research on low-emissions fuels and automobiles.

National Environmental Policy Act (1969): Requires a systematic analysis of major federal actions. Includes a consideration of all reasonable alternatives as well as an analysis of short-term and long-term, irretrievable, irreversible, and unavoidable impacts.

Clean Air Act (1970): Established new primary and secondary standards for ambient air quality, set new limits on emissions from stationary and mobile sources to be enforced by both state and federal governments, and increased funds for air pollution research. It was soon discovered that the deadlines set were overly ambitious (especially those for auto emissions). To reach these standards in such a short period of time, the auto industry faced serious economic limitations and seemingly insurmountable technological challenges. Over the next decade, the legislation was once again amended to extend these deadlines and to mandate states to revise their implementation plans. Congress did not amend the Clean Air Act during the decade of the 1980s, in part because governmental policies placed economic goals ahead of environmental goals.

Montreal Protocol (1989): An agreement among nations requiring the phaseout of chemicals that damage the ozone layer.

Clean Air Act (1990): Addressed five main areas: air quality standards, motor vehicle emissions and alternative fuels, toxic air pollutants, acid rain, and stratospheric ozone depletion.

Pollution Prevention Act (1990): Requires industries to reduce pollution at its source. Reduction can be in terms of volume and/or toxicity.

Kyoto Protocol (1997): An agreement among 150 nations requiring greenhouse gas reductions.

NOISE POLLUTION

Noise pollution is unwanted human-created sound that disrupts the environment. The dominant form of noise pollution is from transportation sources, principally motor vehicles, aircraft noise, and rail transport noise. Besides transportation noise, other prominent sources are office equipment, factory machinery, appliances, power tools, and audio entertainment systems. Noise regulation by governmental agencies effectively began in the United States with the 1972 Federal Noise Control Act.

Effects

Normal hearing depends on the health of the inner, middle, and outer ear. Three kinds of hearing loss occur: conductive, sensory, and neural. Sensory hearing loss is caused by damage to the inner ear and is the most common form associated with noise.

In addition to contributing to hearing loss, too much noise can affect health in other ways too. Immediate effects may be temporary or may become permanent. They may include cardiovascular problems with an accelerated heartbeat and high blood pressure, gastric-intestinal problems, a decrease in alertness and ability to memorize, nervousness, pupil dilation, and a decrease in the visual field. Effects that may be longer lasting include insomnia, nervousness, bulimia, chronically high blood pressure, anxiety, depression, and sexual dysfunction.

Control Measures

Roadway noise can be reduced through the use of noise barriers, limitations on vehicle speed, newer roadway surface technologies, limiting times for heavy-duty vehicles, computer-controlled traffic flow devices that reduce braking and acceleration, and changes in tire design. Aircraft noise can be reduced through developing quieter jet engines and rescheduling takeoff and landing times. Industrial noise can be reduced through new technologies in industrial equipment and installation of noise barriers in the workplace. Residential noise such as power tools, garden equipment, and loud radios can be controlled through local laws and enforcement.

RELEVANT LAW

Noise Control Act (1972): Establishes a national policy to promote an environment for all Americans free from noise that jeopardizes their health and welfare. To accomplish this, the act establishes a means for the coordination of federal research and activities in noise control, authorizes the establishment of federal noise emissions standards for products distributed in commerce, and provides information to the public respecting the noise emission and noise reduction characteristics of such products.

WATER POLLUTION

Water pollution can originate from either a point or a nonpoint source. A point source occurs when harmful substances are emitted directly into a body of water. An example of a point source of water pollution is a pipe from an industrial facility discharging effluent directly into a river. Point source pollution is usually monitored and regulated in developed countries.

Nonpoint sources deliver pollutants indirectly through transport or environmental change. An example of a nonpoint source of water pollution is when fertilizer from a farm field is carried into a stream by rain (run off). Nonpoint sources are much more difficult to monitor and control, and they account for the majority of contaminants in streams and lakes.

Sources of Water Pollution

The following sections describe the varied sources of water pollution. They include air pollution, chemicals, microbiological sources, mining, noise, nutrients, oxygen-depleting substances, suspended matter, and thermal sources.

AIR POLLUTION

Pollutants like mercury, sulfur dioxide, nitric oxides, and ammonia fall out of the air and into the water. They can then cause mercury contamination in fish and acidification and eutrophication of lakes. The oceans have absorbed enough carbon dioxide to have already caused a slight increase in ocean acidification. This may be causing the carbonate structures of corals, algae, and marine plankton to dissolve. These organisms form the base of the food pyramid in the ocean.

CHEMICALS

A variety of chemicals from industrial and agricultural sources can cause water pollution. Examples include metals, solvents and oils, detergents, and pesticides. These can accumulate in fish and shellfish, poisoning the people, animals, and birds that eat them. On a square-foot basis, homeowners apply more chemicals to their lawns than farmers do to their fields. Each year, road runoff and other nonspill sources impart an amount of oil to the oceans that is more than 5 times greater than the *Exxon Valdez* spill—about 21 million barrels. Discharge of oily wastes and oil-contaminated ballast water and wash water are all significant sources of marine pollution. Drilling and extraction operations for oil and gas can also contaminate coastal waters and groundwater. The EPA estimates that about 100,000 gasoline storage tanks are leaking chemicals into groundwater. In Santa Monica, California, wells supplying half the city's water have been closed because of dangerously high levels of the gasoline additive MTBE. New evidence strongly suggests that components of crude oil, called polycyclic aromatic hydrocarbons (PAHs), persist in the marine environment for years and are toxic to marine life at concentrations in the low parts per billion (ppb) range. Chronic exposure to PAHs can affect the development of marine organisms, increase susceptibility to disease, and jeopardize normal reproductive cycles in many marine species.

Studies have shown that up to 90% of drug prescriptions pass through the human body unaltered. Animal farming operations that use growth hormones and antibiotics also send large quantities of these chemicals into the water. Most wastewater treatment facilities are not equipped to filter out personal care products, household products, or pharmaceuticals. As a result, a large portion of these chemicals pass directly into local waterways. Study of the effects of these chemicals have discovered fragrance molecules inside fish tissues, ingredients from birth control pills causing gender-bending hormonal effects in frogs and fish, and the chemical nonylphenol, a remnant of detergent, disrupting fish reproduction and growth.

MICROBIOLOGICAL SOURCES

Disease-causing (pathogenic) microorganisms such as bacteria, viruses, and protozoa can result in swimmers getting sick and fish and shellfish becoming contaminated. Examples of waterborne diseases include cholera, typhoid, shigella, polio, meningitis, and hepatitis. In developing countries, an estimated 90% of the wastewater is discharged directly into rivers and streams without treatment. In the United States, 850 billion gallons (3 trillion L) of raw sewage are dumped into rivers, lakes, and bays each year by leaking sewer systems and inadequate combined sewer/storm systems that overflow during heavy rains. Leaking septic tanks and other sources of sewage can also cause groundwater and stream contamination. Beaches suffer the effects of water pollution from sewage. About 25% of all beaches in the United States annually have water pollution advisories or are closed each year due to bacterial buildup caused by sewage.

MINING

Mining causes water pollution in a number of ways. The mining process exposes heavy metals and sulfur compounds that were previously locked away in Earth. Rainwater leaches these compounds out of the exposed Earth, resulting in acid mine drainage and heavy-metal pollution that can continue long after the mining operations have ceased. Second, the action of rainwater on piles of mining waste (tailings) transfers pollution to freshwater supplies. In the case of gold mining, cyanide is intentionally poured on piles of mined rock (a leach heap) to extract the gold from the ore chemically. Some of the cyanide ultimately finds its way into nearby water. Additionally, huge pools of mining waste slurry are often stored behind containment dams that often leak or infiltrate ground water supplies. Fourth, mining companies in developing countries often dump mining waste directly into rivers or other bodies of water as a method of disposal.

The U.S. government in 2003 reclassified mining waste from mountaintop removal (a type of coal mining) so it could be dumped directly into valleys and burying streams altogether. The Iron Mountain mine in California has been closed since 1963 but continues to drain sulfuric acid and heavy metals (such as cadmium and zinc) into the Sacramento River. The river's bright-orange water is completely devoid of life and has a pH that is 10,000 times more acidic than battery acid. Experts say the pollution may continue for another 3,000 years.

NOISE

Many marine organisms, including marine mammals, sea turtles, and fish, use sound to communicate, navigate, and hunt. Because of this oceanic water noise pollution, some species may have a harder time hunting or detecting predators. They may also not being able to navigate properly.

NUTRIENTS

Phosphorus and nitrogen are necessary for plant growth and are plentiful in untreated wastewater. When added to lakes and streams, they can cause the growth of aquatic weeds that block waterways as well as algal blooms. If the source is from humans, it is called cultural eutrophication. Deposition of atmospheric nitrogen (from nitrogen oxides) also causes nutrient-type water pollution. Nutrient pollution is also a problem in estuaries and deltas, where the runoff that was aggregated by watersheds is finally dumped at the mouths of major rivers.

OXYGEN-DEPLETING SUBSTANCES

Biodegradable wastes are used as nutrients by bacteria and other microorganisms. Excessive biodegradable wastes can cause oxygen depletion in receiving waters. This can result in increases in anaerobic bacteria that produce ammonia, amines, sulfides, and methane (swamp gas) and decreases of aerobic organisms such as fish.

SUSPENDED MATTER

Suspended wastes eventually settle out of water and form silt or mud at the bottom. Toxic materials can also accumulate in the sediment and affect organisms throughout the food web. When forests are clear-cut, the root systems that previously held soil in place die and the sediment is free to run off into nearby streams, rivers, and lakes.

Plastics and other plastic-like substances (such as nylon from fishing nets and lines) can entangle fish, sea turtles, and marine mammals, causing injury and death. Plastic that has broken down into microparticles is now being ingested by tiny marine organisms and is moving up the marine food chain. Plastic remains in the ecosystem and continues to harm marine organisms far into the future.

THERMAL SOURCES

Produced by industry and power plants. Heat reduces the ability of water to hold oxygen and causes death to organisms that cannot tolerate heat and/or low oxygen levels. Global warming is also imparting additional heat to the oceans, rivers, and streams with unknown consequences.

CASE STUDIES

Minamata disease: Twenty-seven tons of mercury-containing compounds from industrial processes were dumped into Minamata Bay in Japan between 1932 and 1968. The mercury collected in fish and shellfish caught from the bay. Symptoms included blurred vision, hearing loss, loss of muscular coordination, and reproductive disorders.

***Exxon Valdez* (1989):** In 1989, the oil tanker *Exxon Valdez* spilled 11 to 30 million gallons (42 to 110 million L) of crude oil into Prince William Sound. As a result, 250,000 sea birds, 3,000 otters, 300 seals, 300 bald eagles, and 22 whales died along with billions of salmon and herring eggs. The oil also destroyed the majority of the plankton in the sound.

Cultural Eutrophication

Cultural eutrophication is defined as the process whereby human activity increases the amount of nutrients entering surface waters. The two most important nutrients that cause cultural eutrophication are nitrates (NO_3^-) and phosphates (PO_4^{3-}) that come from fertilizer, sewage discharge, and animal wastes.

Nitrates are water soluble. Nitrates found in fertilizers can remain on fields and accumulate, leach into groundwater, end up in surface runoff, and/or volatize and enter the atmosphere where they contribute to acid precipitation. Nitrates cause nitrate poisoning in water supplies, reduce the effectiveness of hemoglobin, and may be responsible for worldwide declines in amphibians.

Phosphates are also a component of inorganic fertilizers. However, they are not water soluble and adhere to soil particles. Soil erosion contributes to the buildup of phosphates in water supplies. Phosphate levels in water supplies are 75% higher than they were during preindustrial times. Phosphate buildup is more damaging in freshwater systems. In contrast, nitrate pollution is more damaging in wetlands where nitrogen is the limiting factor.

Nitrates and phosphates are algal nutrients. Increased concentrations of these nutrients increase the carrying capacity of lakes and streams. Explosions in the amount of algae as a result of cultural eutrophication are called algal blooms. The steps involved in algal bloom include the following:

- Increased algae due to increased nitrate and/or phosphate concentrations result in decreased light penetration, killing off deeper plants and their supply of oxygen to water.
- Oxygen concentration decreases in the water due to the consequences of increased material for decomposers.
- Lower oxygen concentrations cause fish and other aquatic organisms to die and contaminate the water at a high rate.
- Decaying fish and algae produce toxins in the water.

Several methods can control cultural eutrophication.. Planting vegetation (buffer zones) along streambeds slows erosion and absorbs some of the nutrients. Controlling the application and timing of applying fertilizer, controlling runoff from feedlots, and using biological controls such as denitrifying bacteria that convert nitrates into atmospheric nitrogen are other methods.

Groundwater Pollution

About 50% of the people in the United States depend on groundwater for their water supplies. In some countries, it may reach as high as 95%. Almost half of the water used for agriculture in the United States comes from groundwater. The Environmental Protection Agency (EPA) estimates that each day, 4.5 trillion liters of contaminated water seep into groundwater supplies in the United States. In the United States, 34 billion liters per year (60%) of the most hazardous liquid waste solvents, heavy metals, and radioactive materials are injected directly into deep groundwater via thousands of injection wells. Although the EPA requires that these effluents be injected below the deepest source of drinking water, some pollutants have already entered underground water supplies in Florida, Texas, Ohio, and Oklahoma.

Water entering an aquifer remains there for approximately 1,400 years compared with 16 days for water entering a river system. Once an aquifer is contaminated, it is practically impossible to remove the pollutants. For example, in Denver, Colorado, just 80 liters of organic solvents contaminated 4.5 trillion liters of groundwater. Initial cleanup of contaminated groundwater locations in the United States could cost up to $1 trillion over the next 30 years.

Maintaining Water Quality and Water Purification

DRINKING WATER TREATMENT METHODS

- **Adsorption**—Contaminants stick to the surface of granular or powdered activated charcoal.
- **Disinfection**—Chlorine, chloramines, chlorine dioxide, ozone, and UV radiation.
- **Filtration**—Removes clays, silts, natural organic matter, and precipitants from the treatment process. Filtration clarifies water and enhances the effectiveness of disinfection.
- **Flocculation-Sedimentation**—Process that combines small particles into larger particles that then settle out of the water as sediment. Alum, iron salts, or synthetic organic polymers are generally used to promote coagulation.
- **Ion Exchange**—Removes inorganic constituents. It can be used to remove arsenic, chromium, excess fluoride, nitrates, radium, and uranium.

WATER TREATMENT REMEDIATION TECHNOLOGIES

- **Adsorption/absorption**—Solutes concentrate at the surface of a sorbent (an absorbing surface), thereby reducing their concentration.
- **Aeration**—Bubbling air through water increases rates of oxidation.
- **Air stripping**—VOCs are separated from groundwater by exposing water to air (the VOCs evaporate due to their high vapor pressure).
- **Bioreactors**—Groundwater is acted upon by microorganisms.
- **Constructed wetlands**—Uses natural geochemical and biological processes that parallel natural wetlands. Also known as living machines.
- **Deep-well injection**—Uses injection wells to place treated or untreated liquid waste into geologic formations that do not pose a potential risk to groundwater.
- **Enhanced bioremediation**—The natural rate of bioremediation is enhanced by adding oxygen and nutrients into groundwater.
- **Fluid-vapor extraction**—A vacuum system is applied to low-permeable soil to remove liquids and gases.
- **Granulated activated carbon** (GAC)—Groundwater is pumped through a series of columns containing activated carbon.
- **Hot water or steam flushing**—Steam or hot water is forced into an aquifer to vaporize volatile contaminants and is then treated through fluid-vapor extraction.
- **In-well air stripping**—Air is injected into wells—the air picks up various contaminants, particularly VOCs. Vapors are drawn off by vapor extraction.
- **Ion exchange**—Involves exchange of one ion for another.
- **Phytoremediation**—Uses plants to remove contamination.
- **UV oxidation**—Uses ultraviolet light, ozone, or hydrogen peroxide to destroy microbiological contaminants.

Sewage Treatment/Septic Systems

Sewage treatment incorporates physical, chemical, and biological processes to remove contaminants from wastewater. There are three stages of wastewater treatment.

A septic system consists of a tank and a drain field. Wastewater enters the tank, where solids settle. Anaerobic digestion using bacteria treats the settled solids and reduces their volume. Excess liquid leaves the tank and moves through a pipe with holes in it to a leach field where the water then percolates into the soil.

Some pollutants, especially nitrogen, do not decompose in a septic system and may contaminate the groundwater. Approximately 25% of Americans rely on septic systems.

PRIMARY TREATMENT

Primary treatment is to reduce oils, grease, fats, sand, grit, and coarse solids. Specific steps include sand catchers, screens, and sedimentation. This is a physical method of cleaning.

SECONDARY TREATMENT

Secondary treatment is designed to degrade substantially the biological content of the sewage derived from human waste, food waste, soaps, and detergent. Specific steps include filters, activated sludge, filter (oxidizing) beds, trickling filter beds using plastic media, and secondary sedimentation. This is a biological method of cleaning.

TERTIARY TREATMENT

Tertiary treatment provides a final stage to raise the effluent quality to the standard required before it is discharged to the receiving environment (sea, river, lake, or ground). Specific steps may include sand filtration, lagooning, constructed wetlands, nutrient removal through biological or chemical precipitation, denitrification using bacteria, phosphorous removal using bacteria, microfiltration, and disinfection using UV light, chlorine, or ozone.

FACTOIDS

- Water treatment accounts for 15% of the cost of water.
- 44% of all lakes, 37% of all rivers, and 32% of all estuaries are unsafe in the United States for recreational activities.
- 2.5 billion people on Earth lack adequate sanitation.
- Over 1 billion people on Earth do not have access to clean drinking water.
- 1.5 million Americans each year become ill due to fecal contamination in drinking water supplies.
- Less-developed countries discharge 95% of all sewage into local rivers, lakes, and nearby oceans.

RELEVANT LAWS

Federal Water Pollution Control Act (1948): Created comprehensive programs for eliminating or reducing the pollution of interstate water and improving the sanitary condition of surface and underground water supplies.

Water Quality Act (1965): Established water purity standards with states retaining initial responsibility for water purity.

Clean Water Act (1972): Established the basic structure for regulating discharges of pollutants into the waters of the United States. It gave the EPA the authority to implement pollution control programs such as setting wastewater standards for industry. The Clean Water Act also continued requirements to set water quality standards for all contaminants in surface waters. The act made it unlawful for any person to discharge any pollutant from a point source into navigable waters unless a permit was obtained under its provisions. It also funded the construction of sewage treatment plants under the construction grants program and recognized the need for planning to address the critical problems posed by nonpoint source pollution.

Safe Drinking Water Act (1974): Established standards for safe drinking water in the United States.

Ocean Dumping Ban Act (1988): Made it unlawful for any person to dump or transport for the purpose of dumping sewage, sludge, or industrial wastes into the ocean.

Oil Spill Prevention and Liability Act (1990): Strengthened the EPA's ability to prevent and respond to catastrophic oil spills.

Source Water Assessment Program—SWAP (1996): Required states to identify sources of public drinking water supplies and assess susceptibility to contamination.

Source Water Protection Program—SWPP (1996): Encouraged states to adopt a community-based approach to preventing water pollution.

Surface Water Treatment Rule—SWTR (1996): Addressed control of microbial pathogens, including cryptosporidium.

TIP

Reference relevant laws in your essay answers whenever possible. They will substantiate your thoughts and provide a historical framework for the issue.

SOLID WASTE

Types of solid waste include:

- **Organic**—Kitchen wastes, vegetables, flowers, leaves, or fruits. Usually decomposes within 2 weeks. Wood can take 10 to 15 years to decompose.
- **Radioactive**—Spent fuel rods and fire alarms. Radioactive wastes can take hundreds of thousands of years to decompose.
- **Recyclable**—Paper, glass, metals, or plastics. Paper decomposes in 10 to 30 days. Glass does not decompose. Metals decompose in 100 to 500 years. Plastics can take up to 1 million years to decompose.
- **Soiled**—Hospital wastes. Cotton and cloth can take 2 to 5 months to decompose.
- **Toxic**—Paints, chemicals, pesticides, and so on. Toxic wastes can take hundreds of years to decompose.

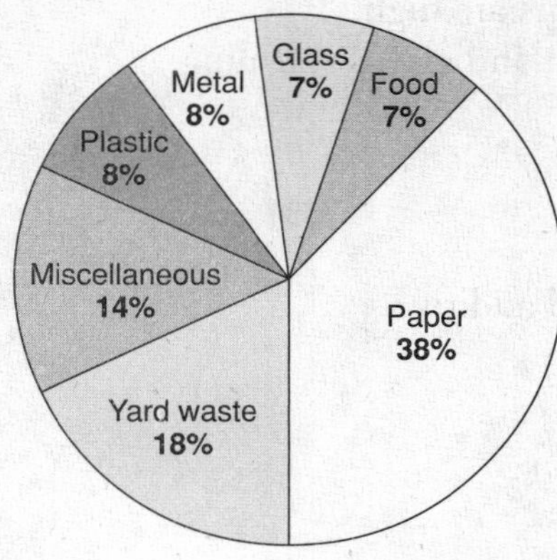

Figure 9.2 Amounts and types of municipal solid wastes (MSW) in the U.S.

FACTOIDS

- Each year, 220 million tons of municipal wastes are generated in the United States.
- Each American generates approximately 4.5 pounds of waste per day.
- The United States, with 5% of the world's population, generates 33% of the world's total waste.
- Waste from all categories in the United States works out to be 10,000 pounds per person.
- 6 billion metric tons of solid agricultural wastes are generated each year in the United States.
- One-third of all solid wastes are mine tailings and smelter residues.

Disposal and Reduction

BURNING, INCINERATION, OR ENERGY RECOVERY

Pros

- Heat can be used to supplement energy requirements.
- Reduces impact on landfills.
- Mass burning is inexpensive.
- What is left is 10% to 20% of original volume.
- U.S. incinerates 15% of its wastes.
- France, Japan, Sweden, and Switzerland incinerate > 40% of their wastes and use the heat to generate electricity.

Cons

- Air pollution including lead, mercury, NO_x, cadmium, SO_2, HCl, and dioxins.
- Sorting out batteries, plastics, etc. is expensive.
- No way of knowing toxic consequences.
- Ash is more concentrated with toxic materials.

- Initial costs of incinerators are high.
- Add to acid precipitation and global warming.

COMPOSTING

Pros

- Creates nutrient-rich soil additive.
- Aids in water retention.
- Slows down soil erosion.
- No major toxic issues.

Cons

- Public reaction to odor, vermin, and insects.
- Not in my backyard (NIMBY).

REMANUFACTURING

Pros

- Recovers materials that would have been discarded.
- Beneficial to inner cities as an industry because material is available and jobs are needed.

Cons

- Toxic materials may be present (CFCs, heavy metals, toxic chemicals, and so on).

DETOXIFYING

Pros

- Reduces impact on the environment.

Cons

- Expensive.

EXPORTING

Pros

- Gets rid of problem immediately.
- Source of income for poor countries.

Cons

- Garbage imperialism or environmental racism.
- Long-term effects not known.
- Expensive to transport.

LAND DISPOSAL—SANITARY LANDFILLS

Pros

- Waste is covered each day with dirt to help prevent insects and rodents.
- Plastic liners, drainage systems, and other methods help control leaching material into groundwater.

- Geologic studies and environmental impact studies are performed prior to building.
- Collection of methane and use of fuel cells to supplement energy demand.
- Use of anaerobic methane generators reduces dependence on other energy sources.

Cons

- Rising land prices. Current costs are $1 million per hectare.
- Transportation costs to the landfill.
- High cost of running and monitoring landfill.
- Legal liability.
- Suitable areas are limited.
- NIMBY.
- Degradable plastics do not decompose completely.

LAND DISPOSAL—OPEN DUMPING

Pros

- Inexpensive.
- Provides a source of income to the poor by providing recyclable products to sell.

Cons

- Trash blows away in the wind.
- Vermin and disease.
- Leaching of toxic materials into the soil.
- Aesthetics.

OCEAN DUMPING

Pros

- Inexpensive.
- Illegal in the United States.

Cons

- Debris floats to unintended areas.
- Marine organisms and food webs are impacted.

FACTOIDS

- 1 gallon (4 L) of gasoline or oil can pollute 1 million gallons (4 million L) of freshwater.
- Two-thirds of all aluminum cans are recycled.
- Minnesota has a 46% recycling rate—highest in the United States.
- 2 million trees a day are harvested in the United States for paper products.
- Of all heavy-metal toxic wastes, 90% come from consumer electronics and batteries.
- 200 million liters of waste motor oil are dumped into sewers or onto land each year in the United States—five times as much oil that leaked out of the *Exxon Valdez.*
- 60 million metric tons of hazardous wastes are generated each year in the United States.

RECYCLING

Pros

- Turns waste into an inexpensive resource.
- Reduces impact on landfills.
- Reduces need for raw materials and the costs associated with it.
- Reduces energy requirements to produce product. For example: recycling aluminum cuts energy use 95% and producing steel from scrap reduces energy requirements 75%.
- Reduces dependence on foreign oil.
- Reduces air and water pollution.
- Bottle bills provide economic incentive to recycle.

Cons

- Poor regulation.
- Fluctuations in market price.
- Throwaway packaging is more popular.
- Current policies and regulations favor extraction of raw materials. Energy, water, and raw materials are sold below real costs to stimulate new jobs and the economy.

REUSE

Pros

- Most efficient method of reclaiming materials.
- Industry models already in place—auto salvage yards, building materials, and so on.
- Refillable glass bottles can be reused 15 times.
- Cloth diapers do not impact landfills.

Cons

- Cost of collecting materials on a large scale is expensive.
- Cost of washing and decontaminating containers is expensive.
- Only when items are expensive and labor is cheap is reuse economical.

RELEVANT LAWS

Solid Waste Disposal Act (1965): First federal law that required environmentally sound methods for disposal of household, municipal, commercial, and industrial wastes.

Marine Plastic Pollution Research and Control Act (1987): Discharge of plastics into water is prohibited. Food waste, paper, rags, glass, and metal cannot be discharged into navigable water or within 12 miles (19 km) of land.

Medical Waste Tracking Act (1988): Required tracking systems, stringent management standards, packaging, and labeling of medical supplies.

Waste Reduction Act (1990): Requires the EPA to develop and coordinate a pollution prevention strategy and develop source reduction models. Requires owners and operators of manufacturing facilities to report annually on source reduction and recycling activities.

MULTIPLE-CHOICE QUESTIONS

1. __________contributes to the formation of __________ and thereby compounds the problem of __________.

   (A) ozone, carbon dioxide, acid rain
   (B) carbon dioxide, carbon monoxide, ozone depletion
   (C) sulfur dioxide, acid deposition, global warming
   (D) nitrous oxide, ozone, industrial smog
   (E) nitric oxide, ozone, photochemical smog

2. Photochemical smog does NOT require the presence of

   (A) nitrogen oxides
   (B) ultraviolet radiation
   (C) peroxyacyl nitrates
   (D) volatile organic compounds
   (E) ozone

3. Which of the following is a natural component of the atmosphere, comprises about 0.036% by volume of the atmosphere and is produced by the decay of vegetation, volcanic eruptions, exhalation of animals, burning of fossil fuels and deforestation?

   (A) Carbon monoxide
   (B) Carbon dioxide
   (C) Nitrous oxide
   (D) Nitrogen dioxide
   (E) Methane

4. Which of the following steps is NOT involved in the production of industrial smog?

   (A) $C + O_2 \rightarrow CO_2$
   (B) $C + O_2 \rightarrow CO$
   (C) $S + O_2 \rightarrow SO_2$
   (D) $NO_2 \rightarrow NO + O$
   (E) $SO_2 + O_2 \rightarrow SO_3$

5. Household water is most likely to be contaminated with radon in homes that

   (A) are served by public water systems that use a groundwater source
   (B) are served by public water systems that use a surface water source
   (C) have private wells
   (D) use bottled water
   (E) are served by water agencies that use ozone to disinfect the water

6. Which reaction is NOT involved in the formation of acid deposition?

(A) $O_3 + C_xH_y \rightarrow$ PANS
(B) $SO_2 + H_2O \rightarrow H_2SO_3$
(C) $H_2SO_3 + ½O_2 \rightarrow H_2SO_4$
(D) $NO + \rightarrow ½O_2 \rightarrow NO_2$
(E) $2NO_2 + H_2O \rightarrow HNO_2 + HNO_3$

7. Normal rainfall has a pH of about

(A) 2.3
(B) 5.6
(C) 7.0
(D) 7.6
(E) 8.3

8. According to the Environmental Protection Agency, about __________ of all commercial buildings in the United States are classified as sick.

(A) 5%
(B) 15%
(C) 50%
(D) 75%
(E) 100%

9. In developing countries, the most likely cause of respiratory disease would be

(A) photochemical smog
(B) industrial smog
(C) smoking
(D) PM_{10}
(E) asbestos

10. Humans LEAST susceptible to the effects of air pollution are

(A) newborns
(B) children between the age of 2 and 10
(C) teenagers
(D) adult males
(E) the elderly

11. Acid precipitation, leaching out the metal __________ , causes fish and other aquatic organisms to die from acid shock.

(A) Al
(B) Pb
(C) Hg
(D) Cd
(E) Fe

12. Which pollutant best illustrates the effectiveness of legislation?

(A) NO_2
(B) SO_2
(C) CO_2
(D) O_3
(E) Pb

13. The diagram below shows the range of organisms found within certain sections of a river in an industrial area. Which section of the river most likely has the LOWEST level of dissolved oxygen?

Effects of Sewage Discharge in a River
Organisms commonly found in discharge zone

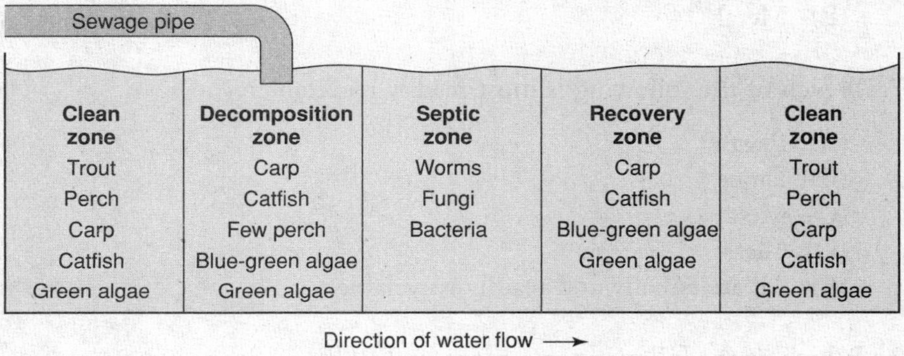

Sewage pipe

Clean zone	**Decomposition zone**	**Septic zone**	**Recovery zone**	**Clean zone**
Trout	Carp	Worms	Carp	Trout
Perch	Catfish	Fungi	Catfish	Perch
Carp	Few perch	Bacteria	Blue-green algae	Carp
Catfish	Blue-green algae		Green algae	Catfish
Green algae	Green algae			Green algae

Direction of water flow ⟶

(A) Clean zone
(B) Decomposition zone
(C) Septic zone
(D) Recovery zone
(E) None of the above

14. Which one of the following statements is TRUE?

(A) The United States generates approximately 230 million tons of municipal solid waste a year, about 4.6 pounds per person per day.
(B) Food waste is the third largest component of generated waste (after yard waste and corrugated boxes) and the second largest component of discarded waste.
(C) It takes 3 to 12 months to produce compost, depending on the type of waste.
(D) The average American college student produces 640 pounds of solid waste each year, including 500 disposable cups and 320 pounds of paper.
(E) All statements are true.

15. The major source of solid waste in the United States comes from what source?

    (A) Homes
    (B) Factories
    (C) Agriculture
    (D) Petroleum refining
    (E) Mining wastes

16. What is the largest type of domestic solid waste in the United States?

    (A) Yard wastes
    (B) Paper
    (C) Plastic
    (D) Glass
    (E) Metal

17. Which of the following is most readily recyclable?

    (A) Plastic
    (B) Paper
    (C) Metal
    (D) Glass
    (E) All are equally and readily recyclable

18. Which of the following statements is TRUE?

    (A) Recycling is more expensive than trash collection and disposal.
    (B) Landfills and incinerators are more cost effective and environmentally sound than recycling options.
    (C) The marketplace works best in solving solid waste management problems; no public sector intervention is needed.
    (D) Landfills are significant job generators for rural communities.
    (E) None of the above are true.

19. In the 1970s, houses were built over a toxic chemical waste disposal site. This case study is known as

    (A) Love Canal
    (B) Bet Trang
    (C) Bhopal
    (D) Brownfield
    (E) Chernobyl

20. Which of the following methods of handling solid wastes is against the law in the United States?

    (A) Incineration
    (B) Dumping it in open landfills
    (C) Burying it underground
    (D) Exporting the material to foreign countries
    (E) Dumping the material in the open ocean

FREE-RESPONSE QUESTION

By Dr. Ian Kelleher
Brooks School
North Andover, MA

(a) Study the following graph, which shows projected trends in annual carbon dioxide emissions, and then answer the following questions.

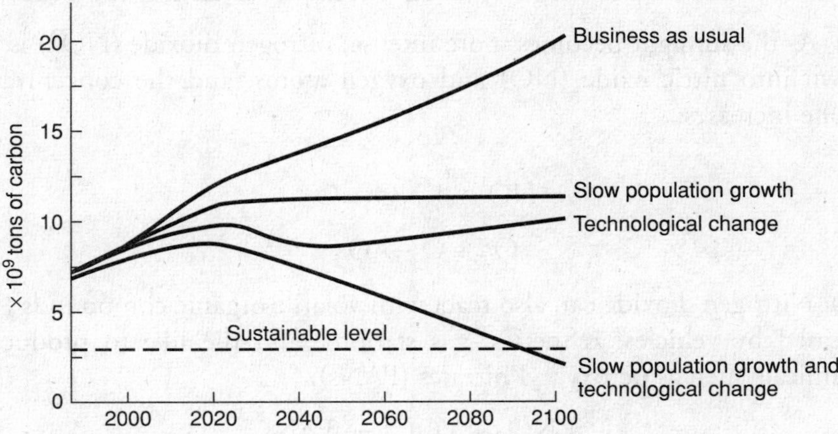

(i) In the "business as usual" model, what factors do you think might contribute to the increase in carbon dioxide emissions?

(ii) Given the shape of the graph, what do you think is meant in this case by "technological change"?

(iii) What is meant by a "sustainable level" of carbon dioxide emissions? According to these predictions, what needs to happen for this level to be brought about?

(b) Use the example of acid deposition to illustrate the difference between remediation and alleviation of an environmental problem.

(c) Look at the graph of CFC production and account for the trends you observe.

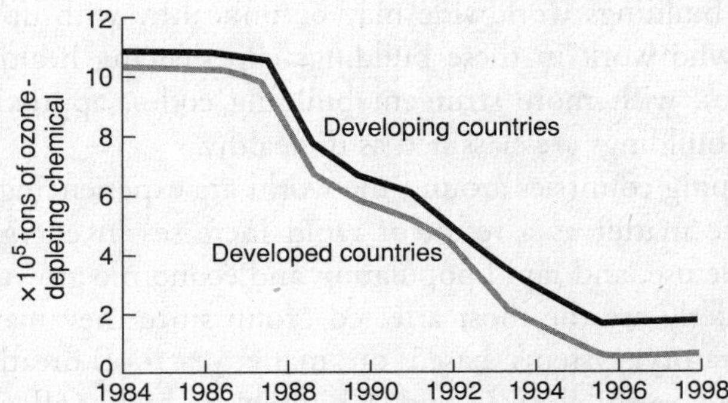

(d) In 1992, the World Bank designated indoor air pollution in developing countries as one of the four most critical global environmental problems. Use examples of major sources of indoor pollutants to illustrate how the issue of indoor air quality in developing countries differs from that in developed countries.

MULTIPLE-CHOICE ANSWERS AND EXPLANATIONS

1. **(E)** As the sunlight becomes more intense, nitrogen dioxide (NO_2) is broken down into nitric oxide (NO) and oxygen atoms, and the concentration of ozone increases:

$$NO_2 \xrightarrow{uv} NO + O$$

$$O_2 + O \rightarrow O_3$$

2. **(C)** Nitrogen dioxide can also react with volatile organic compounds (VOCs) released by vehicles, refineries, gas stations, and the like to produce toxic chemicals such as peroxyacyl nitrates (PANs):

$$NO_2 + VOCs \rightarrow PANs$$

3. **(B)** Exhalation of animals was the key in this question. Carbon dioxide is a product of cellular respiration:

$$C_6H_{12}O_6 + 6O_2 \rightarrow 6CO_2 + 6H_2O$$

4. **(D)** This step is present in the formation of photochemical smog.
5. **(C)** Radon is a naturally occurring radioactive gas that causes cancer and may be found in drinking water and indoor air. The Safe Drinking Water Act requires the EPA to develop regulations to reduce radon in drinking water.
6. **(A)** Review the reactions required to form acid deposition.
7. **(B)** Rain is naturally acidic due to the presence of carbon dioxide in the air, which forms carbonic acid with rainwater.
8. **(B)** The World Health Organization (WHO) estimates that up to 30% of commercial buildings worldwide may be unhealthy with up to one-third of the people who work in these buildings experiencing health effects. In the United States, with more stringent building codes, approximately 15% of commercial buildings are classified as unhealthy.
9. **(D)** Developing countries around the world are experiencing increased levels of particulate matter as a result of rapid increases in energy consumption, motor vehicle use, and rapid population and economic growth.
10. **(D)** The elderly are the most affected group since they may have compromised respiratory systems based on many years of breathing pollutants. Newborns are a very close second. Children are highly affected due to high activity levels and increased cases of asthma.
11. **(A)** Aluminum reduces the ion exchange through the gills and causes fish to excrete excess mucus, which reduces oxygen uptake.

12. (E) From 1976 to 1980, the average amount of lead in children was 15 μg/dL. From 1991 to 1994, the average level had been reduced to 2.7 μg/dL.
13. (C) The level of dissolved oxygen in the septic zone is too low to support organisms that live by aerobic respiration. In this region, anaerobic bacteria flourish and produce waste products such as hydrogen sulfide (H_2S) and methane (CH_4).
14. (E) No explanation needed.
15. (E) Mining wastes, along with oil and gas production, constitute about 75% of all solid wastes generated in the United States.
16. (B) Paper accounts for approximately 35% of domestic solid waste in the United States.
17. (B) Paper can be broken down into fibers and reused generally without requiring a sorting process. For some refined paper products, recycled paper must be sorted based on ink types.
18. (E) In choice (A), when designed properly, recycling programs are cost-competitive with trash collection and disposal. In choice (B), properly designed recycling programs are cost competitive with landfills and incinerators. In choice (C), solid waste systems generally operate under public sector guidelines. In choice (D), recycling creates many more jobs in rural and urban communities than landfill employment.
19. (A) Love Canal is an abandoned canal in New York where a huge amount of toxic waste was buried. The waste was composed of at least 300 different chemicals, totaling an estimated 20,000 metric tons.
20. (E) The Ocean Dumping Act prohibits dumping of wastes into territorial ocean waters.

FREE-RESPONSE ANSWER

(a) (i) Increased carbon dioxide emissions primarily come from increased burning of fossil fuels worldwide. The "business as usual" model would include an increasing global population, which would mean an increase in the demand for energy. In the "business as usual" model, this need would be met primarily by increased use of fossil fuels, by far the most common source of energy in the world today. Increasing rates of development, particularly of developing countries in this model, would also lead to an increase in fossil fuel use.

Notable points:

Cover all of the relevant aspects of "business as usual," such as increasing size and development of populations.

"Technological change" is a vague phrase that could mean many things. In this case, the curve thus labeled shows a dramatic reduction in carbon dioxide emissions, so the question is asking about technologies that reduce fossil fuel use.

(a) (ii) "Technological change" reduces carbon dioxide emissions on the graph. Increased use of alternative sources of energy such as nuclear power, hydroelectric, solar, and tidal energy would all decrease fossil fuel emissions. Technologies that increase energy efficiency and save power, such as more efficient car engines, would also decrease the use of fossil fuels and thus carbon dioxide emissions.

(a) (iii) A "sustainable level" of carbon dioxide emissions is one in which the amount absorbed by Earth's natural systems, such as oceans and plants, equals the amount released into the atmosphere. A combination of both technological change and slowed population growth is needed to bring carbon dioxide emissions to sustainable levels. As the graph shows, neither factor can produce sustainable levels on its own.

Explaining the difference between these two similar terms would be a good way to start. These are basic terms that you should know. Starting like this should lessen the chance of mixing them up.

Thus, cause and effect must both be covered in the answer, and you must take care to distinguish the two. Acid deposition is a secondary pollutant. Explaining the sequence of events that leads to its formation might be helpful in distinguishing cause from effect.

(b) Alleviation of an environmental problem means stopping or lessening its cause. Remediation means cleaning up the effects of the problem.

Acid deposition forms in the atmosphere mainly from sulfur oxides (SO_x) and nitrogen oxides (NO_x). Both are present primarily as a consequence of the combustion of fossil fuels.

Methods of alleviation may concentrate on decreasing the amount of sulfur and nitrogen oxides released into the atmosphere. Any method of reducing the rate of consumption of fossil fuels, such as increased use of nuclear power and alternative energy sources, or laws and education to help conserve energy would decrease the amount of sulfur and nitrogen oxides released. Clean-fuel technologies, such as fluidized-bed combustion of coal, would result in less of these gases being produced on combustion. Scrubbers in smokestacks can remove much of what is produced. Perhaps the most important step in reducing emissions in the United States, however, was the passing of the Clean Air Act in 1963 and its subsequent amendments.

Methods of remediation may concentrate on neutralizing the acid deposited in an environment. Acids can be neutralized by adding a base. Since an abundant, low-cost, and nontoxic material is often needed, limestone, $CaCO_3$, is commonly used. For example, limestone might be added to a lake to increase its pH. Powdered limestone could also be spread over agricultural lands to increase the pH of the soil. Many nutrients are more soluble in acidic soils and therefore might be washed away by rain. As a result, the addition of fertilizer is also required. Deforestation caused by acid deposition may be addressed by treating the soil and then replanting.

Anything that reduces the use of fossil fuels or limits the emissions of sulfur and nitrogen oxides would be an example of alleviation. Anything that neutralizes the effects of acid in the environment would be an example of remediation. Be careful in giving just an appropriate level of detail.

(c) Between 1984 and 1988, both developed and developing countries used ozone-depleting chemicals at a fairly consistent rate of approximately 1 million tons per year. After 1988, the amount used by both developed and developing countries decreased sharply and at a fairly constant rate for the next seven years. By 1996, developed countries used 50,000 tons per year and developing countries used 150,000 tons. Use remained at approximately these levels for the next two years.

The reduction is likely to be due to countries implementing technologies to comply with the Montreal Protocol (1989), which set limits for the emission of chemicals that cause depletion of the ozone layer. The biggest reduction has come from using alternatives to chlorofluorocarbons (CFCs)—the principal ozone-depleting agents. Alternatives include HFCs and HCFCs.

When asked to explain the trends shown in a graph, a good starting point is to describe what they are. Writing this down first should also help you compose your explanation.

This sample essay may provide more detail than necessary. The important point is that the amounts for both developed and developing countries fell dramatically at the same time and basically at the same rate. This suggests that the drops come as a result of the implementation of new laws. Since we are dealing with a global situation, it is likely to be in the form of an international agreement.

The wording of the World Bank designation suggests that the question is referring to the general masses of the population in developing countries rather than the small, technologically developed percentage. The answer is thus strongly focused on this difference in lifestyle.

(d) Homes are perhaps the most important factor when considering indoor air quality since people tend to spend more time there than anywhere else. The majority of people in developing countries live in homes with much simpler technologies. Many of the pollutants found in houses in developed countries are not present. The major source of indoor air pollutants in developing countries, therefore, is the combustion of poor-quality fuels for heating, lighting, and cooking. Such dirty fuels include animal wastes, kerosene, and low-grade coal that may release large amounts of particulates, carbon monoxide, sulfur oxides, and other toxins on combustion. As fires are often burned indoors in places with inadequate ventilation, the levels of these pollutants and carbon monoxide can be greatly concentrated.

Energy sources are much more technologically advanced in developed countries, so this is not such an important source of indoor air pollution. For example, poorly maintained furnaces may produce some carbon monoxide. However, this problem is not nearly as widespread or generally as serious as the energy issue in developing countries. Instead, major sources of air pollution include lead (from lead paints), asbestos, and fumes from volatile organic compounds (VOCs) in paints, glues, plastics, and furniture. These things will be less common in houses in developing countries. Houses in developed countries are often tightly sealed, more so than in developing countries. This leads to increased levels of concentration. For example, radon gas, which occurs naturally from radioactive decay in certain rocks, might accumulate to dangerous levels in a modern air-conditioned house but not in a simpler hut with no glass in the windows.

CHAPTER 10

Impacts on the Environment and Human Health

Health, learning and virtue will ensure your happiness; they will give you a quiet conscience, private esteem and public honour.

—Thomas Jefferson

HAZARDS TO HUMAN HEALTH

Environmental risk analysis is the comparing of the risk of a situation to its related benefits. It is the overall process that allows one to evaluate and deal with the consequences of events, based on their probability. There are four classes of risk:

1. High risk—such as smoking or driving while intoxicated
2. Low risk—infrequent events that may have a large consequence, such as an earthquake on the East Coast of the United States
3. Very low risk—events that have never occurred in recorded history, such as a major meteor striking the North American continent
4. Mixed risk—outcomes that increase in frequency against a background of occurrences, such as additional cases of cancer beyond that normally expected

Understanding how people accept risk requires an understanding of how preferences are accepted and measured. The three types of preferences are revealed, expressed, and natural standards. Revealed preferences are observations on the risks people actually take. Expressed preferences are often measured through public opinion polls. Natural standards are levels of risk humans have lived with in the past.

Risk estimation is a scientific question, while acceptability of a given level of risk is a political question. We can see this distinction by comparing the risk of smoking with working in a coal mine. If the United States spent as much money on premature deaths and illnesses caused by smoking tobacco as we do on coal mine safety, there would be little money left in the United States for any other purpose. This is the political reality of risk acceptance that goes beyond risk estimation.

Three problems arise in risk analysis. First, lack of information leads to uncertainty and is known as a default option. Next, complexity of the information often leads to confusion. Finally, failure to interpret uncertainty and complexity occurs in meaningful ways.

External influences are factored into decisions regarding risk. These influences include public concern, economic interest, and legislative actions that affect the possible choices available.

Risk analysis is divided into risk assessment and risk management. Risk assessment is an objective estimation of risk. It includes the identification of hazards, dose-response assessment, exposure assessment, and risk characterization.

Risk management is the process of determining what to do about risk. This includes risk identification and use of mitigating measures to reduce risk. Risk management has a scientific basis and uses inference to the extent possible that is free of policy implications. Once a risk has been characterized, risk management can be used either to prevent or to mitigate the risk. Risk prevention focuses on not doing or not allowing the risk to occur. In contrast, risk mitigation focuses on subsequent activity to limit the consequences. Risk management should take into consideration societal, economic, and political factors when weighing strategies for risk management. It is this dimension that many times causes tension between the scientific community and the decision makers.

RISK MANAGEMENT STRATEGY

- **Market-based method**—Relies on market forces to provide indirect controls. Usually the response from industry.
- **Hierarchical method**—Relies on explicit controls and top-down management styles (government, laws, etc.). Usually the response from lawmakers.
- **Sectarian method**—Relies on emotions. Usually the response from citizens.
- **Rational method**—Relies on logic and facts in decision making. Usually the response from researchers.

Acute and Chronic Effects

Acute health effects are characterized by sudden and severe exposure and rapid absorption of the substance. Normally, a single large exposure is involved. Acute health effects are often reversible, such as carbon monoxide poisoning.

Chronic health effects are characterized by prolonged or repeated exposures over many days, months, or years. Symptoms may not be immediately apparent. Chronic health effects are often irreversible. Examples include lead or mercury poisoning, asbestosis, or cancer.

Dose-Response Relationships

Dose-response relationships describe the change in effect on an organism or a population caused by differing levels of exposures to a substance. These relationships are used to determine whether various environmental risks are safe or hazardous.

A dose-response curve is a graph that relates the amount of drug or toxin given (plotted on the x-axis and usually the logarithm of the dose) compared with the response (plotted on the y-axis and usually provided in percentages). The point on the graph where the response is first observed is known as the threshold dose. For most drugs, the desired effect is found slightly above the threshold dose. When past this point, negative side effects begin to appear.

LD_{50} (lethal dose, 50%) is the median lethal dose of a pollutant or drug that kills half the members of a tested population within 14 days and is the most common indicator of toxicity. EC_{50} is the concentration of a compound where 50% of its effect is observed. IC_{50} represents the concentration of a pollutant or drug that is required for obtaining inhibition.

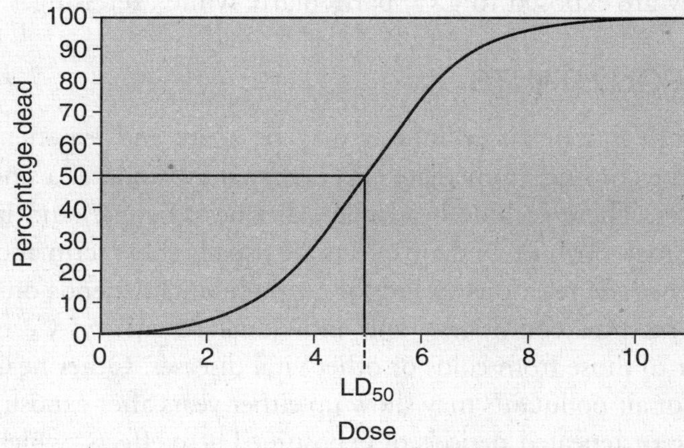

Figure 10.1 A dose-response curve

Air Pollutants and Their Effect on Human Health

AIR TOXICS

Air toxics are a group of air pollutants that are known or suspected to cause serious health problems. Examples of air toxics include asbestos, benzene, chloroform, formaldehyde, lead, mercury and nickel compounds, and perchloroethylene. People exposed to air toxics at sufficient concentrations and durations may have an increased chance of developing cancer or other serious health problems. These include damage to the immune system as well as neurological, reproductive (reduced fertility), developmental, and respiratory problems.

ASBESTOS

Studies of people who were exposed to asbestos in factories and shipyards have shown that breathing high levels of asbestos fibers can lead to an increased risk of lung cancer, mesothelioma (a cancer of the lining of the chest and the abdominal cavity), and asbestosis (the lungs become scarred with fibrous tissue). Researchers have not yet determined a safe level of exposure. However, they know that the greater and longer the exposure, the greater the risk of contracting an asbestos-related disease. Risks of lung cancer and mesothelioma increase with the number of fibers inhaled. The risk of lung cancer from inhaling asbestos fibers is also greater for smokers. People who develop asbestosis have usually been exposed to high levels of asbestos for a long time. The symptoms of these diseases do not usually appear until about 20 to 30 years after the first exposure.

CARBON MONOXIDE (CO)

Carbon monoxide enters the bloodstream through the lungs and binds chemically to hemoglobin, the substance in blood that carries oxygen to cells. In this way, CO interferes with the ability of the blood to transport oxygen to organs and tissue throughout the body. This can cause slower reflexes, confusion, and drowsiness. It can also reduce visual perception and coordination and can decrease the ability to learn. People with cardiovascular disease, such as angina, are most at risk from exposure to CO. These individuals may experience chest pain and other cardiovascular symptoms if they are exposed to CO, particularly while exercising.

INDOOR AIR POLLUTANTS

REMEMBER

Environmental problems have a cultural and social context. Understanding the role of cultural, social, and economic factors is vital to the development of solutions.

Health effects from indoor air pollutants may be acute and experienced soon after exposure or may be chronic. Immediate effects may show up after a single exposure or repeated exposures. These include headaches, dizziness, fatigue, and irritation of the eyes, nose, and throat. Such immediate effects are usually short term and treatable. The likelihood of immediate reactions to indoor air pollutants depends on several factors: age, preexisting medical conditions, and individual sensitivity. Certain immediate effects are similar to those from colds or other viral diseases. Other health effects from exposure to indoor air pollutants may show up either years after exposure has occurred or only after long or repeated periods of exposure. These effects, which include some respiratory diseases, heart disease, and cancer, can be severely debilitating or fatal.

LEAD (Pb)

Exposure to lead can occur through inhalation of air and ingestion of lead in food, water, soil, or dust. Excessive lead exposure can cause seizures, brain and kidney damage, mental retardation, and/or behavioral disorders. Children six and under are most at risk because their bodies are growing quickly.

NITROGEN DIOXIDE (NO_2)

Health effects of exposure to nitrogen dioxide include coughing, wheezing, and shortness of breath in children and adults with respiratory disease such as asthma. Even short exposures to nitrogen dioxide can affect lung function and may cause permanent structural changes in the lungs.

OZONE (O_3)

The reactivity of ozone causes health problems because it damages lung tissue, reduces lung function, and sensitizes the lungs to other irritants. Exposure to ozone for several hours at relatively low concentrations has been found to reduce lung function significantly and to induce respiratory inflammation in normal, healthy people during exercise. This decrease in lung function is generally accompanied by symptoms including chest pain and pulmonary congestion.

PM_{10}

Coarse particles can aggravate respiratory conditions such as asthma. Exposure to fine particles is associated with several serious health effects, including premature death. When exposed to PM_{10}, people with existing heart or lung diseases such as asthma, chronic obstructive pulmonary disease, and congestive or ischemic heart disease are at increased risk of premature death. Older persons are especially sensitive to PM_{10} exposure. When exposed to PM_{10}, children and people with existing lung disease may not be able to breathe as deeply or as vigorously as they normally would. They may also experience symptoms such as coughing and shortness of breath. PM_{10} can increase susceptibility to respiratory infections and can aggravate existing respiratory diseases, such as asthma and chronic bronchitis.

SULFUR DIOXIDE (SO_2)

High concentrations of sulfur dioxide affect breathing and may aggravate existing respiratory and cardiovascular disease. Sensitive populations include asthmatics, individuals with bronchitis or emphysema, children, and the elderly.

Smoking and Other Risks

Cigarette smoke contains over 4,700 chemical compounds including 60 known carcinogens. No threshold level of exposure to cigarette smoke has been defined. However, conclusive evidence indicates that long-term smoking greatly increases the likelihood of developing numerous fatal conditions.

Cigarette smoking is responsible for more than 85% of lung cancers and is also associated with cancers of the mouth, pharynx, larynx, esophagus, stomach, pancreas, kidney, bladder, and colon. Cigarette smoking has also been linked to leukemia. Apart from the carcinogenic aspects of cigarette smoking, links to increased risks of cardiovascular diseases (including stroke), sudden death, cardiac arrest, peripheral vascular disease, and aortic aneurysm have also been established. Many components of cigarette smoke have also been characterized as ciliotoxic materials. These irritate the lining of the respiratory system, resulting in increased bronchial mucus secretion and chronic decreases in pulmonary and mucociliary function.

FACTOIDS

- Cancer is the second leading cause of death and was among the first diseases linked to smoking.
- One in four U.S. high school students smokes; however, smoking rates among students in grades 9–12 have been declining since 1997.
- 90% of all adult smokers first lit up as teenagers.
- Lung cancer is the leading cause of cancer death, and cigarette smoking causes most cases.
- Smoking causes about 90% of lung cancer deaths in men and almost 80% in women.
- When compared with nonsmokers, men who smoke are about 23 times more likely to develop lung cancer and women who smoke are about 13 times more likely.
- In 2003, an estimated 171,900 new cases of lung cancer occurred, and approximately 157,200 people died from lung cancer.
- Carcinogens in tobacco smoke damage important genes that control the growth of cells, causing them to grow abnormally or to reproduce too rapidly.

RELEVANT LAWS

Federal Hazardous Substances Act (1960): Required that certain hazardous household products bear cautionary labels to alert consumers to the potential hazards of these products.

Federal Environmental Pesticides Control Act (1972): Required registration of all pesticides in the United States.

Hazardous Materials Transportation Act (HAZMAT) (1975): Governs the transportation of hazardous materials and wastes. Covers containers, labeling, and marking standards.

Toxic Substances Control Act (1976): Gives the Environmental Protection Agency (EPA) the ability to track the 75,000 industrial chemicals currently produced in or imported into the United States.

Comprehensive Environmental Response, Compensation and Liability (CERCLA or SUPERFUND) (1980): Established federal authority for emergency response and cleanup of hazardous substances that have been spilled, improperly disposed of, or released into the environment.

Low-Level Radioactive Policy Act (1980): Made states responsible for disposing of their own low-level radioactive wastes.

Nuclear Waste Policy Act (1982): Established a study to find a suitable site for disposal of spent fuel from nuclear reactors. Yucca Mountain, Nevada, seems most feasible at this time.

HAZARDOUS CHEMICALS IN THE ENVIRONMENT

A hazardous waste is a waste with properties that make it dangerous or potentially harmful to human health or the environment. Hazardous wastes can be liquids, solids, contained gases, or sludges. The Environmental Protection Agency has separated hazardous wastes into the following categories:

CORROSIVE

Corrosive wastes are strong acids or strong bases that are capable of corroding metal containers, such as storage tanks, drums, or barrels. Battery acid is an example.

DISCARDED COMMERCIAL PRODUCTS

These are specific commercial chemical products in an unused form. Some pesticides and some pharmaceutical products become hazardous wastes when discarded.

IGNITABLE

Ignitable wastes can create fires under certain conditions, are spontaneously combustible, or have a flash point less than 140°F (60°C). Examples include waste oils and used solvents.

NONSPECIFIC SOURCE

These include wastes from common manufacturing and industrial processes, such as solvents that have been used in cleaning or degreasing operations.

REACTIVE

Reactive wastes are unstable under normal conditions. They can cause explosions, toxic fumes, gases, or vapors when heated, compressed, or mixed with water. Examples include lithium batteries and explosives.

SOURCE SPECIFIC

These are wastes from specific industries, such as petroleum refining or pesticide manufacturing. Examples include certain sludges and wastewaters from treatment and production processes.

TOXIC

Toxic wastes are harmful or fatal when ingested or absorbed. They may contain mercury or lead for example. When toxic wastes are land disposed, contaminated liquid may leach from the waste and pollute groundwater.

Treatment, Disposal, and Cleanup of Contaminated Sites

Reduction and cleanup of hazardous wastes can occur by producing less waste, converting the hazardous material to less hazardous or nonhazardous substances, and placing the toxic material into perpetual storage.

PRODUCE LESS WASTE

- **Recycle**—Improved technology seeks ways of collecting hazardous wastes and using them as raw materials for new products.
- **Reduce or eliminate toxicity**—Improved technology seeks substitutes for hazardous chemicals. For example, Puron replaced Freon.

CONVERSION TO LESS HAZARDOUS OR NONHAZARDOUS SUBSTANCES

Chemical, physical, and biological treatment

Bioremediation is the use of bacteria and enzymes to break down hazardous materials. *Phytoremediation* involves *rhizofiltration* (using sunflowers to absorb radioactive wastes), *phytostabilization* (using willow and poplar trees to absorb organic contaminants), *phytodegradation* (using poplars to absorb and break down contaminants), and *phytoextraction* (using Indian mustard and brake ferns to absorb inorganic metal contaminants). Pros: inexpensive, low energy use, little to no air pollution, easy to build. Cons: slow, effective only as far down as roots will reach, some toxic materials can evaporate through plants. Plants would be toxic and need to be properly disposed. Chemical methods involve use of cyclodextrin.

Incineration

Can release air pollutants and toxic ash (such as lead, mercury, and dioxins).

Thermal Treatment

Plasma arcs. Pros: small, mobile, no toxic ash. Cons: expensive and can release particulates, chlorine gas, toxic metals, and radioactive wastes.

PERPETUAL STORAGE

Arid Region Unsaturated Zone

The unsaturated zone is the subsurface between the land surface and underlying aquifers. It includes sites in the arid western United States that are being relied upon to isolate a significant portion of the nation's radioactive and other hazardous wastes for thousands of years.

Landfill

Pros: inexpensive. Cons: groundwater seepage and contamination.

Salt Formations

Toxic wastes are deposited in deep salt formations. The absence of flowing water within natural salt formations prevents dissolution and subsequent spreading of the waste products. Rooms and caverns in the salt close with time, thus isolating the waste from the biosphere.

Surface Impoundments

Excavated ponds, pits, or lagoons. Pros: low cost, low operating cost, built quickly, wastes can be retrieved and, if lined, can store wastes for long periods. Cons: groundwater contamination, VOC pollution, overflow if flooding occurs, earthquake issues, promotes waste production.

Underground Injection

Pros: low cost, wastes can be retrieved, simple technology. Cons: leaks, earthquake issues, groundwater contamination.

Waste Piles

Storage of toxic materials in drums, underground vaults, or above-ground buildings. Pros: easy to identify leaks. Cons: shipping of materials to facilities results in accidents.

CASE STUDY

The first synthesized chlorinated organic pesticide was DDT. It appeared to have low toxicity and was broad spectrum. It did not break down, so it did not have to be reapplied often. Crop production increased and mosquitoes decreased. In 1962, Rachel Carson published *Silent Spring*, which made the connection between DDT and nontarget organisms by direct and indirect toxicity. DDT persisted in the environment through bioaccumulation (an increase in concentration up the food chain) and biomagnification (the tendency for a compound to accumulate in tissues). DDT was found to decrease the eggshell thickness of various species of birds, nearly wiping out bald eagles and peregrine falcons. DDT was beginning to show up in native people of the Arctic, seals, and human breast milk. It was pulled off the U.S. market and is now being manufactured in Indonesia.

Biomagnification

Biomagnification is the increase in concentration of a pollutant from one link in a food chain to another. In order for biomagnification to occur, the pollutant must be long-lived, mobile, soluble in fats, and biologically active. If a pollutant is short-lived, it will be broken down before it can become dangerous. If it is not mobile, it will stay in one place and be less likely to be taken up by many organisms. If the pollutant is soluble in water, it will be excreted by the organism. Pollutants that dissolve in fats, however, may be retained for a long time. It is traditional to measure the amount of pollutants in fatty tissues of organisms such as fish. In mammals, milk produced by females is often tested since the milk is high in fat and because the young are often more susceptible to damage from toxins.

ECONOMIC IMPACTS

A cost-benefit analysis is a technique for deciding whether to make a change. To use the technique, one adds up the value of the benefits of a course of action and subtracts the costs associated with it.

Costs are either one time or ongoing. Benefits are most often received over time. Time is factored into a cost-benefit analysis by calculating a payback period—the time for the benefits of a change to repay its costs.

In its simple form, a cost-benefit analysis is carried out using only financial costs and benefits. For example, a simple cost-benefit analysis of a new road would measure the cost of building the road and subtract this from the economic benefit of improving transport links. It would not measure either the cost of environmental damage or the benefit of quicker and easier travel to work. A more sophisticated approach is to attempt to put a financial value on intangible costs and benefits, which is highly subjective.

Framework of Cost-Benefit Analysis

Step	Description
Cost-benefit	Determine an action and levels of action that achieve the greatest net economic benefit. Exploring options and determining incremental levels of remediation provide the most benefit for the least cost.
Cost-effectiveness	Implementing a specific environmental, health, or safety objective at the least cost. Emphasis is on achieving the objective. Flexible regulatory guidelines are adapted to find the lowest cost to solve a problem.
Health or environmental protection standards	Reducing risk to the public whatever the cost.
Risk-benefit	Balancing health or environmental protection with the costs of providing the protection.
Technology	To achieve results that are predictable and certain.

A cost-benefit analysis applies to three economic situations: First, it can help judge whether public services provided by the private sector are adequate. Second, it can be used when judging and assessing inefficiencies (market failures) in the private sector and their impact on the health, safety, and environmental needs of the country. Third, it helps in determining how to meet societal needs in a cost-effective manner in areas that only government can address. These include defense, preservation of scenic areas, environmental protection, and so on.

A cost-benefit analysis can be used for evaluating policy alternatives, shaping regulatory strategies, and evaluating specific regulations. A cost-benefit analysis requires:

1. Gathering all information and data about a public issue, including history and background
2. Defining the possible solutions to solving the issue
3. Brainstorming the possible environmental and societal consequences of the alternatives
4. Quantifying the benefits and the costs
5. Making decisions and balancing concerns

Externalities and Marginal Costs

Externality is a general term for a wide variety of costs and benefits that are not included in prices or the effects of an action on people who were not a part of the process. One example of environmental externalities is the cost of environmental damage associated with production and transportation activities such as species die-off or poisonings. Another example is the cost of flooding of low-elevation areas and the increase in home and property damage from extreme weather associated with climate change. A third sample is air pollution generated by a local industry that affects others who had no choice or say in the matter.

If there are externalities, then there is overproduction of a good. The total cost of a good to society (called the social cost) includes the costs of production incurred by the industry as well as the external costs. The marginal cost is reached when the change in the total cost of a product changes with the production of just one more item. For example, it may cost a company the same price to produce 5 widgets as it does to produce 10 widgets. However, if the company has to pay more (and consequently charge more) to produce the 11th widget, then it has exceeded its marginal cost. The optimal level of an externality is found by equating the marginal benefit to the marginal cost. Negative externalities can be handled in three ways:

- Regulation prohibiting or limiting the activity
- Making the activity more costly through taxation—by setting the tax equal to the cost to others, the externality has been internalized
- A hybrid approach that includes tradable licenses, with the number of licenses set by regulation and the market determining the use of the licenses

Sustainability

Sustainability deals with the continuity of the economic, social, and institutional aspects of human society while at the same time preserving biodiversity and the environment. Sustainable activities seek to provide the best outcomes for both human societies and natural ecosystems. Several issues are common to both interests:

- Consideration of risk, uncertainty, and irreversibility
- Ensuring appropriate valuation, appreciation, and restoration of nature
- Integration of environmental, social, and economic goals in policies and activities
- Equal opportunity and community participation
- Conservation of biodiversity and ecological integrity
- A commitment to best practice
- No net loss to either human or natural capital
- Continuous improvement
- The need for good governance

Unlimited economic and population growth puts many demands on natural resources. Its effects on pollution and the carrying capacity of Earth are factors that impact sustainability. These are all analyzed using life cycle assessments and ecological footprint analyses.

MULTIPLE-CHOICE QUESTIONS

1. About 70% of U.S. hazardous wastes come from

   (A) agricultural pesticides
   (B) smelting, mining, and metal manufacturing
   (C) nuclear power plants
   (D) chemical and petroleum industries
   (E) households

2. Which act established federal authority for emergency response and cleanup of hazardous substances that have been spilled, improperly disposed of, or released into the environment?

   (A) Resource Recovery Act
   (B) Resource Conservation and Recovery Act
   (C) Solid Waste Disposal Act
   (D) Superfund
   (E) Hazardous Materials Transportation Act

3. Effects produced from a long-term, low-level exposure are called

   (A) acute
   (B) chronic
   (C) pathological
   (D) symptomatic
   (E) synergistic

4. Which of the following techniques is NOT an example of bioremediation?

    (A) land farming
    (B) composting
    (C) rhizofiltration
    (D) phytoremediation
    (E) all are examples of bioremediation

5. The following dose-response curve shows that

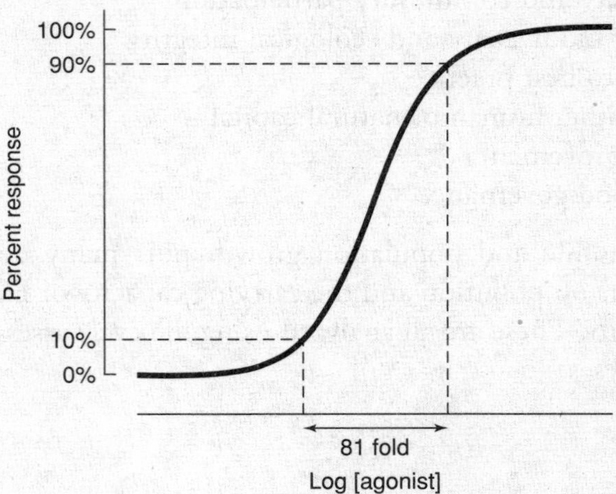

    (A) larger amounts of agonist produce a corresponding increase in response
    (B) 80% more agonist is required to achieve 80% more response
    (C) 81 times more agonist is needed to achieve a 90% response than a 10% response
    (D) an 80% high response is achieved with a tenfold increase in agonist
    (E) none of the above are true

6. A dose that is represented as LD_{50}

    (A) shows a response in 50% of the population
    (B) kills half of the study group
    (C) is a dose that has an acceptable risk level of 50%
    (D) is a dose that has a threshold response of 50%
    (E) is a dose that is administered to 50% of the population

7. Problem(s) associated with risk management include

    (A) people making a risk assessment vary in their conclusions of long-term versus short-term risks and benefits
    (B) some technologies benefit some groups and harm others
    (C) there is consideration of the cumulative impacts of various risks rather than consideration of each impact separately
    (D) there may be conflict of interest in those carrying out the risk assessment and review of the results
    (E) all of the above are true

8. Currently, the single-most significant threat to human health is

    (A) toxic chemicals
    (B) accidents
    (C) pathogenic organisms
    (D) pollution
    (E) nontransmissible diseases such as cancer and cardiovascular disease

9. The accumulation of DDT by peregrine falcons, brown pelicans, and other predatory birds during the 1960s is an example of

    (A) bioaccumulation
    (B) bioremediation
    (C) acute exposure
    (D) biomagnification
    (E) a case-controlled study

10. A concentration of 30 ppm would be equivalent to

    (A) 0.3%
    (B) 0.03%
    (C) 0.003%
    (D) 0.0003%
    (E) 0.00003%

11. Preserving the value of a resource for the future is a(n)

    (A) aesthetic value
    (B) cultural value
    (C) existence value
    (D) use value
    (E) option value

12. It costs a copper smelter \$200 to reduce emissions by 1 ton and \$250 for each additional ton. It costs an electric company \$100 to reduce its emissions by 1 ton and \$150 for each additional ton. What is the least expensive way of reducing total emissions by 2 tons?
   (A) Legislate that both firms must reduce emissions by 1 ton
   (B) Charge both firms \$251 for every ton they emit
   (C) Allow each firm to buy a \$151 permit to pollute
   (D) File an injunction to halt production until the firms reduce emissions by 2 tons
   (E) None of the above

13. In economic analysis, the optimum level of pollution
   (A) is always zero
   (B) is where the marginal benefits from further reduction equals the marginal cost of further reduction
   (C) occurs when demand crosses the private cost supply curve
   (D) should be determined by the private market without any government intervention
   (E) none of the above

14. Externalized costs of nuclear power include all of the following EXCEPT
   (A) disposing of nuclear wastes
   (B) government subsidies
   (C) costs associated with Three Mile Island
   (D) Price-Anderson Indemnity Act
   (E) all are external costs

15. Which of the following is NOT part of a cost-benefit analysis?
   (A) Judging whether public services provided by the private sector are adequate
   (B) Judging and assessing inefficiencies in the private sector and their impact on health, safety, and environmental need
   (C) Determining external costs to society
   (D) Meeting societal needs in a cost-effective manner
   (E) All are part of a cost-benefit analysis

16. A federal tax on a pack of cigarettes is an example of

    (A) full-cost pricing
    (B) an internal cost
    (C) an external cost
    (D) a marginal cost
    (E) a life cycle cost

17. The threshold dose-response model

    (A) cannot be used to determine how toxic a substance is
    (B) implies a risk associated with all doses
    (C) implies that no detectable harmful effects can occur below a certain dose
    (D) is similar to a linear dose-response model
    (E) all of the above are true

18. How many people aged 15–24 are newly infected worldwide with the AIDS virus each day?

    (A) 60
    (B) 600
    (C) 6,000
    (D) 60,000
    (E) 600,000

19. Which of the following is NOT part of risk assessment?

    (A) Determining the probability of a particular hazard
    (B) Determining types of hazards
    (C) Coming up with an estimate on the chances of how many people could be exposed to a particular risk
    (D) Informing the public about the chances of risks
    (E) All are part of risk assessment

20. Which factor listed below is generally considered to be the primary cause of reduced human life span?

    (A) AIDS
    (B) Infectious disease
    (C) Cancer
    (D) Poverty
    (E) Heart disease

FREE-RESPONSE QUESTION

By: Annaliese Berry, B.A.
California Institute of Technology
Williams College
A.P. Environmental Science Teacher
Marin Academy, San Rafael, California

Part (a): 2 points
Part (b): 2 points
Part (c): 4 points
Part (d): 2 points

Total: 10 points

In 1989, the oil tanker *Exxon Valdez* struck an offshore reef in Alaska and spilled more than 11 million gallons of crude oil into Prince William Sound. It was the largest spill in U.S. history. Although Exxon spent $2.2 billion in partially dispersing and removing the oil, many seabirds and mammals died as a result of this catastrophe.

Ironically, environmental accidents such as the *Exxon Valdez* oil spill actually stimulate our nation's economy by causing an increase in gross domestic product and gross national product. Using this information, answer the following questions.

(a) Explain what gross domestic product (GDP) or gross national product (GNP) is a measure of and give examples of a possible cause for an increase in the GDP or GNP following the *Exxon Valdez* oil spill.
(b) Give at least one criticism of GDP and GNP as progress indicators, and mention an alternative progress indicator. What does that alternative attempt to measure?
(c) Name and explain one internal cost and one external cost that might be associated with the oil industry. Include in your answer an explanation of what internal and external costs are.
(d) How would internalizing external costs (sometimes called full-cost analysis or true-cost analysis) affect the pricing and economic competitiveness of petroleum products?

TIP

For Free Response essays, be sure to practice making good arguments in favor of, or against, a particular position on an environmental issue. Make sure you understand that restating the question is a waste of time in a timed test, and doing so will never earn you any points.

MULTIPLE-CHOICE ANSWERS AND EXPLANATIONS

1. (D) In 1999, over 20,000 hazardous waste generators produced over 40 million tons of hazardous wastes.
2. (D) The Superfund program began in 1980 to locate, investigate, and clean up the most polluted sites nationwide.
3. (B) Chronic diseases are of long duration with slow progress. Examples would include congestive heart failure, Parkinson's disease, and cerebral palsy.
4. (E) Land farming, composting, rhizofiltration, and phytoremediation all involve the use of bacteria and enzymes to break down hazardous materials.
5. (C) "81 fold" means 81 more times. An agonist is a drug or other chemical that produces a reaction typical of a naturally occurring substance.

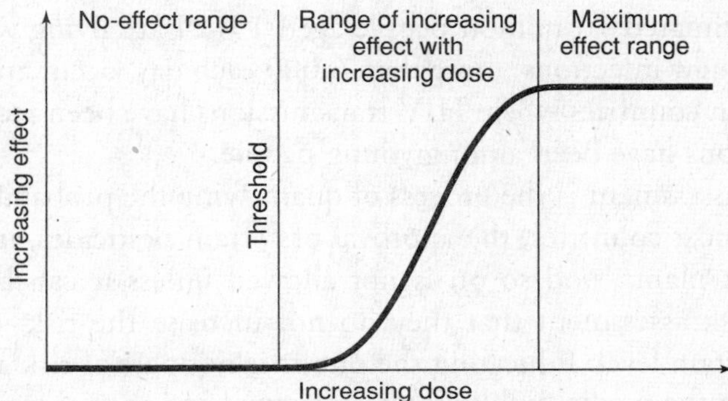

6. (B) The LD_{50} value is typically expressed in milligrams of material per kilogram of body weight. It indicates the quantity of material that will cause 50% of the subjects to perish.
7. (E) Risk assessment is the process of evaluating the likelihood of an adverse health effect. Risk assessment does NOT determine the level of allowable or acceptable risk—that is risk management.
8. (E) For most of human history, pathogenic (disease-causing) organisms were the greatest threat to health. Today cardiovascular, cancer, and other noninfectious diseases have become the major killers.
9. (D) Biomagnification is the increase in the concentration of toxic substances as one moves up the food chain. Animals at the top of the food chain receive the highest concentration of toxins and experience the worst effects.
10. (C) To change ppm to a percentage, move the decimal place four places to the left and add a % sign.
11. (E) An option value is the value that people place on having the option to enjoy something in the future.
12. (C) The copper smelter would pay $450 to emit 2 tons of pollutants. The electric company would pay $250 to emit 2 tons of pollutants. Clearly, $151 would be the least expensive option available.
13. (B) The marginal benefit of pollution control declines as the level of environmental quality goes up. At the same time, the marginal cost of pollution tends to increase. The optimal level of pollution from an economic standpoint is where the marginal benefit equals the marginal cost. As long as the marginal benefit exceeds the marginal cost, there is economic incentive for cleaning up.

14. (E) External costs are the costs that are borne by people other than the producer of a good.
15. (C) Cost-benefit analysis is a technique for deciding whether to make a change and requires adding up the value of the benefits of a course of action and subtracting the costs associated with it.
16. (A) Full-cost pricing accounts for the cost of a good when its internal costs and its estimated short- and long-term external costs are included in its market price. Washington State increased its cigarette tax to $1.425 a pack on Jan. 1, 2002, making it the nation's highest. Virginia has the lowest state sales tax of 2.5 cents per pack. There is a 39-cent per pack federal excise tax.
17. (C) The threshold dose-response model has long been recognized as the most important tool in understanding the risk assessment processes used by regulatory and public health agencies worldwide.
18. (C) An estimated 10 million people aged 15–24 are living with HIV/AIDS. Half of all new infections, more than 6,000 each day, occur among the young. However, in countries where HIV transmissions have been reduced, the greatest reductions have been among young people.
19. (D) Risk assessment is the process of quantifying the probability of a harmful effect. In most countries, the approval of certain pesticides, industrial chemicals, power plants, and so on is not allowed unless it can be demonstrated through risk assessment that they do not increase the risk, death, or illness above a certain level. Educating the public is not part of risk assessment. That is left up to the media, public interest groups, and special interest groups.
20. (D) Poverty is the fundamental issue that affects human life span. People who are without financial resources suffer from malnutrition, exposure, and disease. Those with financial resources are able to obtain proper nutrition, have shelter, and are able to seek medical treatment.

FREE-RESPONSE ANSWER

Notable points:

Explanation of GDP or GNP.

Example of how cleaning up the spill or replacing lost equipment/oil causes goods or services to be transacted.

(a) Gross domestic product (GDP) is a measure of the market value of goods and services transacted within a country during a year, so it increases any time products or services are paid for. This means that the additional expense that Exxon (now Exxon-Mobil) incurred by controlling the spill ($2.2 billion) increased the GDP. For example, money was spent to hire boats with skimmers in an attempt to capture and contain floating oil.

(b) GDP and GNP have been criticized as progress indicators because economic growth is only indirectly related to quality of life. The GDP and GNP are often stimulated by events that directly lower quality of life, such as the *Exxon Valdez* oil spill, the Oklahoma City bombing, Hurricane Katrina in New Orleans, or the events of 9/11 in New York City. Alternative progress indicators attempt to capture quality of life or standard of living, although these are often harder to define and measure than the flow of dollars. One such indicator is the net national product (NNP), which accounts for depletion and destruction of natural resources along with changes in GNP.

GDP and GNP are economic measures, not direct quality of life measures.

Example of alternative indicator. Examples include the NNP, ISEW, GPI, or NEW. A general alternative is to take into account any standard of living data (literacy rate, percent of population below poverty line, and so on).

Explanation of what example indicator measures.

(c) An internal cost is a cost that is paid for by the organization producing a product. This cost is typically passed along to the consumers when they purchase the product. One internal cost of the petroleum industry could be the cost of extracting petroleum using oil drills and platforms.

Naming of an internal cost. Costs may be expenses related to cleanup, ship or oil replacement, cost of labor, and so on.

Explanation that internal costs are costs paid for by the organization producing the product.

Naming of an external cost such as water pollution left behind after a spill, air or water pollution from operating tankers, or an increase of CO_2 in the atmosphere when the petroleum products are consumed.

Explanation that external costs are harmful effects of the product that are not paid for by the organization.

Price of product would increase as external costs are paid for by the producer and passed along to the consumer.

Petroleum products would become less economically competitive.

(d) Internalizing external costs would mean that organizations producing a product would pay for any harmful effects (the external costs) of their products. This would allow the market price of a product to reflect the full cost of producing and cleaning up the product. It would also increase the price of any product that has external costs. A product with high external costs would become substantially more expensive and less economically competitive when this happened. In the energy industry, sources of power such as oil and gas have relative high external costs and could become more expensive than lower-impact sources of energy such as wind power.

UNIT VII: GLOBAL CHANGE (10–15%)

Areas on Which You Will Be Tested

A. **Stratospheric Ozone**—formation of stratospheric ozone, ultraviolet radiation, causes of ozone depletion, effects of ozone depletion, strategies for reducing ozone depletion, and relevant laws and treaties.

B. **Global Warming**—greenhouse gases and the greenhouse effect, impacts and consequences of global warming, reducing climate change, and relevant laws and treaties.

C. **Loss of Biodiversity**
   1. Habitat loss—overuse, pollution, introduced species, and endangered and extinct species.
   2. Maintenance through conservation.
   3. Relevant laws and treaties.

Stratospheric Ozone and Global Warming

CHAPTER 11

I believe that global warming is a myth. And so, therefore, I have no conscience problems at all and I'm going to buy a Suburban next time.

—Rev. Jerry Falwell

STRATOSPHERIC OZONE

The stratosphere contains approximately 97% of the ozone in the atmosphere, and most of it lies between 9 and 25 miles (15 to 40 km) above Earth's surface. Most ozone is formed over the tropics. However, slow circulation currents carry the majority of it to the poles, resulting in the thickest layers over the poles and the thinnest layer above the tropics. It also varies somewhat due to season, being somewhat thicker in the spring and thinner during the autumn. The increase in temperature with height in the stratosphere occurs because of absorption of ultraviolet radiation by ozone.

Ozone is formed in the stratosphere by the reaction of ultraviolet radiation striking an oxygen molecule. This results in the oxygen molecule splitting apart and forming atomic oxygen: $O_2 + UV \rightarrow O + O$. Atomic oxygen can now react with molecular oxygen to form ozone: $O + O_2 \rightarrow O_3$. The reverse reaction also occurs when ultraviolet radiation strikes an ozone molecule, causing it to form atomic oxygen and molecular oxygen: $O_3 + UV \rightarrow O + O_2$. Atomic oxygen can also react with ozone to produce oxygen gas: $O + O_3 \rightarrow O_2 + O_2$. Generally, these reactions balance each other so that the concentration of ozone in the stratosphere remains fairly constant, which keeps the amount of UV radiation reaching Earth also constant. Various forms of life on Earth have evolved over millions of years with fairly constant amounts of UV radiation.

Ultraviolet Radiation

The sun emits a wide variety of electromagnetic radiation, including infrared, visible, and ultraviolet. Ultraviolet radiation can be subdivided into three forms: UVA, UVB, and UVC.

UVA

320 to 400 nm wavelength. Closest to blue light in the visible spectrum. The form that usually causes skin tanning. UVA radiation is 1,000 times less effective than UVB in producing skin redness, but more of it reaches Earth's surface than UVB. Birds, reptiles, and bees can see UVA. Many fruits, flowers, and seeds also stand out more strongly from the background in ultraviolet wavelengths. Many birds have patterns in their plumage that are not visible in the normal spectrum (white light) but become visible in ultraviolet. Urine of some animals is also visible only in the UVA spectrum.

UVB

290 to 320 nm wavelength. Causes blistering sunburns and is associated with skin cancer.

UVC

10 to 290 nm. Found only in the stratosphere and largely responsible for the formation of ozone.

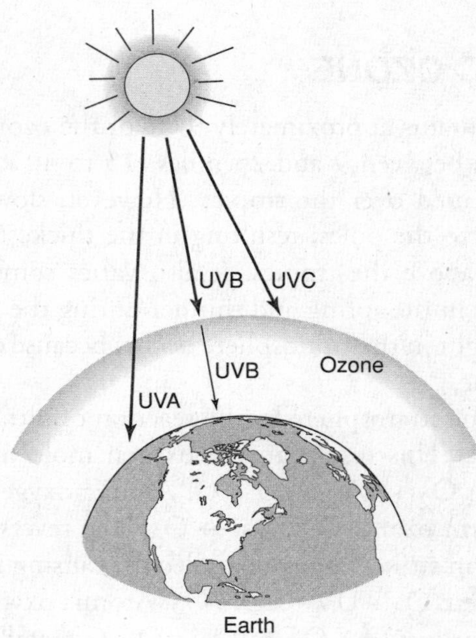

Figure 11.1 Ultraviolet radiation reaching Earth

Causes of Ozone Depletion

Thinning of the ozone layer was first discovered in 1985. It occurs seasonally and is due to the presence of human-made compounds containing halogens (chlorine, bromine, fluorine, or iodine). Measurements indicate that the ozone over the Antarctic has decreased as much as 60% since the late 1970s with an average net loss of about 3% per year worldwide.

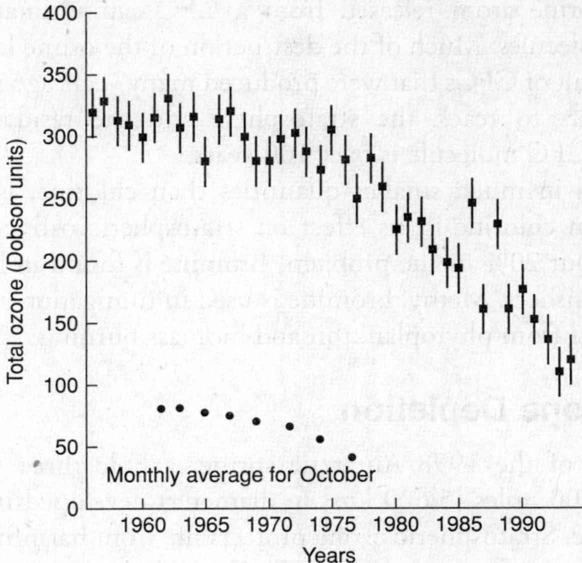

Figure 11.2 Ozone concentration over Antarctica (1955–1995)

The main culprits in the depletion of the ozone layer are compounds known as CFCs (chlorofluorocarbons). First manufactured during the 1920s, they are used as refrigerants (for example, Freon), aerosol propellants, electrical part cleaning solvents, and in the manufacture of foam products and insulation. By 1974, nearly 1 million tons of CFC gases were produced each year, and the chemicals were generating $8 billion worth of business. The largest single source of CFCs to the atmosphere is leakage from air conditioners. The average residence time of CFCs in the environment is 200 years.

When a CFC molecule enters the stratosphere, ultraviolet radiation causes it to decompose and produce atomic chlorine:

$$F{-}C(Cl)_2{-}F \xrightarrow{\text{photon of UV radiation}} F{-}\dot{C}(Cl){-}F + \cdot Cl$$

This atomic chlorine then reacts with the ozone in the stratosphere to produce chlorine monoxide (ClO):

$$Cl + O_3 \rightarrow ClO + O_2$$

The chlorine monoxide then reacts with more ozone to produce even more atomic chlorine in what becomes essentially a chain reaction:

$$ClO + O_3 \rightarrow Cl + 2O_2$$

Thus, one chlorine atom released from a CFC can ultimately destroy over 100,000 ozone molecules. Much of the destruction of the ozone layer that is occurring now is the result of CFCs that were produced many years ago since a CFC molecule takes 8 years to reach the stratosphere and the residence time in the stratosphere for a CFC molecule is over 100 years.

Bromine, found in much smaller quantities than chlorine, is about 50 times more effective than chlorine in its effect on stratospheric ozone depletion and is responsible for about 20% of the problem. Bromine is found in halons, which are used in fire extinguishers. Methyl bromide is used in fumigation and agriculture. It is naturally released from phytoplankton and biomass burning.

Effects of Ozone Depletion

During the onset of the 1998 Antarctic spring, a hole three times the size of Australia (over 3,500 miles [5,600 km] in diameter) developed in the ozone layer over the South Pole. Stratospheric ozone protects life from harmful ultraviolet radiation. Harmful effects of increased UV radiation include:

- Increases in skin cancer
- Increases in sunburns and damage to the skin
- Increases in cataracts of the eye
- Reduction in crop production
- Deleterious effects to animals (they don't wear sunglasses or sunscreen)
- Reduction in the growth of phytoplankton and the cumulative effect on food webs
- Increases in mutations since UV radiation causes changes in DNA structure
- Cooling of the stratosphere
- Reduction in the body's immune system
- Climatic change

Strategies for Reducing Ozone Depletion

Although most developed countries have phased out ozone-destroying chemicals, they are still legal in developing nations. There are several alternatives to CFC use. First, HCFC replaces chlorine with hydrogen. Unfortunately, it is still capable of destroying ozone albeit less effectively because it breaks down more readily in the troposphere. Second, alternatives to halons can be used in fire extinguishers. Third, helium, ammonia, propane, or butane can be used as a coolant. Helium-cooled refrigerators use 50% less electricity than those using CFCs or HCFCs. Individuals can use pump sprays instead of aerosol spray cans when possible, comply with disposal requirements of the Clean Air Act for old refrigerators and air conditioners, read labels and, when choices are available, use ozone friendly products, and support legislation that reduces ozone-destroying products.

RELEVANT LAWS AND TREATIES

Montreal Protocol (1987): Designed to protect the stratospheric ozone layer. The treaty was originally signed in 1987 and substantially amended in 1990 and 1992. The Montreal Protocol stipulated that the production and consumption of compounds that deplete ozone in the stratosphere—chlorofluorocarbons (CFCs), halons, carbon tetrachloride, and methyl chloroform—were to be phased out by 2000 (2005 for methyl chloroform).

London (1990): The countries that signed the Montreal Protocol met again in London and decided that a total phaseout of CFCs was necessary. They agreed that this could be achieved by the year 2000. Control measures were also adopted for carbon tetrachloride and methyl chloroform.

Copenhagen (1992): The phaseout schedule for CFCs was again accelerated, with the industrialized countries agreeing to stop production by 1996. This goal had already been prescribed in the United States in the 1990 amendments to the Clean Air Act. In 1994, the European Community decided that a phaseout could be achieved in Europe by 1995.

GLOBAL WARMING

When sunlight strikes Earth's surface, some of it is reflected back toward space as infrared radiation (heat). Greenhouse gases absorb this infrared radiation and trap the heat in the atmosphere.

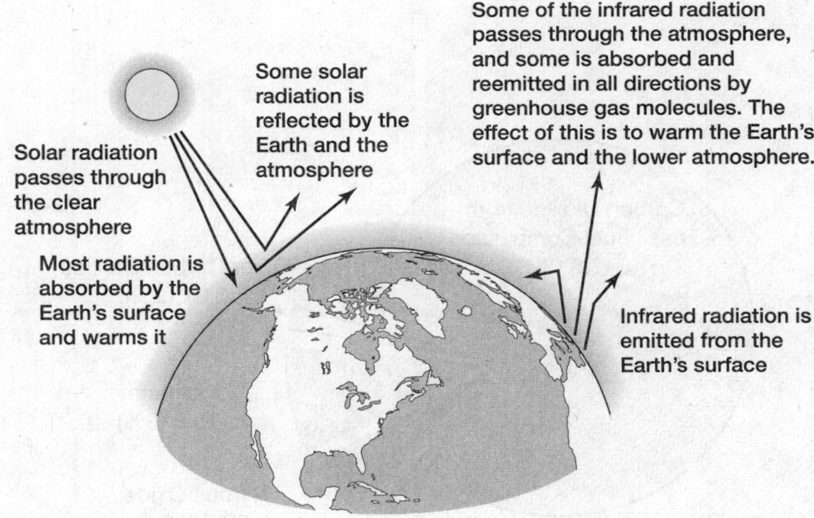

Figure 11.3 The greenhouse effect

Greenhouse Gases

Greenhouse Gas	Average Time in Troposphere (Years)	Relative Warming Potential (CO_2 = 1)	Source
Carbon dioxide (CO_2)	100	1	Burning oil, coal, plants, deforestation, cellular respiration
Carbon tetrachloride (CCl_4)	45	1,500	Cleaning solvent
Chlorofluorocarbons (CFCs)	15 (100 in stratosphere)	1,000–8,000	Air conditioners, refrigerators, foam products, insulation
Halons	65	6,000	Fire extinguishers
Hydrochlorofluorocarbons (HCFCs)	10–400	500–2,000	Air conditioners, refrigerators, foam products, insulation
Hydrofluorocarbons (HFCs)	15–400	150–13,000	Air conditioners, refrigerators, foam products, insulation
Methane (CH_4)	15	25	Rice cultivation, enteric fermentation, production of coal, natural gas leaks
Nitrous oxide (N_2O)	115	300	Burning fossil fuels, fertilizers, livestock wastes, plastic manufacturing
Sulfur hexafluoride (SF_6)	3,200	24,000	Electrical industry as a replacement for PCBs

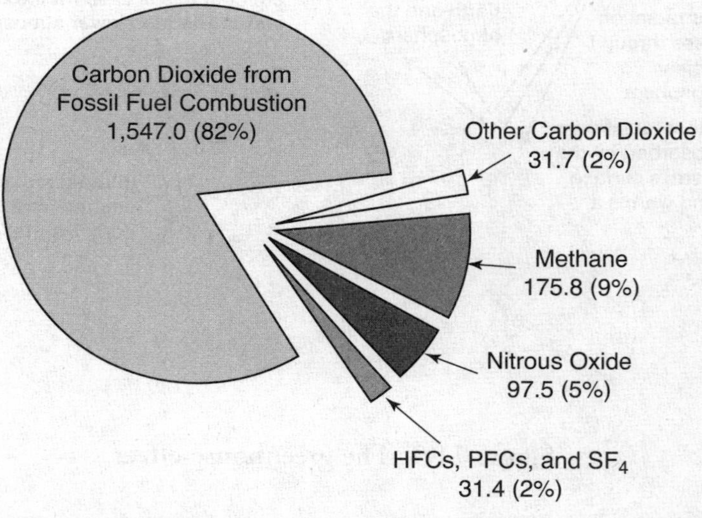

Figure 11.4 U.S. anthropogenic greenhouse gas emissions 2001
(Million Metric Tons of Carbon Equivalent)

Levels of several important greenhouse gases have increased by about 25% since large-scale industrialization began around 150 years ago. During the past 20 years, about three-quarters of human-made carbon dioxide emissions were from burning fossil fuels. In the United States, greenhouse gas emissions come mostly from energy use. These are driven largely by economic growth, fuel used for electricity generation, and weather patterns affecting heating and cooling needs. Energy-related carbon dioxide emissions, resulting from petroleum and natural gas, represent 82% of total U.S. anthropogenic (caused by humans) greenhouse gas emissions.

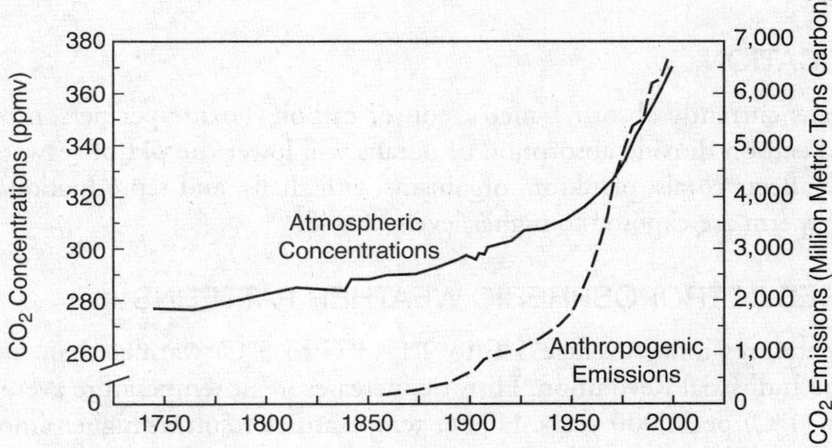

Figure 11.5 Trends in atmospheric concentrations

Source: Oak Ridge National Laboratory

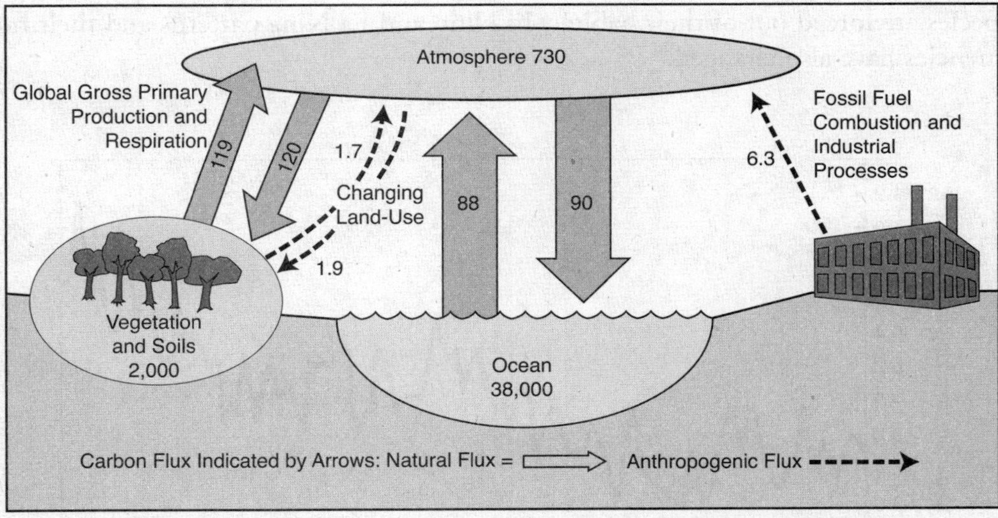

Figure 11.6 Global carbon cycle (billion metric tons carbon)

Source: Intergovernmental Panel on Climate Change

Concentrations of carbon dioxide in the atmosphere are naturally regulated by numerous processes collectively known as the carbon cycle. The movement (flux) of carbon between the atmosphere and the land and oceans is dominated by natural processes, such as plant photosynthesis. Although these natural processes can absorb some of the net 6.1 billion metric tons of anthropogenic carbon dioxide emissions produced each year (measured in carbon equivalent terms), an estimated 3.2 billion metric tons is added to the atmosphere annually. Earth's positive imbalance between emissions and absorption results in the continuing growth in greenhouse gases in the atmosphere.

Impacts and Consequences of Global Warming

Global warming affects the weather, the economy, and numerous other aspects of life.

ACIDIFICATION

The oceans currently absorb 1 metric ton of carbon dioxide per person per year. Increased carbon dioxide absorption of oceans will lower the pH of seawater. This adversely affects corals, plankton, organisms with shells, and reproduction rates as eggs and sperm are exposed to higher levels of acid.

CHANGES IN TROPOSPHERIC WEATHER PATTERNS

Air temperatures today average 5°F to 9°F (3°C to 5°C) warmer than they were before the Industrial Revolution. Historic increases in air temperature averaged less than 2°F (1°C) per 1,000 years. Higher temperatures result in higher amounts of rainfall due to higher rates of evaporation. Worldwide, hurricanes of category 4 or 5 have risen from 20% of all hurricanes in the 1970s to 35% in the 1990s. Precipitation due to hurricanes in the United States has increased 7% during the 20th century. More rainfall increases erosion, which then leads to higher rates of desertification due to deforestation. This then leads to losses in biodiversity as some species are forced out of their habitat. El Niño and La Niña patterns and their frequencies have also changed.

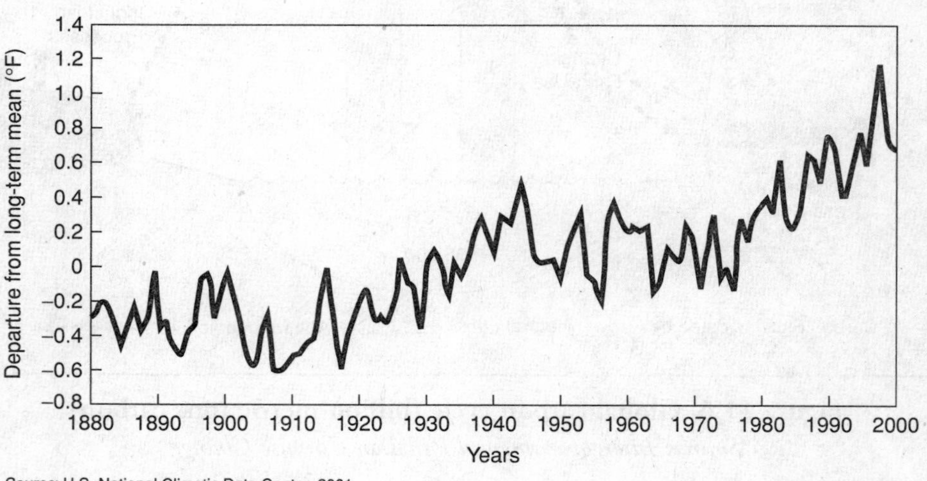

Figure 11.7 Global temperature changes (1880–2000)

DISPLACEMENT OF PEOPLE

The United Nations estimates that by the year 2050, 150 million people will need to be relocated worldwide. This will occur due to the effects of coastal flooding, shoreline erosion, and agricultural disruption.

ECOLOGICAL PRODUCTIVITY

Satellite photos have shown that productivity in the Northern Hemisphere has increased since 1982. However, biomass increases due to warmer temperatures reaches a certain point—the point where limiting factors of water and nutrients curb future productivity increases. In the tropics, plants increase productivity more so than trees (which are carbon sinks). With higher percentages of plants due to increased temperatures and carbon dioxide concentrations, the rates of decomposition increase because plants are shorter lived. As a result, more carbon enters the carbon cycle.

FOREST FIRES

Boreal forest fires in North America used to average 2.5 million acres (10,000 sq km). They now average 7 million acres (28,000 sq km). Forest management practices may also be contributing to the increase.

GLACIER MELTING

Total surface area of glaciers worldwide has decreased by 50% since the end of the 19th century. Temperatures of the Antarctic Southern Ocean rose 0.31°F (0.17°C) between the 1950s and the 1980s. Glacier melting causes landslides, flash floods, glacial lake overflow, and increased variation in water flows into rivers. Hindu Kush and Himalayan glacier melts are reliable water sources for many people in China, India, and much of Asia. Global warming initially increases water flow, causing flooding and disease. Flow will then decrease as the glacier volume dwindles, resulting in drought. Eventual decreases in glacial melt will also affect hydroelectric production.

INCREASED HEALTH AND BEHAVIORAL EFFECTS

Higher temperatures result in higher incidences of heat-related deaths. Estimates indicate that a temperature increase of just 2°F (1°C) will result in approximately 25,000 additional homicides in the United States due to stress and resulting rage.

INCREASE IN DISEASE

Rates of malaria (due to increase in mosquitoes), cholera, and other waterborne diseases will increase. Remediation and mitigation efforts will end up costing more.

INCREASED PROPERTY LOSS

Weather-related disasters have increased 3-fold since the 1960s. Insurance payouts have increased 15-fold (adjusted for inflation) during this same time period. Much of this can be attributed to people moving to vulnerable coastal areas.

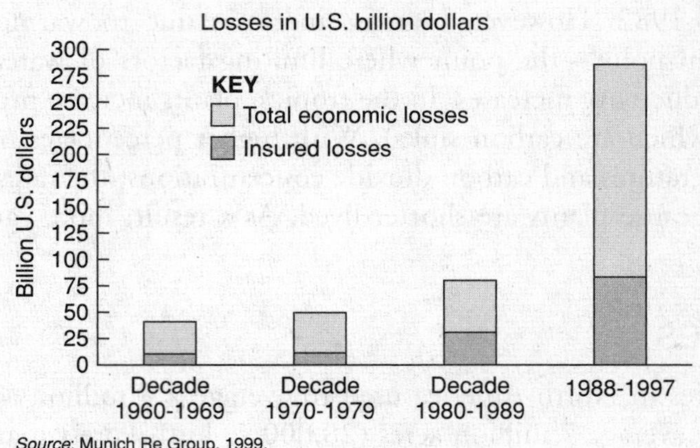

Figure 11.8 Weather and flood catastrophes since 1960

LOSS IN ECONOMIC DEVELOPMENT

Money that was designed to increase education, improve health care, reduce hunger, and improve sanitation and fresh water supplies, will instead be spent on mitigating the effects of global warming.

LOSS OF BIODIVERSITY

Arctic fauna will be most affected. The food webs of polar bears that depend on ice flows, birds, and marine mammals will be negatively impacted. Many refugee species currently have shifted their ranges toward the poles, averaging 4 miles (6 km) per decade. Bird migrations are averaging over two days earlier per decade. Grasses have become established in Antarctica for the first time. Many species of fish and krill that require cooler waters will be negatively impacted. Decreased glacier melt will impact migratory fish, such as salmon, that need sufficient river flow.

RELEASES OF METHANE FROM HYDRATES IN COASTAL SEDIMENTS

Methane hydrate (methane clathrate) is a form of water ice that contains methane within its crystal structure. Extremely large deposits of this resource have been discovered in ocean sediments.

RELEASES OF METHANE FROM THAWING PERMAFROST REGIONS

Thawing of permafrost would increase bacterial levels in the soil and eventually lead to higher releases of methane. Estimates of melting of permafrost peat bogs in Siberia could release as much as 70,000 million metric tons of methane (a greenhouse gas) within the next few decades.

RISE IN SEA LEVEL

Sea levels have risen 400 feet (120 m) since the peak of the last ice age (18,000 years ago). From 3,000 years ago to the start of the Industrial Revolution, the rate of sea level rise averaged 0.1 to 0.2 mm per year. Since 1900, sea level has risen about 3 mm per year (over a 10-fold increase). An increase in global temperatures of 3°F to 8°F (1.5°C to 4.5°C) is estimated to lead to an increase of 6 to 37 inches (15 cm to 95 cm) in sea level. If all glaciers, ice caps, and ice sheets melted on Earth, the sea level would rise about 225 feet (69 m). Rises in sea level would:

- Increase coastal erosion
- Create higher storm surge flooding with coastal inundation
- Increase loss of property and coastal habitats
- Cause losses in fish and shellfish catches
- Cause loss of cultural resources and values such as tourism and recreation
- Cause losses in agriculture and aquaculture due to diminishing soil and water quality
- Result in the intrusion of salt water in aquifers.

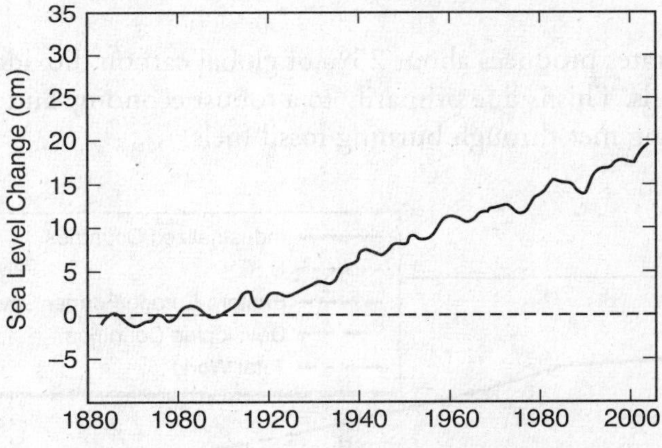

Figure 11.9 Sea level rise (1880–2000)

SLOWING OR SHUTDOWN OF THERMOHALINE CIRCULATION

Melting of the glaciers in Greenland would shift the salt water–freshwater balance in the North Atlantic. This would result in a decrease of heavier saline waters sinking than in traditional ocean circulation patterns. This would have significant effects on the fishing industry. Localized cooling in the North Atlantic brought about through the reduction of thermohaline circulation currents (North Atlantic drift) could result in much colder temperatures in Great Britain and Scandinavia.

Reducing Climate Change

World carbon dioxide emissions are expected to increase by 2% annually between 2001 and 2025. Much of the increase in these emissions is expected to occur in the developing world, such as China and India, where emerging economies fuel economic development with fossil energy. Developing countries' emissions are expected to grow by 3% annually between 2001 and 2025 and surpass emissions of industrialized countries by 2018.

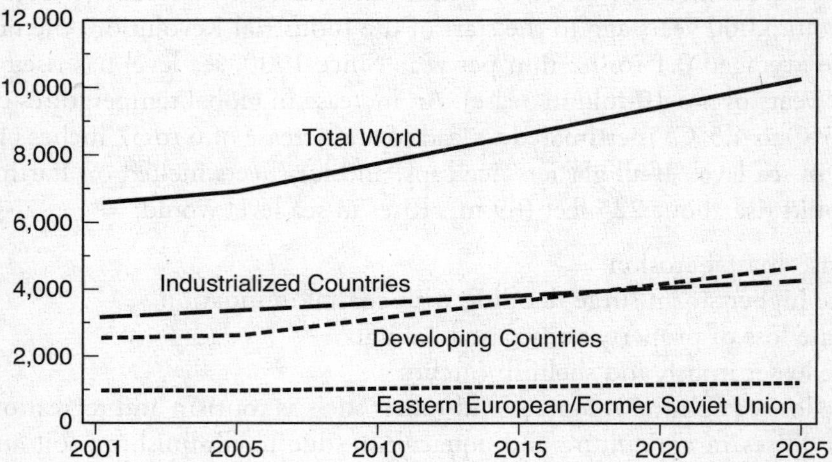

Figure 11.10 World carbon dioxide emissions by region, 2001–2025 (million metric tons of carbon equivalent)

Source: Energy Information Administration

The United States produces about 25% of global carbon dioxide emissions from burning fossil fuels. This is due primarily to a robust economy and 85% of the U.S. energy needs being met through burning fossil fuels.

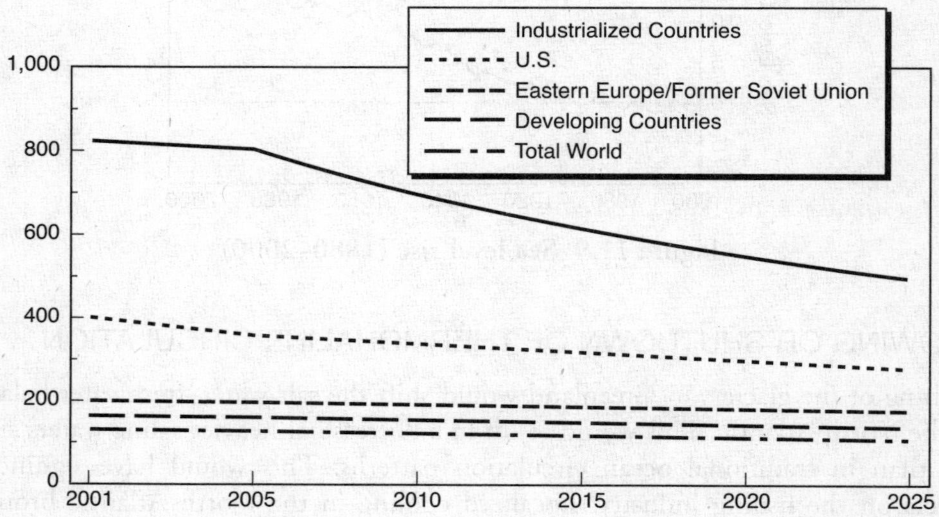

Figure 11.11 Carbon intensity by region, 2001–2025 (metric tons of carbon equivalent per million)

Source: Energy Information Administration

To stabilize the current global warming crisis would require:

1. A decrease in methane emissions by 8%
2. A decrease nitrous oxide emissions by 50%
3. A decrease in carbon dioxide emissions by up to 80%

Six methods can be used to reduce dramatic climatic changes brought on by global warming. The first is increasing the efficiency of cars, which would then reduce the dependence on oil and other fossil fuels. Cars currently represent 25% of the carbon dioxide emissions in the United States. The second method is to use energy more efficiently and move to renewable energy sources (wind, solar, geothermal, bioenergy, and so on). The third is finding chemical substitutes that do not impact global warming and banning chemicals that do. The fourth method is to slow down the rate of deforestation and encourage reforestation. The fifth is reducing dependence on inorganic, nitrogen-based fertilizers and substituting conservation tillage techniques. The sixth is to support treaties, protocols, and legislation that require a reduction in greenhouse gases.

LOSS OF BIODIVERSITY

Information on habitat destruction needs to be compared with past habitat conditions. Long-term history of habitat conditions can be obtained from fossil records, ice core samples, tree ring analysis, and the analysis of pollen. Organisms can cope with habitat destruction by migration, adaptation, and acclimatization. Migration depends on the magnitude of degradation, rate of degradation, organism's ability to migrate, access routes or corridors, and proximity and availability of suitable new habitats. Adaptation is the ability to survive due to changing environmental conditions. Adaptation depends on the magnitude of degradation, rate of degradation, birth rate, length of generation, population size, genetic variability, and gene flow between populations as a function of variation. Acclimatization is the ability to adjust to environmental changes on an individual or population level. Acclimatization depends on the magnitude and rate of degradation and the physiological and/or behavioral limitations of the species.

Plants are initially more susceptible to habitat loss than animals. This occurs for several reasons. Plants cannot migrate. Dispersal rates of seeds are slow events. For example, spruce trees can increase their range about 1 mile (1.6 km) every 100 years. Plants cannot seek nutrients or water. Seedlings must survive and grow in degraded conditions. Stressed plants become prone to disease and infestation.

FACTOIDS

- More than 50% of coastal wetlands (including mangrove swamps and salt marshes) have been lost worldwide.
- In the Philippines, 90% of the mangrove swamps have disappeared in the last 40 years.
- Of the world's beaches, 70% are eroding at a faster-than-natural rate due to coastal construction and mining.

Overuse

Worldwide, more than 3 million fishing boats remove 70 million to 90 million tons of fish and shellfish from the oceans each year. Of the world's marine fish stocks, 70% are fully exploited, overfished, or depleted. Twenty-million metric tons of bycatch (fish and shellfish not sought after) are destroyed each year and represents one-fourth of all catch taken from the sea. Around 80,000 whales, dolphins, seals, and other marine mammals perish as bycatch each year. Shrimp trawlers throw back 5.2 pounds (2.4 kg) of marine life that will eventually die for every pound (0.5 kg) of shrimp they catch. Overall, this amounts to almost 3 billion pounds (1.4 billion kg) of fish destroyed annually by the shrimp industry.

Pollution

Pollution results from human activities such as urban sprawl, transportation, and industry. More significantly in terms of wildlife and biodiversity, human activity expands the effects of pollution into areas where environmental quality may not yet have been severely compromised. Industrial and transportation sources release pollutants into the air, water, and soil that are toxic to most species and degrade their habitats. In addition, light and noise associated with human activity often have negative effects on wildlife behavior and natural cycles.

Pollutants have devastating effects on wildlife when they are released into the soil and water. They affect an animal's endocrine, reproductive, and immune systems. Wildlife can be exposed to pollutants through skin absorption, the animals or plants they eat, and the water they drink. Amphibians and other aquatic species are especially vulnerable to pollutants in streams, rivers, and lakes.

Plants or wildlife (especially species with thin, moist skin such as amphibians) can inhale or absorb air pollutants, causing health problems. Air pollution also contaminates water and soil through acid rain. Sulfur dioxide, a main component of acid rain, can damage plant communities by impairing photosynthesis processes. Plants weakened by acid rain may become vulnerable to root rot, insect damage, and disease, causing serious damage to plant communities and forest ecosystems.

Noise can affect wildlife behavior and physiology. Noise can also cause chronic stress, increased heart rates, metabolism and hormone imbalance, and nervous system stimulation. These responses may cause energy loss, food intake reduction, an avoidance and abandonment of habitat, and even injury or death. Noise will often flush birds from their nests, causing broken eggs, injured young, or exposure to predators. Finally, birdsongs and calls that are used to mate and establish territories may be disrupted by excessive human noise pollution.

Developed areas produce artificial light that causes problems for wildlife that depend on the light from the moon and stars to navigate. Streetlights, neon signs, parking lots, and late-night golf driving ranges may potentially keep wildlife from moving, nesting, or foraging near urban areas.

Introduced Species

Invasive or exotic species are animals and plants that are transported to any area where they do not naturally live. The spread of nonnative species has emerged in recent years as one of the most serious threats to biodiversity. It has undermined the ecological integrity of many native habitats and pushed some rare species to the edge of extinction. Some introduced species simply outcompete native plants and animals for space, food, or water. Others may also fundamentally alter natural disturbance regimes and other ecological processes, making it difficult or impossible for native species to survive. About 15% of the estimated 6,000 nonnative plant and animal species in the United States cause severe economic or ecological impacts. Exotic species have been implicated in the decline of approximately 40% of the species listed for protection under the federal Endangered Species Act.

A major marine source is marine ballast. Every 60 seconds, ships discharge 40,000 gallons (150,000 L) of ballast water that contains foreign plant and animal species into U.S. harbors. Often these plants and animals grow at uncontrolled rates because they have no natural predators in these new locations. These introduced species often become pests and crowd out native plants and animals. Over 250 invasive species of plants and animals are found in San Francisco Bay alone. Common examples of invasive species include zebra mussels, tumbleweed, kudzu, the mongoose, and gypsy moths.

FACTOIDS

- More than 200 species of plants and 71 species of vertebrates have gone extinct in North America since European settlement.
- More than 3,000 species await listing under the federal Endangered Species Act. Since the act was enacted in 1973, only a handful of species have recovered to the point of being completely removed from the federal endangered species list.
- A third of all native freshwater fish species in the United States are threatened or endangered, as are two-thirds of all freshwater mussels and nearly three-quarters of freshwater crayfish.
- Among anadromous fish, 214 salmon and steelhead fish stocks in the Pacific Northwest are now threatened and 101 of these are near extinction. Eastern anadromous fish species (the Atlantic salmon and shad) are gone from most of their original spawning grounds because of dams, development, and pollution.
- Nearly a third of North America's 86 species of frogs and toads may be endangered or extinct.
- Grizzly bears, wolves, cougars, and other large carnivores have virtually disappeared from many areas where they were formerly abundant.
- Some of the most endangered ecosystems in the United States include Southern California coastal sage scrub, southeastern longleaf pine forests and savannas, midwestern tall grass prairies, Hawaiian dry forests, old-growth forests of the Pacific Northwest, and midwestern wetlands.

Endangered and Extinct Species

Since 1500 C.E., 816 species have become extinct, 103 of them since 1800—a rate 50 times greater than the natural background rate. In the next 25 years, extinction rates are expected to rise as high as 25%. According to the Nature Conservancy, about one-third of all U.S. plant and animal species are at risk of becoming extinct. Mammals listed as "endangered" rose from 484 in 1996 to 520 in 2000, with primates increasing from 13 to 19. Endangered birds increased from 403 species to 503 species. Endangered freshwater fish more than doubled—from 10 species to 24 species in four years. One-fourth of all mammals and reptiles, one-fifth of all amphibians, one-eighth of all birds, and one-sixth of all conifers are in some manner endangered or are threatened to the point of extinction.

Maintenance Through Conservation

Maintaining and protecting wildlife consists of three major approaches:

- **Species approach**—Protecting *endangered species* through legislation.
- **Ecosystem approach**—Preserving balanced *ecosystems*.
- **Wildlife management approach**—Managing *game species* for sustained yield through international treaties to protect migrating species, improving wildlife habitats, regulating hunting and fishing, creating harvest quotas, and developing population management plants.

Biodiversity can be protected by:

1. Properly designing and updating laws that legally protect endangered and threatened species (such as the Endangered Species Act)
2. Protecting the habitats of endangered species through private or governmental land trusts
3. Reintroducing species into suitable habitats
4. Managing habitats and monitoring land use
5. Establishing breeding programs for endangered or threatened species
6. Creating and expanding wildlife sanctuaries
7. Restoring compromised ecosystems
8. Reducing nonnative and invasive species

RELEVANT LAWS AND TREATIES

Lacey Act (1900): Aided in restoring birds in parts of the United States where they had become scarce.

Migratory Bird Treaty Act (1918): Implemented the protection of migratory birds.

Multiple-Use Act (1960): Directed that the national forests be managed for timber, watershed, range, outdoor recreation, wildlife, and fish purposes.

The Convention on International Trade in Endangered Species of Wild Fauna and Flora (CITES) (1963): An international agreement between governments to ensure that international trade in wild animals and plants did not threaten their survival.

Fur Seal Act (1966): Prohibited the taking of fur seals.

National Wildlife Refuge System Act (1966): Provided for the administration and management of the National Wildlife Refuge System.

Marine Mammal Protection Act (1972): Established federal responsibility to conserve marine mammals.

Marine Protection and Sanctuaries Act (1972): Established national marine sanctuaries.

Endangered Species Act (1973): Provided a program for the conservation of threatened and endangered plants and animals and the habitats in which they are found.

Fishery Conservation and Management Act—Magnuson-Stevens Fisher Act (1976): Established a 200-mile (320-km) zone off the coast of the United States that is off-limits for foreign fishing.

Alaska National Interest Lands Conservation Act (1980): Designated large areas of Alaska as national parks, wildlife refuges, and national forests.

Convention for the Conservation of Antarctic Marine Living Resources (1980): A treaty that required management of all southern ocean fisheries that have potential negative effects on the Antarctic ecosystem.

Nonindigenous Aquatic Nuisance Prevention and Control Act (1990): Federal program to prevent introduction of and to control the spread of aquatic invasive species.

MULTIPLE-CHOICE QUESTIONS

1. Which of the following represents the greatest contribution of methane emission to the atmosphere?

   (A) Enteric fermentation or flatulence from animals
   (B) Coal mining
   (C) Landfills
   (D) Rice cultivation
   (E) Burning of biomass

2. Ozone depletion reactions that occur in the stratosphere are facilitated by

   (A) NO_2^-
   (B) NH_3
   (C) NH_4^+
   (D) N_2O
   (E) NO_3^-

3. In addition to absorbing harmful solar rays, how do ozone molecules help to stabilize the upper atmosphere?

   (A) They release heat to the surroundings.
   (B) They create a buoyant lid on the atmosphere.
   (C) They create a warm layer of atmosphere that keeps the lower atmosphere from mixing with space.
   (D) All of the above are correct.
   (E) None of the above are correct.

4. The bromine released from methyl bromide is about 40 times more damaging to the ozone layer than chlorine on a molecule of chlorofluorocarbon. Under the Montreal Protocol, methyl bromide in developing countries will be phased out in 2005. What is methyl bromide used for?

   (A) Sterilizing soil in fields and greenhouses
   (B) Killing pests on fruits, vegetables, and grain before export
   (C) Fumigating soils
   (D) Killing termites in buildings
   (E) All of the above

5. Rising sea levels due to global warming would be responsible for all of the following EXCEPT

   (A) destruction of coastal wetlands
   (B) beach erosion
   (C) increased damage due to storms and floods
   (D) increased salinity of estuaries and aquifers
   (E) all of the above

6. The Intergovernmental Panel on Climate Change (IPCC) is a United Nations-sponsored group of 2,500 scientists around the world and serves as a resource and authority on global warming issues. What percentage of the world's population does this organization believe will be at risk of infectious disease due to global warming?

   (A) 0%
   (B) 20%
   (C) 40%
   (D) 65%
   (E) 100%

7. One of the reasons the vortex winds in the Antarctic are important in the formation of the hole in the ozone layer is

   (A) they prevent warm, ozone-rich air from mixing with cold, ozone-depleted air
   (B) they quickly mix ozone-depleted air with ozone-rich air
   (C) they harbor vast quantities of N_2O
   (D) they bring in moist, warm air, which accelerates the ozone-forming process
   (E) none of the above

8. Although the Montreal Protocol curtailed production of ozone-depleting substances, the peak concentration of chemicals in the stratosphere is only now being reached. When do scientists expect a recovery in the ozone levels in the stratosphere to occur?

   (A) Immediately
   (B) Within 5 years
   (C) Within 20 years
   (D) Within 50 years
   (E) In about 100 years

9. Over the past 1 million years of Earth's history, the average trend in Earth's temperature would best be described as:

   (A) very large fluctuations, from periods of extreme heat to periods of extreme cold
   (B) very little change, if any, fairly uniform and constant temperature
   (C) a general cooling trend
   (D) a general warming trend
   (E) a number of fluctuations varying by a few degrees

10. The agency responsible for the identification and listing of endangered species is the

    (A) Forest Service
    (B) Agriculture Department
    (C) National Park Service
    (D) Fish and Wildlife Service
    (E) Endangered Species Department of the Interior

11. A treaty that controls international trades in endangered species is known as (the)

    (A) Endangered Species Act
    (B) CITES
    (C) International Treaty on Endangered Species
    (D) Lacey Act
    (E) Federal Preserve System

12. Managing game species for sustained yields would be consistent with what conservation approach?

    (A) Wildlife management approach
    (B) Species approach
    (C) Ecosystem approach
    (D) Sustainable yield approach
    (E) Holistic approach

13. A certain insect was causing extensive damage to local crops. The farmers, who were environmentally conscious and did not want to use pesticides, decided to introduce another insect into their fields that studies have shown would prey on the pest insect. Before any action is taken, the farmers should consider (the)

    (A) principle of natural balance
    (B) principle of unforeseen events
    (C) Murphy's law
    (D) law of supply and demand
    (E) precautionary principle

14. Which of the following is the biggest threat to wildlife preserves?

    (A) Hunters
    (B) Poachers
    (C) Global warming
    (D) Invasive species
    (E) Tourists

15. A certain species of plant is placed on the threatened species list. Several years later it is placed on the endangered species list. This is an example of

    (A) a negative-negative feedback loop
    (B) a positive-negative feedback loop
    (C) a negative-positive feedback loop
    (D) a negative-feedback loop
    (E) a positive-feedback loop

16. The specific form of radiation largely responsible for the formation of ozone in the stratosphere is

    (A) UVA
    (B) UVB
    (C) UVC
    (D) infrared
    (E) gamma rays

17. Which of the following is NOT an effect caused by increased levels of ultraviolet radiation reaching Earth?

    (A) Warming of the stratosphere
    (B) Cooling of the stratosphere
    (C) Increased genetic damage
    (D) Reduction in immunity
    (E) Reduction in plant productivity

18. Which of the following gases is NOT considered a greenhouse gas, responsible for global warming?

    (A) SO_2
    (B) N_2O
    (C) H_2O
    (D) SF_6
    (E) CH_4

19. Which of the following effects would result from the shutdown or slowdown of the thermohaline circulation pattern?

    (A) A warmer Scandinavia and Great Britain
    (B) Freshwater fish moving into the open ocean
    (C) A colder Scandinavia and Great Britain
    (D) A saltier ocean
    (E) A drier Scandinavia and Great Britain

20. The largest number of species are being exterminated per year in

    (A) grasslands
    (B) deserts
    (C) forests
    (D) tropical rain forests
    (E) the tundra

FREE-RESPONSE QUESTION

Part (a): 3 points
Part (b): 4 points
Part (c): 3 points

Total: 10 points

The following charts show global temperature changes and carbon dioxide emissions.

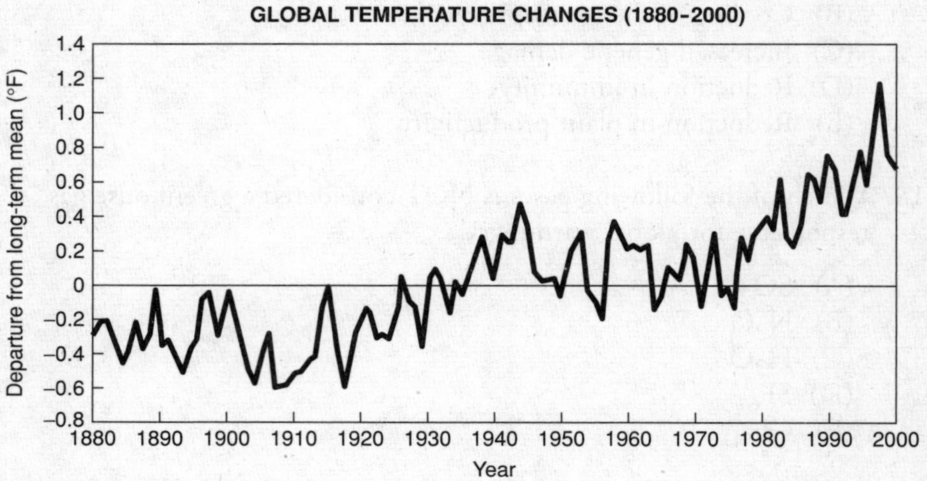

Source: U.S. National Climatic Data Center, 2001

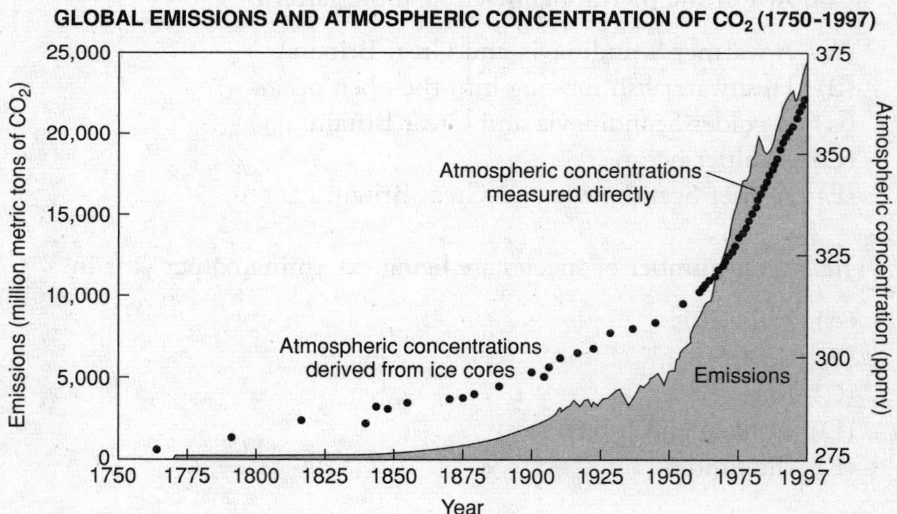

Source: Carbon Dioxide Information Analysis Center, 2001

(a) Describe global warming.
(b) What factors may be increasing global temperatures?
(c) What are the environmental consequences of global warming?

MULTIPLE-CHOICE ANSWERS AND EXPLANATIONS

1. (C) Since 1800, there has been a 150% increase in methane concentrations in the atmosphere compared with a 30% increase in carbon dioxide. Methane is about 30 times more effective in contributing to global warming than carbon dioxide.
2. (D) Nitrous oxide is responsible for about 7% of the anthropogenic greenhouse gases and stays in the atmosphere for about 120 years. The concentration of nitrous oxide in the atmosphere has increased about 15% since the Industrial Revolution. It has an ozone-depleting potential comparable with many HCFCs. The main source of nitrous oxide are inorganic fertilizers and industrial processes. Natural sources of nitrous oxide arise from biological processes within the soil and oceans.
3. (D) Solar energy that is absorbed by ozone molecules and is partly turned into heat creates a warm region in the stratosphere. This creates a stable air mass that resists sinking and mixing with the lower atmosphere, effectively forming a barrier. In the ozone layer, temperature increases with height, creating a stable and buoyant air mass that keeps an effective lid on the lower atmosphere.
4. (E) No explanation needed.
5. (E) No explanation needed.
6. (D) No explanation needed.
7. (A) Ozone depletion follows an annual cycle that corresponds to the amount of light that reaches the Antarctic. The cycle begins every year around June when vortex winds develop in the Antarctic. Cold temperatures produced by these winds create polar stratospheric clouds that capture floating chlorofluorocarbons (CDCs) and other ozone-depleting compounds. For the next two months, a reaction occurs on the cloud surface that frees the chlorine in the CFCs but keeps it contained within the vortex. In September, sunlight returns to the Antarctic and triggers a chemical reaction, causing chlorine to convert ozone to oxygen gas. November brings a breakdown in the vortex and allows the ozone-rich air to combine with the thinning ozone. Wind currents carry this mixture over the southern hemisphere.
8. (E) Concentrations of ozone-depleting chlorofluorocarbons (CFCs) have leveled off in the stratosphere and have actually begun to decline in the lower atmosphere. However, the largest Antarctic ozone hole ever recorded occurred in 2000. A CFC molecule can take about 8 years after being released at ground level to reach the stratosphere. Decades may pass until it is converted by sunlight into a form that depletes ozone.
9. (E) The mean surface temperature of the planet has been warming at a rate of 0.2°C per decade for the past 30 years. The global mean temperature is now within 1°C of the maximum for the past million years. The last time Earth was this warm was about 3 million years ago when sea level was about 25 meters higher than today. If carbon dioxide emissions are not curbed, global temperatures are likely to rise 2 to 3 °C by 2100.

10. (D) The Fish and Wildlife Service, in the Department of the Interior, and the National Oceanic and Atmospheric Administration (NOAA) Fisheries, in the Department of Commerce, share responsibility for administration of the Endangered Species Act. An endangered species is one that is in danger of extinction throughout all or a significant portion of its range. A threatened species is one that is likely to become endangered in the foreseeable future.
11. (B) CITES (the Convention on International Trade in Endangered Species of Wild Fauna and Flora) is an international agreement between governments. Its aim is to ensure that international trade in specimens of wild animals and plants does not threaten their survival.
12. (A) The strength of the traditional wildlife management approach is that it explicitly uses and enhances natural processes to perpetuate populations.
13. (E) The precautionary principle states that if the consequences of an action are unknown but are judged to have some potential for major or irreversible negative consequences, that action should be avoided. The concept includes risk prevention, cost effectiveness, ethical responsibilities toward maintaining the integrity of natural biological systems, and risk assessment. It is not the risk that must be avoided but the potential risk that must be prevented.
14. (C) All choices but (C) can be dealt with on a local level through specific enforcement and management practices. Global warming is an issue that affects all ecosystems and must be solved on a global scale, requiring international cooperation.
15. (E) A positive-feedback loop occurs in a situation in which a change in a certain direction (toward extinction in this case) causes the system to change in the same direction.
16. (C) Ozone is produced by oxygen and sunlight in the UVC wavelength range (< 240 nm). In the atmosphere, this reaction works only at higher altitudes where there is adequate high-energy UV penetration (>10 km). The atomic oxygen so formed can combine with oxygen gas (O_2) to form ozone (O_3).
17. (A) The stratosphere has been cooling over the past three decades. The stratosphere contains the ozone layer, which absorbs sunlight and heats the stratosphere. This long-term cooling trend is generally accepted to result from the loss of the ozone layer as a result of human-made influences. However, the cooling trend is not uniform like ozone loss but, rather, broken into a series of jumps or discontinuities most likely associated with major volcanic eruptions that inject aerosols into the stratosphere. The aerosols also absorb sunlight and heat from the stratosphere, thus temporarily offsetting the cooling trend from ozone loss.
18. (A) Sulfur dioxide (SO_2) contributes to acid rain, not global warming.
19. (C) There is speculation that global warming could melt glaciers in Greenland, increasing the amount of freshwater in the North Sea. This disruption in balance between salt water and freshwater could theoretically slow down or shut down thermohaline circulation that is responsible for the North Atlantic Drift (a section of the Gulf Stream) that currently stabilizes temperatures in Great Britain and Scandinavia.
20. (D) Deforestation in the tropics in order to convert the land to agricultural purposes and cattle grazing is occurring at unprecedented levels. The tropics contain the highest biodiversity anywhere on Earth.

FREE-RESPONSE ANSWER

(a) Energy from the sun drives Earth's weather and climate and heats Earth's surface. In turn, Earth radiates energy back into space. Natural atmospheric greenhouse gases such as water vapor, carbon dioxide, methane, and nitrous oxide trap some of the outgoing energy similar to the glass panels of a greenhouse. Without the greenhouse effect, much of the heat reaching Earth would escape into space. Life as we know it, could not exist. However, Earth's climate is predicted to change due to human activities that have caused these natural gas concentrations to increase dramatically. According to the graph, since the beginning of the Industrial Revolution, atmospheric concentrations of carbon dioxide have increased nearly 30%. In addition, methane concentrations have more than doubled, and nitrous oxide concentrations have increased about 15%. Not all gases contribute to global warming equally. As a reference point, if the global warming potential of carbon dioxide gas is given as a 1, then methane is 20 times more effective in trapping heat energy, and nitrous oxide is over 300 times more effective.

Notable points:

Connection with sun as energy driver.

Energy hitting Earth from the sun is absorbed and reradiated back into space.

Some gases prevent energy from radiating back into space.

Not all gases trap heat equally.

Greenhouse effect is necessary for life on Earth.

Both natural and human-made gases cause the greenhouse effect.

Carbon dioxide is the gas most responsible for the greenhouse effect.

Increase in production of CO_2 is primary factor in greenhouse effect.

How CO_2 is produced.

Comparison of CO_2 levels today as compared with years past.

Mentioning one other greenhouse gas such as methane, water vapor, sulfur hexafluoride, or nitrous oxide.

(b) Scientists generally believe that the combustion of fossil fuels and other human activities are the primary reason for the increased concentration of carbon dioxide. Plant respiration and the decomposition of organic matter release more than 10 times the carbon dioxide released through human activities. However, these releases have generally been in balance during the centuries leading up to the Industrial Revolution, with carbon dioxide being absorbed by plants both terrestrial and oceanic. What has changed in the last few hundred years is the additional release of carbon dioxide by human activities. Fossil fuels burned to run cars and trucks, heat homes and businesses, and power factories are responsible for about 98% of U.S. carbon dioxide emissions, 24% of methane emissions, and 18% of nitrous oxide emissions. Increased agriculture, deforestation, landfills, industrial production, and mining also contribute a significant share of emissions.

In 1997, the United States emitted about one-fifth of total global greenhouse gases. By 2100, in the absence of emissions control policies, carbon dioxide concentrations are projected to be 30%–150% higher than today's levels. Water vapor is the most abundant greenhouse gas. It occurs naturally and makes up about two-thirds of the natural greenhouse effect. Other gases that accelerate the greenhouse effect include nitrous oxide, hydrofluorocarbons, perfluorocarbons, and sulfur hexafluoride. Since preindustrial times, atmospheric concentrations of nitrous oxide have increased by 17%, carbon dioxide by 31%, and methane by 151%. Scientists have confirmed that this is primarily due to human activity.

(c) According to the graph, global mean surface temperatures have increased 0.5°F to 1.0°F since the late 1800s. The 20th century's 10 warmest years all occurred in the last 15 years. The snow cover in the Northern Hemisphere and floating ice in the Arctic Ocean have decreased. Globally, sea level has risen 4 to 8 inches in the last 100 years. Worldwide precipitation over land has increased by about 1% (warmer temperatures usually cause greater amounts of water to evaporate). The frequency of extreme rainfall events has increased throughout much of the United States. Increasing concentrations of greenhouse gases are likely to accelerate the rate of climatic change. Scientists expect that the average global surface temperature could rise 2°F to 10°F in the next century, with significant regional variation. Evaporation will increase as the climate warms, which will increase average global precipitation. Soil moisture is likely to decline in many regions, and intense rainstorms are likely to become more frequent. Sea levels are likely to rise 2 feet along most of the U.S. coasts. A few degrees of warming will increase the chances of more frequent and severe heat waves, which can cause more heat-related death and illness. Greater heat results in increased air pollution as well as damaged crops and depleted water resources. Warming is likely to allow tropical diseases, such as malaria, to spread northward in some areas of the world. It will also intensify Earth's hydrological cycle. Both evaporation and precipitation will increase. Some areas will receive more rain, while other areas will be drier. At the same time, extreme events like floods and droughts are likely to become more frequent. Warming will cause glaciers to melt and oceans to expand. Projections state that sea levels will rise between 4 inches and 3 feet over the next century, threatening low-lying coastal areas.

—Decreased snow cover

—Floating ice

—Rise in sea level

—Increased precipitation due to warmer temperatures

—Increased surface temperatures

—Decreased soil moisture in some areas

—Increase in storm intensities

—Increase in rates of disease

—Increase in heatstroke and other heat-associated effects

—Increased flooding in some areas

—Increased droughts in some areas

PRACTICE EXAMS

Practice Exam 1

With Answers and Analysis

SECTION I (MULTIPLE-CHOICE QUESTIONS)

Time: 90 minutes

100 questions

60% of total grade

No calculators allowed

This section consists of 100 multiple-choice questions. Mark your answers carefully on the answer sheet.

General Instructions

Do not open this booklet until you are told to do so by the proctor.

Be sure to write your answers for Section I on the separate answer sheet. Use the test booklet for your scratch work or notes. Remember, though, that no credit will be given for work, notes, or answers written only in the test booklet. Once you have selected an answer, thoroughly blacken the corresponding circle on the answer sheet. To change an answer, erase your previous mark completely, and then record your new answer. Mark only one answer for each question.

Example	**Sample Answer**
The Pacific is	Ⓐ Ⓑ ● Ⓓ Ⓔ

(A) a river
(B) a lake
(C) an ocean
(D) a sea
(E) a gulf

To discourage haphazard guessing on this section of the exam, a quarter point is subtracted for every wrong answer, but no points are subtracted if you leave the answer blank. Even so, if you can eliminate one or more of the choices for a question, it may be to your advantage to guess.

Because it is not expected that all test takers will complete this section, do not spend too much time on difficult questions. First answer the questions you can answer readily. Then, if you have time, return to the difficult questions later. Do not get stuck on one question. Work quickly but accurately. Use your time effectively.

Answer Sheet 1

PRACTICE EXAM 1

1 Ⓐ Ⓑ Ⓒ Ⓓ Ⓔ
2 Ⓐ Ⓑ Ⓒ Ⓓ Ⓔ
3 Ⓐ Ⓑ Ⓒ Ⓓ Ⓔ
4 Ⓐ Ⓑ Ⓒ Ⓓ Ⓔ
5 Ⓐ Ⓑ Ⓒ Ⓓ Ⓔ
6 Ⓐ Ⓑ Ⓒ Ⓓ Ⓔ
7 Ⓐ Ⓑ Ⓒ Ⓓ Ⓔ
8 Ⓐ Ⓑ Ⓒ Ⓓ Ⓔ
9 Ⓐ Ⓑ Ⓒ Ⓓ Ⓔ
10 Ⓐ Ⓑ Ⓒ Ⓓ Ⓔ
11 Ⓐ Ⓑ Ⓒ Ⓓ Ⓔ
12 Ⓐ Ⓑ Ⓒ Ⓓ Ⓔ
13 Ⓐ Ⓑ Ⓒ Ⓓ Ⓔ
14 Ⓐ Ⓑ Ⓒ Ⓓ Ⓔ
15 Ⓐ Ⓑ Ⓒ Ⓓ Ⓔ
16 Ⓐ Ⓑ Ⓒ Ⓓ Ⓔ
17 Ⓐ Ⓑ Ⓒ Ⓓ Ⓔ
18 Ⓐ Ⓑ Ⓒ Ⓓ Ⓔ
19 Ⓐ Ⓑ Ⓒ Ⓓ Ⓔ
20 Ⓐ Ⓑ Ⓒ Ⓓ Ⓔ
21 Ⓐ Ⓑ Ⓒ Ⓓ Ⓔ
22 Ⓐ Ⓑ Ⓒ Ⓓ Ⓔ
23 Ⓐ Ⓑ Ⓒ Ⓓ Ⓔ
24 Ⓐ Ⓑ Ⓒ Ⓓ Ⓔ
25 Ⓐ Ⓑ Ⓒ Ⓓ Ⓔ
26 Ⓐ Ⓑ Ⓒ Ⓓ Ⓔ
27 Ⓐ Ⓑ Ⓒ Ⓓ Ⓔ
28 Ⓐ Ⓑ Ⓒ Ⓓ Ⓔ
29 Ⓐ Ⓑ Ⓒ Ⓓ Ⓔ
30 Ⓐ Ⓑ Ⓒ Ⓓ Ⓔ
31 Ⓐ Ⓑ Ⓒ Ⓓ Ⓔ
32 Ⓐ Ⓑ Ⓒ Ⓓ Ⓔ
33 Ⓐ Ⓑ Ⓒ Ⓓ Ⓔ
34 Ⓐ Ⓑ Ⓒ Ⓓ Ⓔ
35 Ⓐ Ⓑ Ⓒ Ⓓ Ⓔ
36 Ⓐ Ⓑ Ⓒ Ⓓ Ⓔ
37 Ⓐ Ⓑ Ⓒ Ⓓ Ⓔ
38 Ⓐ Ⓑ Ⓒ Ⓓ Ⓔ
39 Ⓐ Ⓑ Ⓒ Ⓓ Ⓔ
40 Ⓐ Ⓑ Ⓒ Ⓓ Ⓔ
41 Ⓐ Ⓑ Ⓒ Ⓓ Ⓔ
42 Ⓐ Ⓑ Ⓒ Ⓓ Ⓔ
43 Ⓐ Ⓑ Ⓒ Ⓓ Ⓔ
44 Ⓐ Ⓑ Ⓒ Ⓓ Ⓔ
45 Ⓐ Ⓑ Ⓒ Ⓓ Ⓔ
46 Ⓐ Ⓑ Ⓒ Ⓓ Ⓔ
47 Ⓐ Ⓑ Ⓒ Ⓓ Ⓔ
48 Ⓐ Ⓑ Ⓒ Ⓓ Ⓔ
49 Ⓐ Ⓑ Ⓒ Ⓓ Ⓔ
50 Ⓐ Ⓑ Ⓒ Ⓓ Ⓔ
51 Ⓐ Ⓑ Ⓒ Ⓓ Ⓔ
52 Ⓐ Ⓑ Ⓒ Ⓓ Ⓔ
53 Ⓐ Ⓑ Ⓒ Ⓓ Ⓔ
54 Ⓐ Ⓑ Ⓒ Ⓓ Ⓔ
55 Ⓐ Ⓑ Ⓒ Ⓓ Ⓔ
56 Ⓐ Ⓑ Ⓒ Ⓓ Ⓔ
57 Ⓐ Ⓑ Ⓒ Ⓓ Ⓔ
58 Ⓐ Ⓑ Ⓒ Ⓓ Ⓔ
59 Ⓐ Ⓑ Ⓒ Ⓓ Ⓔ
60 Ⓐ Ⓑ Ⓒ Ⓓ Ⓔ
61 Ⓐ Ⓑ Ⓒ Ⓓ Ⓔ
62 Ⓐ Ⓑ Ⓒ Ⓓ Ⓔ
63 Ⓐ Ⓑ Ⓒ Ⓓ Ⓔ
64 Ⓐ Ⓑ Ⓒ Ⓓ Ⓔ
65 Ⓐ Ⓑ Ⓒ Ⓓ Ⓔ
66 Ⓐ Ⓑ Ⓒ Ⓓ Ⓔ
67 Ⓐ Ⓑ Ⓒ Ⓓ Ⓔ
68 Ⓐ Ⓑ Ⓒ Ⓓ Ⓔ
69 Ⓐ Ⓑ Ⓒ Ⓓ Ⓔ
70 Ⓐ Ⓑ Ⓒ Ⓓ Ⓔ
71 Ⓐ Ⓑ Ⓒ Ⓓ Ⓔ
72 Ⓐ Ⓑ Ⓒ Ⓓ Ⓔ
73 Ⓐ Ⓑ Ⓒ Ⓓ Ⓔ
74 Ⓐ Ⓑ Ⓒ Ⓓ Ⓔ
75 Ⓐ Ⓑ Ⓒ Ⓓ Ⓔ
76 Ⓐ Ⓑ Ⓒ Ⓓ Ⓔ
77 Ⓐ Ⓑ Ⓒ Ⓓ Ⓔ
78 Ⓐ Ⓑ Ⓒ Ⓓ Ⓔ
79 Ⓐ Ⓑ Ⓒ Ⓓ Ⓔ
80 Ⓐ Ⓑ Ⓒ Ⓓ Ⓔ
81 Ⓐ Ⓑ Ⓒ Ⓓ Ⓔ
82 Ⓐ Ⓑ Ⓒ Ⓓ Ⓔ
83 Ⓐ Ⓑ Ⓒ Ⓓ Ⓔ
84 Ⓐ Ⓑ Ⓒ Ⓓ Ⓔ
85 Ⓐ Ⓑ Ⓒ Ⓓ Ⓔ
86 Ⓐ Ⓑ Ⓒ Ⓓ Ⓔ
87 Ⓐ Ⓑ Ⓒ Ⓓ Ⓔ
88 Ⓐ Ⓑ Ⓒ Ⓓ Ⓔ
89 Ⓐ Ⓑ Ⓒ Ⓓ Ⓔ
90 Ⓐ Ⓑ Ⓒ Ⓓ Ⓔ
91 Ⓐ Ⓑ Ⓒ Ⓓ Ⓔ
92 Ⓐ Ⓑ Ⓒ Ⓓ Ⓔ
93 Ⓐ Ⓑ Ⓒ Ⓓ Ⓔ
94 Ⓐ Ⓑ Ⓒ Ⓓ Ⓔ
95 Ⓐ Ⓑ Ⓒ Ⓓ Ⓔ
96 Ⓐ Ⓑ Ⓒ Ⓓ Ⓔ
97 Ⓐ Ⓑ Ⓒ Ⓓ Ⓔ
98 Ⓐ Ⓑ Ⓒ Ⓓ Ⓔ
99 Ⓐ Ⓑ Ⓒ Ⓓ Ⓔ
100 Ⓐ Ⓑ Ⓒ Ⓓ Ⓔ

Directions: For each question or statement, select the one lettered choice that is the best answer and fill in the corresponding circle on the answer sheet.

1. Which of the following would be most likely to increase competition among the members of a squirrel population in a given area?

   (A) An epidemic of rabies within the squirrel population
   (B) An increase in the number of hawk predators
   (C) An increase in the reproduction of squirrels
   (D) An increase in temperature
   (E) An increase in the food supply

2. According to the most recent data, approximately how many years ago did life first appear on Earth?

   (A) 6 billion years ago
   (B) 4.6 billion years ago
   (C) 3.5 billion years ago
   (D) 1 billion years ago
   (E) 1 million years ago

3. Which one of the following statements is FALSE?

   (A) The greenhouse effect is a natural process that makes life on Earth possible with 98% of total global greenhouse gas emissions being from natural sources (mostly water vapor) and 2% from human-made sources.
   (B) The United States is the number one contributor to global warming.
   (C) The Kyoto Protocol would have allowed the United States to increase its greenhouse gas emissions—primarily carbon dioxide (CO_2), methane (CH_4), and nitrous oxide (N_2O)—by only 2% per year based on 1990 levels.
   (D) The effects of global warming on weather patterns may lead to adverse human health impacts.
   (E) Global warming may increase the incidence of many infectious diseases.

4. Which of the following would be an external cost?

   (A) The cost of steel in making a refrigerator
   (B) The cost of running a refrigerator for one month
   (C) The cost of labor in producing refrigerators
   (D) The taxes paid by consumers in purchasing refrigerators
   (E) The costs associated with health care when the refrigerator leaks refrigerant into the atmosphere

5. The tick bird–rhinoceros association is an example of

(A) facultative commensalism
(B) obligatory commensalism
(C) obligatory mutualism
(D) facultative mutualism
(E) obligatory parasitism

6. Which of the following is NOT an example of environmental mitigation?

(A) Promoting sound land use planning, based on known hazards
(B) Relocating or elevating structures out of the floodplain
(C) Constructing living snow fences
(D) Organizing a beach cleanup
(E) Developing, adopting, and enforcing effective building codes and standards

7. The upthrusting of a mountain barrier

(A) is an instance of diastrophism
(B) is an example of gradation
(C) stimulates desert formation of the ocean
(D) has no effect on the local climate
(E) has never occurred

8. Multicellular organisms began to appear on land approximately how many years ago?

(A) 4.5 billion years ago
(B) 3.5 billion years ago
(C) 1.5 billion years ago
(D) 700 million years ago
(E) 400 million years ago

9. Which theory explains evolution as a result of long, stable periods of time interrupted by geologically brief periods of major change?

(A) Punctuated equilibrium
(B) Gradualism
(C) Natural selection
(D) Reproductive isolation
(E) Use and disuse

10. Rising sea levels due to global warming would be responsible for all of the following EXCEPT

(A) destruction of coastal wetlands
(B) beach erosion
(C) increases due to storm and flood damage
(D) increased salinity of estuaries and aquifers
(E) all would be the result of rising sea level

11. Branching evolution, producing many branching lines of descent from a single ancestral source, is called

    (A) divergent evolution
    (B) convergent evolution
    (C) straight-line evolution
    (D) macroevolution
    (E) adaptive evolution

12. Place the following economic activities in order, starting with those activities closest to natural resources and ending with those furthest away.

    I. Use raw materials to produce or manufacture something new and more valuable.
    II. Professions that process, administer, and disseminate information (e.g., computer engineers, professors, and lawyers).
    III. Agriculture, fishing, hunting, herding, forestry, and mining.
    IV. Include all activities that amount to doing service for others (e.g., doctors, teachers, and secretaries).

    (A) I, II, III, IV
    (B) I, III, IV, II
    (C) III, I, II, IV
    (D) III, I, IV, II
    (E) II, IV, I, III

13. The first heterotrophs that evolved on the primitive Earth must have been able to obtain energy from

    (A) enzymes present in the environment
    (B) radiant energy of the sun
    (C) inorganic compounds in the atmosphere
    (D) organic compounds present in the environment
    (E) molecules of ATP in the environment

14. Most of Earth's mass is found in this region, which is composed of iron, magnesium, aluminum, and silicon-oxygen compounds. At over 1,800°F (1,000°C), most of this region is solid but the upper third is more plasticlike in nature. Which region is being described?

    (A) Troposphere
    (B) Lithosphere
    (C) Crust
    (D) Mantle
    (E) Core

15. Which of the following statements regarding coral reefs is FALSE?

    (A) Modern reefs can be as much as 2.5 million years old.
    (B) Coral reefs capture about half of all the calcium flowing into the ocean every year, fixing it into calcium carbonate rock at very high rates.
    (C) Coral reefs store large amounts organic carbon and are very effective sinks for carbon dioxide from the atmosphere.
    (D) Coral reefs are among the most biologically diverse ecosystems on the planet.
    (E) Coral reefs are among the most endangered ecosystems on Earth.

16. Which of the following strategies to control pollution would incur the greatest governmental cost?

    (A) Green taxes
    (B) Government subsidies for reducing pollution
    (C) Regulation
    (D) Charging a user fee
    (E) Tradable pollution rights

17. A gene pool is

    (A) the number of recessive genes available in an ecosystem
    (B) the sum total of all genes of an individual
    (C) the genes that are available to a community
    (D) the sum total of all the genes of all the individuals in the population
    (E) the sum total of all the mutated genes of the various species that share an ecosystem

18. In 1989, the *Exxon Valdez* spilled 10.8 million gallons of crude oil into Prince William Sound in Alaska. What happened to most of the oil?

    (A) It was cleaned up by Exxon.
    (B) It eventually evaporated into the air.
    (C) It sank into the ground.
    (D) It biodegraded and photolyzed.
    (E) It dispersed into the water column.

19. Which of the following contributes LEAST to speciation?

    (A) Sexual reproduction
    (B) Asexual reproduction
    (C) Selection
    (D) Variation
    (E) Isolation

20. Which act's primary goal is to protect human health and the environment from the potential hazards of waste disposal and calls for conservation of energy and natural resources, reduction in waste generated, and environmentally sound waste management practices?

    (A) RCRA
    (B) FIFRA
    (C) CERCLA
    (D) OSHA
    (E) FEMA

21. One of the earliest sites that utilized Superfund resources was

    (A) Bhopal, India
    (B) Love Canal, New York
    (C) Prince William Sound, Alaska
    (D) Chernobyl, Ukraine
    (E) Donora, Pennsylvania

22. Which growth curve best represents the effects of environmental resistance acting on a sustainable population?

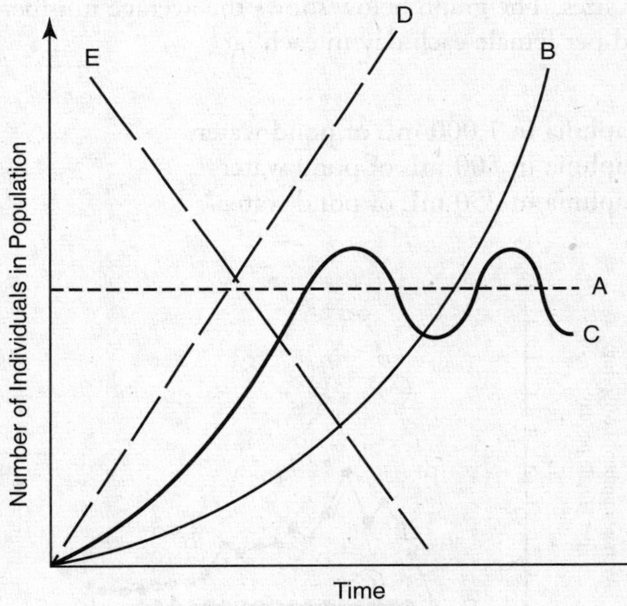

    (A) A
    (B) B
    (C) C
    (D) D
    (E) E

23. The location where two tectonic plates slide apart from each other with the space that was created being filled with molten magma from below, such as the Mid-Atlantic Ridge, the East Pacific Rise, and the East African Great Rift Valley, are known as

 (A) divergent boundaries
 (B) convergent boundaries
 (C) transform boundaries
 (D) tectonic boundaries
 (E) lithospheric boundaries

24. When present in large numbers, the plant most likely to cause a lake to become gradually filled with dead and decaying plants and organisms is the

 (A) lichen
 (B) elodea
 (C) maple tree
 (D) duckweed
 (E) mushroom

25. A group of 100 daphnia was placed into each of three culture jars of three different sizes. The graph below shows the average number of offspring produced per female each day in each jar.

 Key:
 (A) Daphnia in 1,000 mL of pond water
 (B) Daphnia in 500 mL of pond water
 (C) Daphnia in 250 mL of pond water

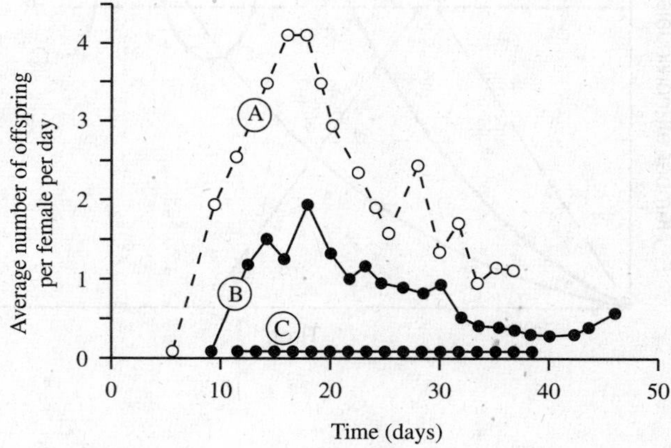

 The information suggests that the population growth rate

 (A) is density-dependent
 (B) is density-independent
 (C) depends on a predator-prey relationship
 (D) increases exponentially after 20 days
 (E) levels off when the carrying capacity is reached

Questions 26–28

Base your answers to questions 26 through 28 on the following diagram.

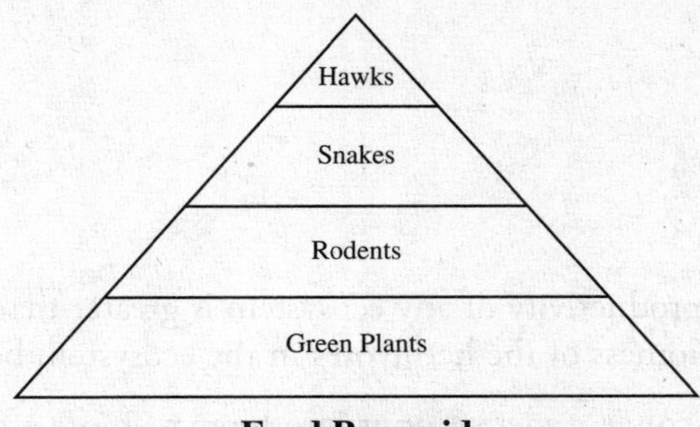

Food Pyramid

26. The greatest amount of energy present in this pyramid is found at the level of the

    (A) hawks
    (B) snakes
    (C) rodents
    (D) green plants
    (E) decomposers

27. The pyramid implies that in order to live and grow, 1,000 kilograms of snakes would require

    (A) less than 1,000 kilograms of green plants
    (B) 1,000 kilograms of rodents
    (C) more than 1,000 kilograms of rodents
    (D) no rodents
    (E) less than 1,000 kilograms of hawks

28. Which of the following represents the food chain shown in the pyramid?

    (A) Green plant, rodent, snake, hawk
    (B) Rodent, green plant, snake, hawk
    (C) Snake, green plant, rodent, hawk
    (D) Hawk, green plant, rodent, snake
    (E) Green plant, snake, hawk, rodent

29. If a city with a population of 100,000 experiences 4,000 births, 3,000 deaths, 500 immigrants, and 200 emigrants within the course of one year, what is net annual percentage growth rate?

    (A) 0.3%
    (B) 1.3%
    (C) 13%
    (D) 101.3%
    (E) 130%

30. The annual productivity of any ecosystem is greater than the annual increase in biomass of the herbivores in the ecosystem because

    (A) plants convert energy input into biomass more efficiently than animals
    (B) there are always more animals than plants in any ecosystem
    (C) plants have a greater longevity than animals
    (D) during each energy transformation, some energy is lost
    (E) animals convert energy input into biomass more efficiently than plants

Questions 31–33

Following is a diagram showing the relationships that exist in an arid ecosystem. Base your answers to questions 31 through 33 on the diagram and your knowledge of environmental science.

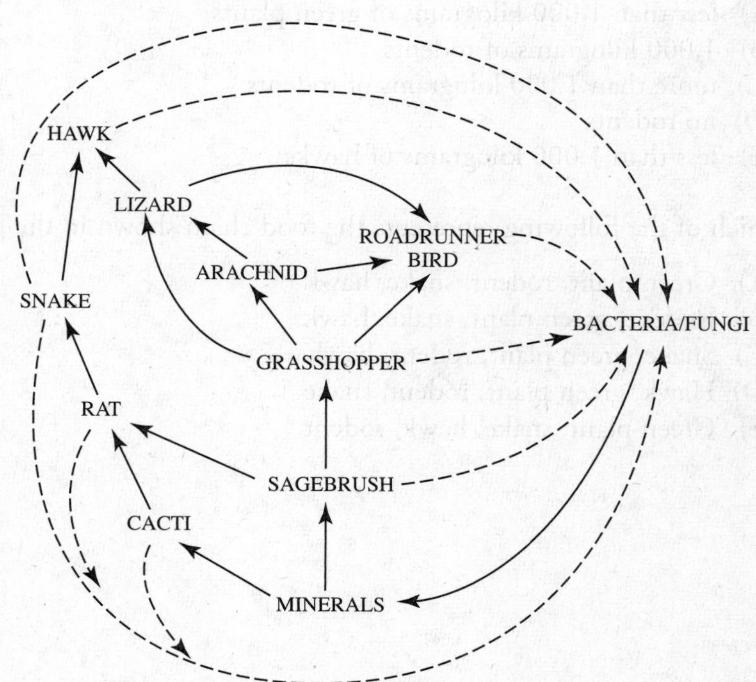

31. The snake is acting as a

    (A) producer
    (B) primary consumer
    (C) secondary consumer
    (D) autotroph
    (E) secondary producer

32. Between which two organisms would there MOST LIKELY be the greatest competition?

    (A) Rat and snake
    (B) Lizard and arachnid
    (C) Cacti and sagebrush
    (D) Grasshopper and bacteria/fungi
    (E) Arachnid and roadrunner

33. In the diagram, which statement correctly describes the role of bacteria/fungi?

    (A) The bacteria/fungi convert radiant energy into chemical energy.
    (B) The bacteria/fungi directly provide a source of nutrition for animals.
    (C) The bacteria/fungi are saprophytic agents restoring inorganic material to the environment.
    (D) The bacteria/fungi convert atmospheric nitrogen into minerals and are found in the nodules of cacti.
    (E) The bacteria/fungi consume live plants and animals.

34. When resources are scarce,

    (A) prices go down
    (B) recycling is not profitable
    (C) investment and potential profits decrease
    (D) there is an impetus for conservation
    (E) there is little competition to discover new or substitute products

35. Examine the three age-structure diagrams below.

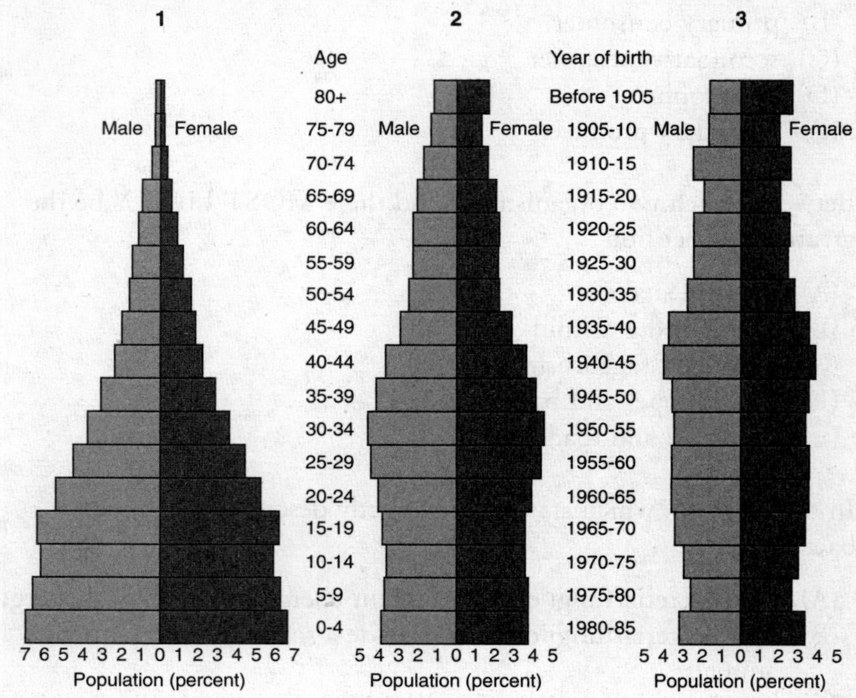

Which of the following is correct?

(A) 1 = United States; 2 = Mexico; 3 = Sweden
(B) 1 = Mexico; 2 = Sweden; 3 = United States
(C) 1 = Sweden; 2 = United States; 3 = Mexico
(D) 1 = Mexico; 2 = United States; 3 = Sweden
(E) 1 = Sweden; 2 = Mexico; 3 = United States

36. Choose the correct order for the passage of a federal environmental bill.

I. Floor debates and vote
II. Referred to the other chamber
III. Presidential action
IV. Bill is introduced, bill is referred to a committee, and a hearing is conducted
V. Committee reports

(A) IV, III, II, I, V
(B) IV, V, I, II, III
(C) I, II, III, IV, V
(D) IV, I, II, III, V
(E) IV, V, I, III, II

37. Approximately five pounds of carbon is released into the atmosphere in the form of carbon dioxide for every gallon of gasoline that is burned. Two cars are making a 500 mile trip. Car A gets 15 miles to the gallon and car B gets 30 miles to the gallon. Approximately how much more carbon (in the form of CO_2) will car A produce than car B?

(A) 5 pounds
(B) 16.7 pounds
(C) 33.3 pounds
(D) 83 pounds
(E) 100 pounds

38. Which factor does NOT significantly affect the amount of solar energy reaching the surface of Earth?

(A) Earth's rotation once every 24 hours
(B) Earth's revolution around the Sun once per year
(C) the tilt of Earth's axis (23.5°)
(D) atmospheric conditions
(E) the distance between Earth and the sun

39. Which of the following statements would NOT be consistent with the deep ecology movement?

(A) The well-being and flourishing of human and nonhuman life on Earth has intrinsic value. These values are independent of the usefulness of the nonhuman world for human purposes.
(B) Richness and diversity of life-forms contribute to the realization of these values and are also values in themselves.
(C) Our success depends on how well we can understand and manage Earth's life-support systems.
(D) The flourishing of human life and cultures is compatible with a substantial decrease in human population. The flourishing of nonhuman life requires such a decrease.
(E) Present human interference with the nonhuman world is excessive, and the situation is rapidly worsening.

40. A chain is dragged across the seafloor, pulling a huge net behind it. A single pass of this net can remove up to a quarter of seafloor life. Repeated passes can remove nearly all seafloor life, including sessile animals and plants plus many species of fish and marine invertebrates. This type of fishing is known as

(A) bottom trawling or dredging
(B) chain-fishing
(C) scraping
(D) scoop netting
(E) drag-lining

41. Which item listed below COULD be placed into a compost pile?

    (A) Chemically treated wood products
    (B) Pet wastes
    (C) Manure
    (D) Pernicious weeds
    (E) Bones

42. Two-thirds of Iceland's energy sources come from clean, renewable hydroelectric and geothermal sources. Research in Iceland is currently underway in developing hydrogen fuel cells. Iceland is an example of a country that is practicing

    (A) sustainability
    (B) remediation
    (C) conservation
    (D) preservation
    (E) mitigation

43. The diagram below represents evolutionary pathways for two hypothetical organisms (*X* and *Y*).

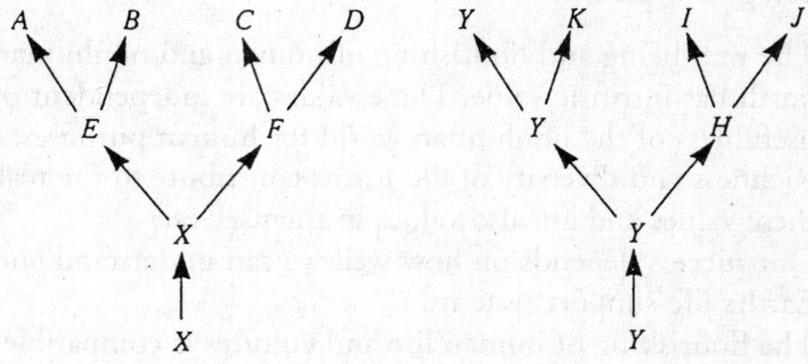

Which species was probably the most successful in the environment over time?

    (A) *X*
    (B) *Y*
    (C) *K*
    (D) *J*
    (E) *C*

44. A group promoting greenbelts would believe in all the following ideas EXCEPT

(A) encouraging smart growth instead of sprawl
(B) designing compact, pedestrian-friendly, transit-oriented neighborhoods
(C) including affordable housing in all city plans for expansion
(D) directing public and private investment into existing communities instead of into wasteful sprawl development
(E) encouraging large retail chains such as Wal-Mart and Costco to locate within their city so that driving distances could be reduced by allowing customers to find everything they need in one location instead of driving all over town

45. As urban environments become denser, noise pollution is becoming a major environmental issue as people deal with noise generated from flight paths, freeways, work environments, and neighborhoods. Which of the following is a TRUE statement?

(A) Loud sound is not dangerous as long as you do not feel any pain in your ears.
(B) Hearing loss after sound exposure is temporary.
(C) Hearing loss is caused mostly by aging.
(D) Loud sound damages only your hearing.
(E) If you have a hearing loss already, you still have to protect your hearing.

46. What is meant by the term "enriched" uranium

(A) pure uranium, with no other elements present
(B) all nuclei are U-238 in the sample
(C) uranium that has been rinsed in heavy water to enrich it
(D) uranium that has a higher proportion of U-235 nuclei than normal
(E) uranium that has had plutonium added to it

47. Due to CITES, elephant populations in Africa

(A) have grown to record numbers
(B) have stabilized
(C) are near the point of collapse
(D) are being moved to safe wildlife sanctuaries
(E) no longer exist

Questions 48–50

Use this information for questions 48 through 50. Two varieties of the same species of voles (meadow mice), albino and red-backed, were used in an experiment. Both varieties were subjected to the predation of a hawk under controlled laboratory conditions. During the experiment, the floor of the test room was covered on alternate days with white ground cover that matched the albino voles and red-brown cover that matched the red-backed voles. The results of 50 trials are shown in the following table:

Number of Voles Captured

Variety	White Cover	Red-Brown Cover	Total
Albino	35	57	92
Red-backed	60	40	100
Total	95	97	192

48. Which result would have been most likely if only red-brown floor covering had been used?

    (A) 97 red-backed voles would have survived.
    (B) 95 albino voles would have survived.
    (C) The survival rate of the albino voles would have decreased markedly.
    (D) A greater number of red-backed voles would not have survived.
    (E) There would have been no change in the results.

49. The results of this experiment lend credence to the concept of

    (A) the female species being more deadly than the male
    (B) use and disuse of organs and tissues
    (C) overproduction as a means of species perpetuation
    (D) adaptation permitting survival of the species
    (E) spontaneous generation of a lethal mutation

50. The purpose of the experiment was to

    (A) measure the visual acuity of hawks
    (B) decrease the vole population
    (C) compare the agility of the several varieties of voles
    (D) determine the basic intelligence of hawks and voles
    (E) compare the effects of protective coloration

Questions 51–52

For questions 51 and 52, choose the term that best fits each symbiotic description.

(A) Commensalism
(B) Mutualism
(C) Parasitism
(D) Saprophytism
(E) Competition

51. Termites eat wood but are unable to digest it without the aid of certain flagellated protozoans that live within the bodies of termites.

52. Certain young clams attach themselves to the gills of a fish. In a short time, each clam becomes surrounded by a capsule formed by cells of the fish. The clam feeds and grows by absorbing nutrients from the fish's body.

53. Natural capitalism is associated with

(A) Paul Hawken
(B) Garrett Hardin
(C) Barry Commoner
(D) Rachel Carson
(E) Aldo Leopold

54. Which agency does NOT manage designated federal wilderness in the United States?

(A) National Park Service
(B) Forest Service
(C) Bureau of Land Management
(D) Fish and Wildlife Service
(E) All share management

Questions 55–56

Answer questions 55 and 56 on the basis of the diagram below. Each circle represents the range for a pair of owls. *X* represents the area where reproduction and nesting take place.

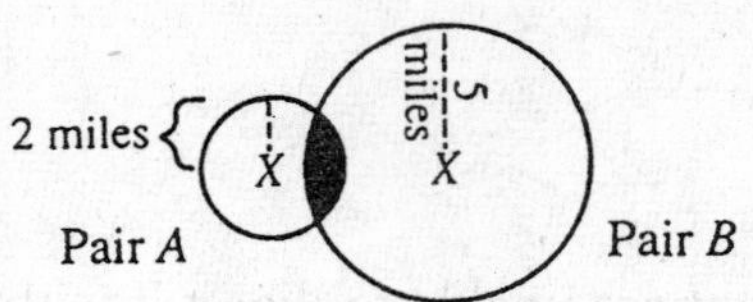

55. The shaded area represents the area where
    (A) most of the owls' predators are located
    (B) most of the owls' food is located
    (C) competition is likely to occur between the pairs
    (D) mating occurs
    (E) there is no competition between the two pairs

56. What is the most likely reason why the range of pair *B* is larger than the range of pair *A*?
    (A) Pair *B* is younger.
    (B) Pair *B* can fly faster than pair *A*.
    (C) Pair *B* has more prey near its nest than pair *A*.
    (D) Pair *B* has less food available near its nest than pair *A*.
    (E) Pair *A* does not need as much living space as pair *B*.

Questions 57–58

In 1940, ranchers introduced cattle into an area. The graph below shows the effect of cattle ranching on the populations of two organisms present in the area before the introduction of the cattle. Base your answers to questions 57 and 58 on the graph and on your knowledge of environmental science.

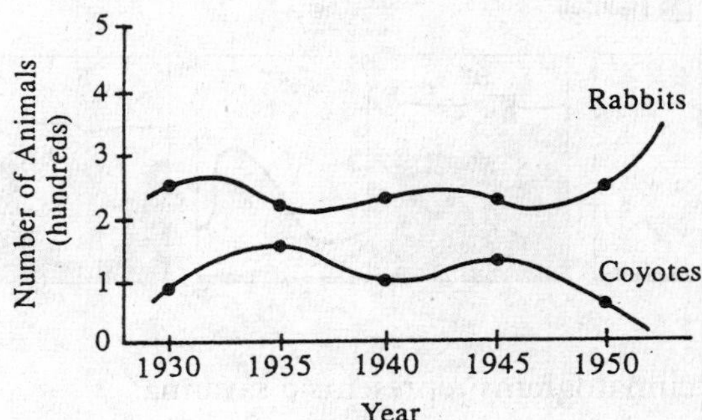

57. The most probable reason for the increase in the rabbit population after 1950 was

   (A) more food became available to the rabbits
   (B) the coyote population declined drastically
   (C) the cattle created a more favorable environment for the rabbits
   (D) the coyotes and cattle competed for the same food
   (E) the coyote population increased

58. If the interrelationship of rabbits and coyotes was once in balance, what is the most probable explanation for the decline of the coyotes?

   (A) Mutations
   (B) Starvation
   (C) Disease
   (D) Increase in reproductive rate
   (E) Removal by human beings

Questions 59–60

Questions 59 and 60 refer to the following graphs of temperature and rainfall for six major ecosystems. A year's temperature from January through December (the line) and rainfall pattern (the shaded area) of each ecosystem are shown.

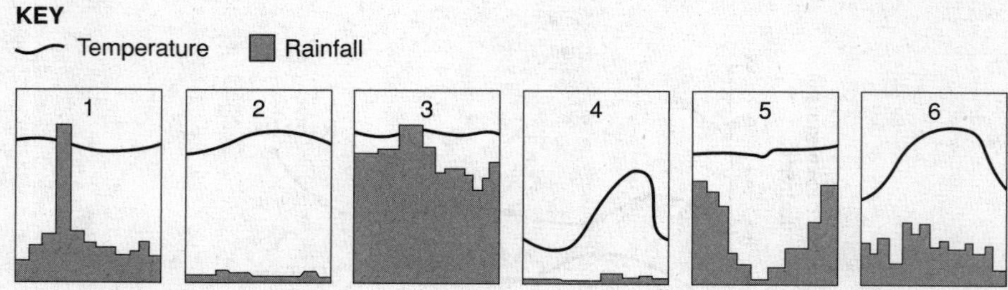

59. Which of the climatograms represents a savanna?

    (A) 1
    (B) 2
    (C) 3
    (D) 5
    (E) 6

60. Which represents a tundra?

    (A) 1
    (B) 2
    (C) 3
    (D) 4
    (E) 5

61. Areas of terrestrial and coastal ecosystems that are internationally recognized within the framework of UNESCO's Man and the Biosphere (MAB) Program are known as

    (A) wilderness preserves
    (B) wildlife sanctuaries
    (C) habitat sanctuaries
    (D) biosphere reserves
    (E) preservation zones

62. Rank the following biomes in order of most productive to least productive, as measured by biomass produced per acre.

 I. Desert
 II. Tropical rain forests
 III. Tundra
 IV. Grassland

(A) I, II, III, IV
(B) IV, III, II, I
(C) III, I, IV, II
(D) II, III, IV, I
(E) II, IV, III, I

63. You work for a company that has violated federal environmental statutes regarding dumping of hazardous wastes. Your company's case would initially be heard in

(A) U.S. District Court
(B) U.S. Claims Court
(C) U.S. Circuit Court of Appeals
(D) State Trial Courts
(E) U.S. Supreme Court

64. Refer to the data below taken from a stream. The Roman numerals indicate collection sites.

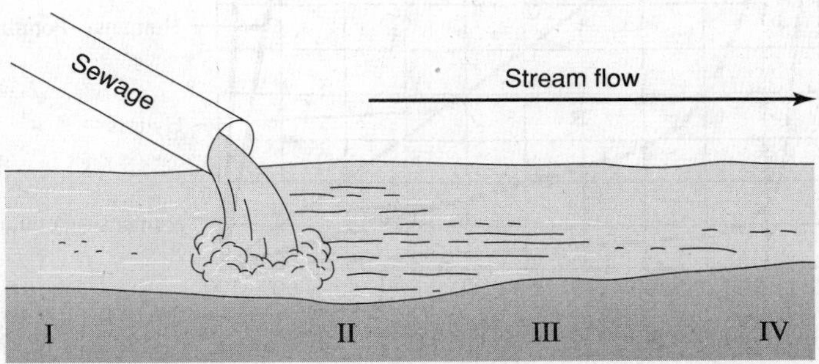

Where would the lowest DO (dissolved oxygen) content be expected?

(A) I
(B) II
(C) III
(D) IV
(E) Dissolved oxygen would not be affected by sewage effluent. Therefore, DO would be equal at all points.

65. The theory that great disasters serve to maintain a population and its food supply balance was initially proposed by

    (A) Darwin
    (B) Wallace
    (C) Hardy and Weinberg
    (D) Malthus
    (E) Lyell

Questions 66–67
The following graph shows survival rates for five animal populations. When survival curves are calculated, the following assumptions are made:

I. All individuals of a given population are the same age.
II. No new individuals enter the population.
III. No individuals leave the population.

These curves show the relationship of the number of individuals in a population to units of physiological life span. Base your answers to questions 66 and 67 on the graph and on your knowledge of environmental science.

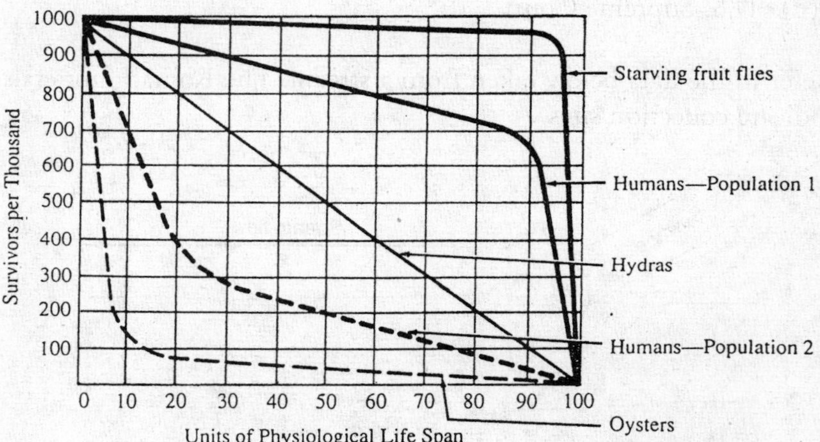

66. According to the data, it can be assumed that

    (A) fruit flies live longer than humans
    (B) oysters outlive fruit flies
    (C) the population of hydras is steadily declining
    (D) the life span of human populations is related to that of oysters
    (E) there is a high mortality rate among young oysters

67. The survival curves indicate that

    (A) starving fruit flies live out their full life span
    (B) human populations are more vulnerable than hydras
    (C) human population 2 has a greater rate of survival than human population 1
    (D) the hydra has a longer life span than the oyster
    (E) fruit flies die directly after pupation

68. The average requirement for drinking water per person per day is approximately

    (A) 1 pint
    (B) 1 quart
    (C) ½ gallon
    (D) 1 gallon
    (E) 3 gallons

69. The average American uses how many gallons of water per day for indoor domestic use?

    (A) Less than 1
    (B) 1–10
    (C) 20–40
    (D) 50–80
    (E) 100–200

70. The world's population in 2000 was approximately 6 billion. If the growth rate were 2%, in what year would the world's population be 12 billion?

    (A) 2035
    (B) 2050
    (C) 2010
    (D) 2100
    (E) 4000

71. You are on a trip visiting Rwanda in Africa. Which of the following would you notice?

    (A) A larger number of people between 30 and 45 than any other age category
    (B) A population that is equally distributed in age categories
    (C) A greater number of people age 65 and older compared with children under 15
    (D) Fewer people age 65 compared with children under 15
    (E) A population wiped out due to AIDS

72. Which of the following is NOT a component of an environmental impact report?

    (A) Alternatives to the proposed action
    (B) A statement of positive and negative environmental impacts of the proposed activities
    (C) A statement of financial liability
    (D) Purpose and need for the project
    (E) The relationship between short-term resources and long-term productivity

73. Which of the following statements is NOT consistent with "The Tragedy of the Commons" by Garret Hardin?

    (A) We will always add one too many cows to the village commons, destroying it.
    (B) The destruction of the commons will not be stopped by shame, moral admonitions, or cultural mores anywhere nearly so effectively as it will be by the will of the people expressed as a protective mandate, in other words, by government.
    (C) The "Tragedy of the Commons" is a modern phenomenon. Humans were not capable of doing too much damage until the population exceeded certain numbers and their technological tools became powerful beyond a certain point.
    (D) A free-market economy, based on capitalism, does not contribute to the tragedy of the commons.
    (E) We will always opt for an immediate benefit at the expense of less-tangible values such as the availability of a resource to future generations.

74. The most common method of disposing of municipal solid wastes in the United States is

    (A) incineration
    (B) ocean dumping
    (C) sanitary landfills
    (D) recycling
    (E) exporting

75. What would be the best environmental solution for beach erosion?

    (A) Bring in sand from other sources to replace sand that has been washed away
    (B) Build seawalls
    (C) Build breakwaters
    (D) Restrict development near beaches
    (E) Restrict access to the beach

76. Which one of the following statements is FALSE?

    (A) Every week, more than 500,000 trees are used to produce the two-thirds of newspapers that are never recycled.
    (B) The equivalent of 10 city blocks of rain forest is destroyed every minute, an area the size of Pennsylvania lost every year.
    (C) Of Earth's dry land surface, 7% is rain forest, home to more than 50% of the world's plants and animals.
    (D) About one-fifth of the United States—442 million acres—is forested.
    (E) Private owners account for 59% of the nation's 490 million acres of commercial forestland; the government owns 27%; and the forest industry owns 14%.

77. The densest populations of most organisms that live in the ocean are found near the surface. The most probable explanation is that

    (A) the surface is less polluted
    (B) the bottom contains radioactive material
    (C) salt water has more minerals than freshwater
    (D) the light intensity that reaches the ocean decreases with increasing depth
    (E) the largest primary consumers are found near the surface

78. Which of the following would be classified as a point source for pollution?

    (A) Sulfur oxides released from an electrical generating plant
    (B) Oil, grease, and toxic chemicals from urban runoff and energy production
    (C) Sediments from improperly managed construction sites, crops, forestlands, and eroding stream banks
    (D) Salt from irrigation practices and acid drainage from abandoned mines
    (E) Bacteria and nutrients from livestock, pet wastes, and faulty septic systems

79. An APES class went on a field trip and saw a field with many small, narrow channels averaging about a foot deep and a foot wide cut into Earth. The class concluded that

    (A) the small channels were most likely due to rill erosion.
    (B) the small channels were most likely due to wind erosion.
    (C) the small channels were most likely due to glacier action.
    (D) the small channels were most likely caused by gully erosion.
    (E) the small channels were due to plate movements of Earth's crust.

80. John Muir was responsible for developing (the)

 (A) Green Peace
 (B) Nature Conservancy
 (C) Audubon Society
 (D) Sierra Club
 (E) Environmental Defense Fund

81. In response to the Great Depression, President Franklin D. Roosevelt created many programs designed to put America back to work. Which agency, listed below, was established in early 1933 with a twofold mission: to reduce unemployment, especially among young men, and to preserve the nation's natural resources?

 (A) WPA
 (B) EPA
 (C) CCC
 (D) ACLU
 (E) TVA

82. Which American president dealt with an energy shortage by establishing a national energy policy, by decontrolling domestic petroleum prices to stimulate production, and expanding the national park system by including protection of 103 million acres of Alaskan lands?

 (A) Jimmy Carter
 (B) Ronald Reagan
 (C) Bill Clinton
 (D) George Bush
 (E) George W. Bush

83. What was the name of the movement, in the late 1970s, that tried to gain land back from the federal government?

 (A) State's Rights Rebellion
 (B) Public Land Rebellion
 (C) Homestead Movement
 (D) Land Rush Rebellion
 (E) Sagebrush Rebellion

84. A country has a population of 700 million people and has an annual rate of growth of 1.75 percent. In one year, what would be the population of this country?

 (A) 12,250,000
 (B) 700,122,500
 (C) 701,225,000
 (D) 712,250,000
 (E) 712,500,000

85. Which legislation was signed into law almost 25 years ago by President Nixon that sets out the conditions under which a species can be listed as endangered or threatened, calls for the development of a plan for the recovery of the species' population, and allows for designating and protecting habitat critical to the species' survival?

(A) National Wildlife Refuge System Act
(B) Species Conservation Act
(C) National Forest Management Act
(D) Endangered Species Act
(E) National Environmental Policy Act

86. One of the greatest successes of the Endangered Species Act has been the

(A) passenger pigeon
(B) whooping crane
(C) bald eagle
(D) condor
(E) auk

87. If global temperature warmed for any reason, atmospheric water vapor would increase due to evaporation. This would increase the greenhouse effect and thereby further raise the temperature. As a result, the ice caps would melt back, making the planet less reflective so that it would warm further. This process is an example of

(A) positive coupling
(B) negative coupling
(C) synergistic feedback
(D) positive-feedback loop
(E) negative-feedback loop

88. Approximately how many people have become infected with the AIDS virus since it was discovered in the early 1980s?

(A) 5 million
(B) 15 million
(C) 60 million
(D) 150 million
(E) 500 million

89. What is the most frequent cause of beach pollution?

(A) Polluted runoff and storm water
(B) Sewage spills from treatment plants
(C) Oil spills
(D) Ships dumping their holding tanks into coastal waters
(E) People leaving their trash on the beach

90. Atmospheric carbon dioxide levels

    (A) are about 10% higher than they were at the time of the Industrial Revolution
    (B) are about 25% higher than they were at the time of the Industrial Revolution
    (C) are about the same as they were at the time of the Industrial Revolution
    (D) are slightly lower than they were at the time of the Industrial Revolution
    (E) are significantly lower than they were at the time of the Industrial Revolution

91. 550 parts per million would be equivalent to

    (A) 5.5 parts per billion
    (B) 55 parts per billion
    (C) 5,500 parts per billion
    (D) 55,000 parts per billion
    (E) 550,000 parts per billion

92. The major sink for phosphorus is

    (A) marine sediments
    (B) atmospheric gases
    (C) seawater
    (D) plants
    (E) animals

93. Certain volcanoes are built almost entirely of fluid lava flows that pour out in all directions from a central summit vent or group of vents. They build a broad, gently sloping, dome-shaped cone. These cones are built up slowly by the accretion of thousands of highly fluid basalt lava flows that spread widely over great distances and then cool as thin, gently dipping sheets. What are these volcanoes called?

    (A) Lava domes
    (B) Composite volcanoes
    (C) Cinder cones
    (D) Shield volcanoes
    (E) Stratovolcanoes

94. Which of the following would NOT be a likely location for seismic activity?

    (A) Along mid-oceanic ridges
    (B) Faults associated with volcanic activity
    (C) Boundaries between oceanic and continental plates
    (D) Interior of continental plates
    (E) Boundaries between continental plates

95. The first reptiles appeared on Earth

(A) during the Cambrian (550–500 million years ago)
(B) during the Devonian (410–360 million years ago)
(C) during the Carboniferous (360–280 million years ago)
(D) during the Permian (280–250 million years ago)
(E) during the Jurassic (210–140 million years ago)

96. Examine the following weather map.

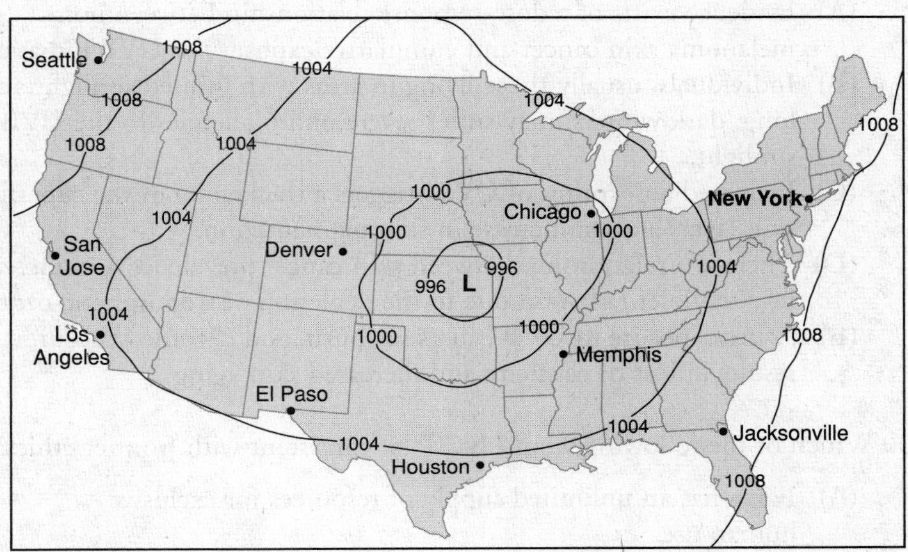

Which of the following would be TRUE?

(A) It is likely to be fair weather in the Midwest.
(B) It is likely to be raining in the Midwest.
(C) It is likely to be raining in the Northeast.
(D) Rain would be expected in the western United States.
(E) Not enough information is provided.

97. An AP Environmental Science class was investigating soil texture. The students each placed 2 teaspoons of soil into their hands and mixed a small amount of water into the soil until it was the consistency of moist putty. The students were able to make a small-sized ball of mud that stayed the shape of a ball after they squeezed it. However, they were unable to make the ball of soil form a ribbon using their thumb and index finger. Most likely the type of soil that they were testing was

(A) gravel
(B) silty clay
(C) sand
(D) clay
(E) loamy sand

98. Causes of sick-building syndrome include all of the following EXCEPT

(A) radon and asbestos
(B) chemical contaminants from indoor sources
(C) chemical contaminants from outdoor sources
(D) biological contaminants
(E) inadequate ventilation

99. Which of the following statements is FALSE?

(A) Evidence exists of a dose-response relationship between non-melanoma skin cancer and cumulative exposure to UVB radiation.
(B) Individuals, usually those living in areas with limited sunlight and long, dark winters, may suffer severe photo-allergies to the UVB in sunlight.
(C) Increased absorption of UVB triggers a thickening of the superficial skin layers and an increase in skin pigmentation.
(D) There is a relationship between skin cancer prevalence and increases in ultraviolet radiation due to the depletion of tropospheric ozone.
(E) Acute exposure to UVB causes sunburn, and chronic exposure results in loss of elasticity and increased skin aging.

100. Which of the following would NOT be consistent with frontier ethics?

(A) Earth has an unlimited supply of resources for exclusive human use.
(B) Humans are a part of nature, not apart from it.
(C) Success comes through domination and control of nature.
(D) Anthropocentrism
(E) Greater value assumption

SECTION II (FREE-RESPONSE QUESTIONS)

Time: 90 minutes

No calculators allowed

4 questions

Directions: Answer all four questions, which are weighted equally. The suggested time is about 22 minutes for answering each question. Write all your answers on the pages following the questions in the pink booklet. Where calculations are required, clearly show how you arrived at your answer. Where explanation or discussion is required, support your answers with relevant information and/or specific examples.

1. The environmental impact of washing a load of clothes in an electric washing machine is different than washing the same clothes by hand. Use the information below to answer the questions that follow. Show your calculations.

Assume the following:

1. All of the clothes can be washed in one load in the washing machine.
2. The water entering the water heater is 60°F.
3. The water leaving the water heater is 130°F.
4. The electric washing machine uses 20 gallons of water. It uses 110 volts of electricity at an average of 1,500 watts for 30 minutes.
5. Washing the clothes by hand requires 35 gallons of hot water.

Other information:

1 gallon of water = 8 pounds
1 Btu = amount of energy required to raise the temperature of 1 pound of water by 1°F
1 kilowatt-hour = 3,400 Btus

(a) Calculate the total amount of energy (in Btus) to wash the clothes using the washing machine.
(b) Calculate the total amount of energy (in Btus) to wash the clothes by hand.
(c) Discuss the environmental impact of washing clothes.

2. An AP Environmental Science class did an investigation on competition. Part I of the investigation focused on intraspecific competition to assess the effect of growth among radish plants at different population densities. Part II of the investigation focused on the relative competitiveness of two species of plants (radish and wheat) when they were planted together. The results are presented below:

Part I: Intraspecific Competition Among Radish Plants

Seeds per Pot	Total Biomass per Pot (g)
1	5.0
10	70.0
20	75.0

Part II: Interspecific Competition Between Radish and Wheat Plants

Seeds per Pot	Total Biomass per Pot (g)
1 radish	5.0
1 wheat	3.0
10 radish	50.0
10 wheat	25.0
20 radish	75.0
20 wheat	40.0

(a) Discuss the results of Part I of the investigation.
  (i) At what population density was the biomass per plant highest?
  (ii) What resources may have been limited?
  (iii) Discuss the results obtained from the class for Part I in terms of biological laws or principles.

(b) Discuss the results of Part II of the investigation.
  (i) Which plant was most affected by the competition between the two species?
  (ii) What are possible reasons why the plant chosen in (i) may have been more successful?
  (iii) Discuss the results obtained from the class for Part II in terms of biological laws or principles.

3. Examine the age-structure diagrams of Sweden and Kenya below, and answer the following questions.

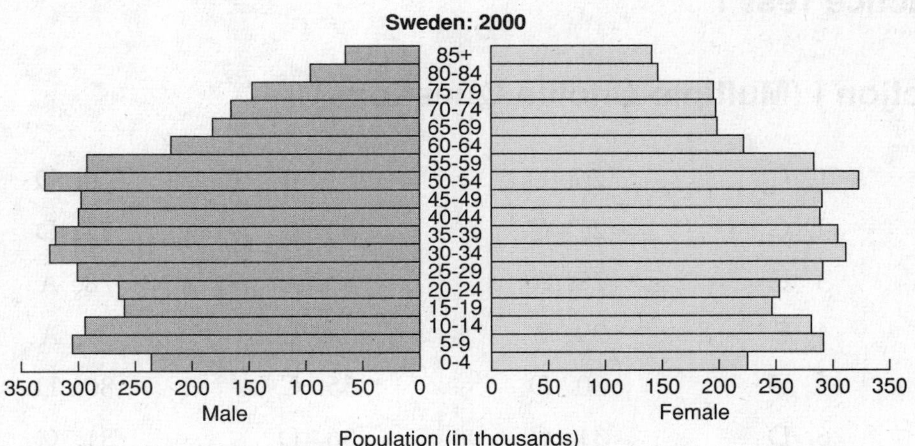

Source: U.S. Census Bureau, International Data Base.

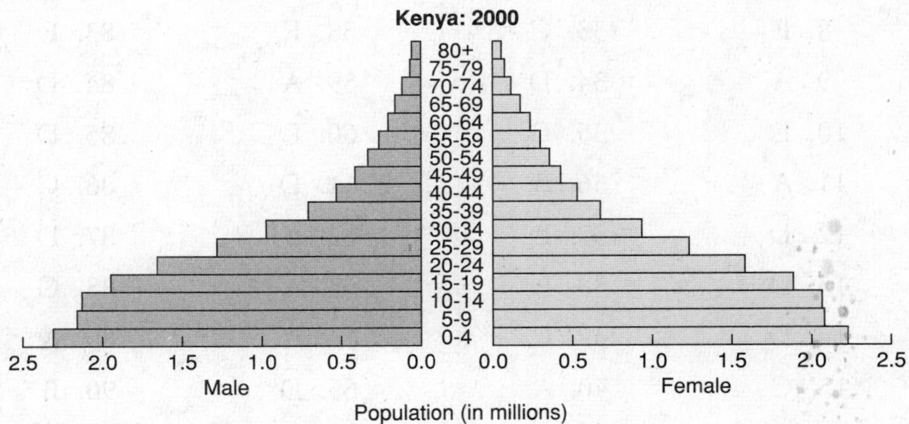

Source: U.S. Census Bureau, International Data Base.

(a) Compare and contrast the two age structure diagrams in terms of two population dynamics—birth rate and death rate.
(b) What factors affect birth rates and death rates?
(c) Discuss methods that have been employed in another country to curb population growth.

4. The use of pesticides has become the most widely used form of pest management and is a controversial topic in environmental science.

(a) Explain what a pesticide is, and describe three major categories of pesticides. For example, insecticide—kills insects and other arthropods. Do NOT use insecticide as one of your categories.
(b) Discuss two positive and two negative effects of pesticide use.
(c) Discuss three alternatives to the use of pesticides.
(d) Name and describe one U.S. federal law or one international treaty that focuses on the use of pesticides.

ANSWER KEY

Practice Test 1

Section I (Multiple-Choice Questions)

1. C
2. C
3. C
4. E
5. D
6. D
7. A
8. E
9. A
10. E
11. A
12. D
13. D
14. D
15. C
16. C
17. D
18. D
19. B
20. A
21. B
22. C
23. A
24. D
25. A
26. D
27. C
28. A
29. B
30. D
31. C
32. B
33. C
34. D
35. D
36. B
37. D
38. E
39. C
40. A
41. C
42. A
43. B
44. E
45. E
46. D
47. B
48. C
49. D
50. E
51. B
52. C
53. A
54. E
55. C
56. D
57. B
58. E
59. A
60. D
61. D
62. E
63. A
64. C
65. D
66. E
67. A
68. C
69. D
70. A
71. D
72. C
73. D
74. C
75. D
76. D
77. D
78. A
79. A
80. D
81. C
82. A
83. E
84. D
85. D
86. C
87. D
88. C
89. A
90. B
91. E
92. A
93. D
94. D
95. C
96. B
97. E
98. A
99. D
100. B

MULTIPLE-CHOICE EXPLANATIONS

1. (C) An increase in the population of squirrels would increase the competition for food and space.
2. (C) Refer to Figure 1.1 (page 4).
3. (C) For (C) to be true, it would have read, "The Kyoto Protocol would have required the United States to *reduce* its greenhouse gas emissions—primarily carbon dioxide (CO_2), methane (CH_4), and nitrous oxide (N_2O)—to 7% below 1990 levels by the year 2012."
4. (E) External costs are the costs that are borne by people other than the producer of a product.
5. (D) The tick bird and the rhinoceros *can* live without each other (facultative or nonobligatory). However, when in close association, they are mutually helpful to each other. Armed with a pointed beak and sharp, curved claws, the tick bird (oxpecker) is able to scavenge parasites from the backs of large, thick-skinned animals such as the rhinoceros to provide food for itself. It rids its host of unwelcome pests and provides early warning of danger to the rhino by calling loudly when alarmed.
6. (D) Mitigation involves taking steps to lessen risk by lowering the probability of a risk event's occurrence or reducing its effect should it occur. Organizing a beach cleanup is remediation—reacting after the beach has been polluted.
7. (A) Diastrophism can be divided into two processes. Orogenesis or mountain building is when rock is forced to buckle under horizontal pressure (folding) and can happen to a degree significant enough to create entire mountain ranges. Epeirogenesis or continent building is when pressures may cause an entire continent to warp or bow.
8. (E) Refer to Figure 1.1 (page 4).
9. (A) Punctuated equilibrium is a theory of evolution that proposes that species are stable and remain unchanged for long periods of time. Instead of a slow, continuous movement, evolution tends to be characterized by long periods of virtual standstill (equilibrium), punctuated by episodes of very fast development of new forms. Speciation occurs when the stability is interrupted by brief periods of major changes.
10. (E) Sea level is rising worldwide and is caused by both natural and human factors. It is predicted that within 100 years, there will be a net loss of 17–43% of coastal wetlands due to rising sea levels.
11. (A) Evolution is the descent from a common ancestor. Divergent evolution is a branching out of species from a single source.
12. (D) Primary economic activities (III) are at the beginning of the production cycle where humans are in closest contact with resources and the environment. Primary economic activities are located at the site of the natural resources being exploited. In many developing nations, approximately 3/4 of the labor force engages in subsistence farming or herding. By contrast, in highly developed countries only a small fraction of the labor force is directly employed in agriculture (less than 3% in the U.S.).

    Secondary economic activities (I) use raw materials to produce or manufacture something new and more valuable. Examples of secondary activities include manufacturing, processing, producing power, and construction. Secondary economic activities are located either at the resource site or in close proximity to the market for the manufactured/processed good. Location depends upon whether the raw material or finished product costs more to ship. The major share of global secondary manufacturing activity is found within a relatively small number of major industrial concentrations. Rather than manufacturing/producing within its own country, richer nations often "set up shop" in developing nations because costs are significantly cheaper.

Tertiary economic activities (IV) include all activities that amount to doing services for others. Doctors, teachers, and secretaries provide personal and professional services. Restaurant staff, store clerks, and hotel personnel provide retail and wholesale services.

Quaternary economic activities (II) are not connected to resources, access to a market, or the environment. Rather, they include professions that process, administer, and disseminate information. Computer engineers, professors, and lawyers serve as examples of white-collar professionals who specialize in the collection and manipulation of information. With vast advancements in technology, quaternary economic activities could potentially exist anywhere. However, typically they exist in nations that have access to research centers, universities, efficient transportation and communication networks, and a pool of highly trained, skilled, flexible workers. Quaternary economic activities loom large in highly advanced, developed societies.

13. (D) Heterotrophs are dependent feeders. This means that they cannot synthesize nutritive material from inorganic compounds but must obtain nutrients from preexisting organic material. The original heterotrophs must have obtained energy from organic compounds that were present in the early environment. The first photosynthesizing autotrophs probably arose in response to organic nutrient depletion—evidence that is found in Precambrian rocks that formed in the presence of free oxygen. These early autotrophs were probably bacterial in nature (cyanobacteria or blue-green algae) with no cell nucleus or specialized organelles with diffuse genetic material, and reproduced through simple body division.
14. (D) Refer to Figure 1.3 (page 4).
15. (C) Coral reefs are among the most ancient of ecosystem types, dating back to the Mesozoic era some 225 million years ago. The release of carbon dioxide from coral reefs is very small (probably less than 100 million tons of carbon per year) relative to emissions due to fossil fuel combustion (about 5.7 billion tons of carbon per year). Coral reefs store very little organic carbon and are not very effective sinks for carbon dioxide from the atmosphere. Forests are more effective sinks for atmospheric carbon. Although tropical rain forests contain more species than coral reefs, reefs contain more phyla than rain forests. Covering less than 0.2% of the ocean floor, coral reefs contain perhaps 1/4 of all marine species. Despite their limited area, coral reefs may be home to up to 25% of the fish catch of developing countries or 10% of the total amount of fish caught globally for human consumption as food. Coral reefs in 93 of the 109 countries containing them have been damaged or destroyed by human activities. In addition, human impacts may have directly or indirectly caused the death of 5–10% of the world's living reefs. If the pace of destruction is maintained, another 60% could be lost in the next 20 to 40 years. The most important short-term threats to coral reefs are sedimentation (from poor land

use such as clear-cutting on steep slopes and other activities such as dredging, eutrophication (overfertilization and sewage pollution), and destructive fishing methods (dynamiting and overharvesting).

16. (C) Green taxes are taxes levied by the government on industries for each unit of pollution and are a source of governmental income. Government subsidies for reducing pollution are limited when industries surpass a break-even point or optimum level of pollution—the point at which cleanup costs exceed the harmful costs of pollution. Regulation is a command-and-control governmental approach that incurs costs to enact and enforce laws, set standards, regulate and monitor potentially harmful activities, and prosecute violators. Furthermore, regulation often focuses on cleanup instead of prevention, discourages innovation by mandating prescribed pollution control strategies, and is often unrealistic with the realities of a competitive global business environment. Charging a user fee provides income to the government by charging industries to utilize a natural resource. Trading pollution or resource use rights occurs between companies. Permits are allocated by the government for certain levels of pollution, and companies are free to trade unused pollution allocations. Once permits are sold or auctioned by the government, further financial impacts occur between the companies, not between the companies and the government. Permits often allow the largest and most financially secure companies to pollute the most, may concentrate pollutants at the most-polluting sites, and may not create economic incentives to reduce pollution since pollution levels are simply moved, not reduced.

17. (D) A gene pool contains all of the genes of a given population. All the genes in a gene pool are capable of being passed from one generation to the next.

18. (D) According to a 1992 study by the National Oceanographic and Atmospheric Administration, 50% of the spilled oil underwent biodegradation and photolysis (chemical decomposition by the action of radiant electromagnetic energy, especially light). Cleanup crews recovered about 14% of the oil, and approximately 13% sunk to the seafloor. About 2% (some 216,000 gallons) remained on the beaches.

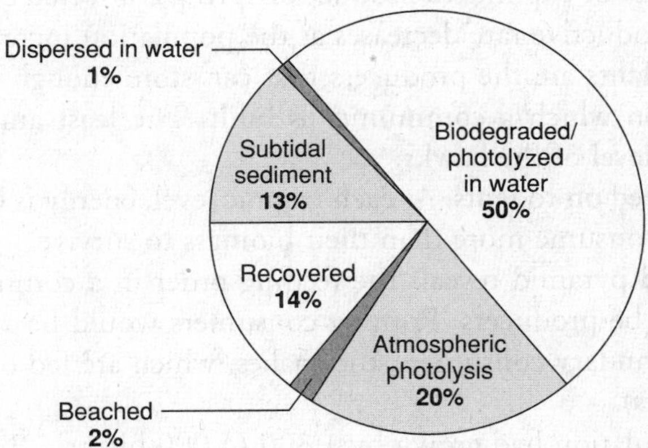

19. (B) Asexual reproduction produces organisms that are genetically identical to the parent. Speciation is the result of genetic variability. There are several types of asexual reproduction. Fission is the simplest form of asexual reproduction and involves the division of a single organism into two complete organisms,

each genetically identical to the other and to the parent. A similar form of asexual reproduction is regeneration, in which an entire organism may be generated from a part of its parent. Budding occurs when a group of self-supportive cells sprouts from and then detaches from the parent organism.

20. **(A)** The Resource Conservation and Recovery Act (RCRA) of 1976 addresses comprehensive management of nonhazardous and hazardous solid waste, sets minimal standards for all waste disposal facilities and for hazardous wastes, and regulates treatment, storage, and transport.
21. **(B)** In 1980, Congress passed the Superfund statute providing broad authorities both to respond to chemical emergencies and to clean up toxic waste sites for long-term protection.
22. **(C)** Environmental resistance or limiting factors consists of all factors in an environment that limit a population's ability to increase in numbers. Predation competition for resources, disease, limited food and space, etc., all work to hold a population in check. Environmental resistance can be either density-dependent or density-independent. Factors that are density-dependent are stronger when a population has a higher density (more crowded). Examples would include predation, parasitism, disease, and competition for space or food. Density-independent factors (usually abiotic) will kill organisms—whether they are crowded or not. Examples would include floods, storms, earthquakes, fire, etc. In nature, biotic potential and environmental resistance work together to level out population numbers to an amount that can be supported by the environment. When environmental resistance "pushes down" on J curve growth, the curve levels into what is known as an S curve (or sigmoid curve) and reflects a balanced community.
23. **(A)** Refer to "Plate Tectonics" in Chapter 1.
24. **(D)** The duckweed (family Lemnaceae) can cover the surface of lakes. These plants grow floating in still or slow-moving freshwater except in the coldest regions. The growth of these high-protein plants can be extremely rapid. The blockage of light results in the death and decay of subsurface plants, and the lake gradually becomes filled with dead and decaying plants and organisms.
25. **(A)** The ability of daphnia to produce offspring is affected by population density. The reproductive rate decreases as the population increases.
26. **(D)** Green plants are the producers that can store enough energy to provide the basis upon which a community is built. The least amount of energy is found at the level of the hawks.
27. **(C)** Snakes feed on rodents. At each trophic level, energy is lost. Therefore the snakes must consume more than their biomass to survive.
28. **(A)** The food pyramid reveals the feeding order in a community. The green plants would be producers. Primary consumers would be rodents, which are fed on by secondary consumers, the snakes, which are fed on by tertiary consumers (hawks).
29. **(B)** The population had grown by 1,300 (4,000 births – 3,000 deaths + 500 immigrants – 200 emigrants = 1,300).

$$\text{Net annual growth rate} = \frac{1{,}300}{100{,}000} \times 100\% = 1.3\%$$

30. (D) Less energy is available at each trophic level because energy is lost by organisms through respiration and incomplete digestion of food sources. Therefore, fewer herbivores can be supported by the vegetative material.

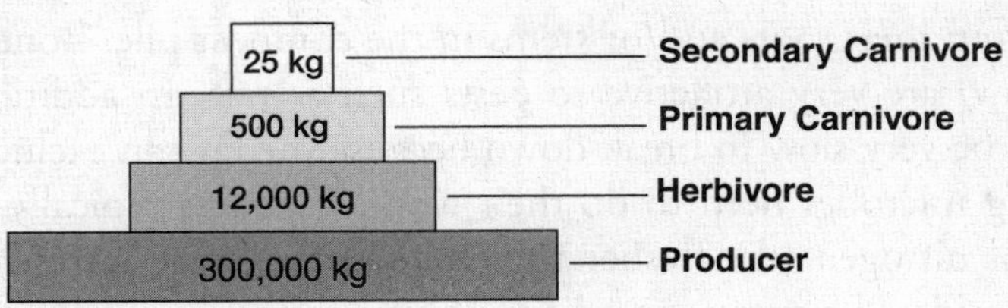

31. (C) Secondary consumers are carnivores that feed upon the flesh of other animals (snake eating a rat).
32. (B) Both lizards and arachnids eat grasshoppers. Competition is the struggle between different species for the same resources.
33. (C) Saprophytic agents gain nourishment from dead organic matter. They are the organisms of decay. Bacteria and fungi are saprophytes.
34. (D) When resources are scarce, the price goes up (law of supply and demand). The increase in prices causes people to conserve resources and use less because they are saving money.
35. (D) An investigation of this figure shows that Mexico has a large percentage of its population in the pre-reproductive stage, indicating a continued high growth rate for many years to come. Sweden, on the other hand, has no such "bulge" in its age structure diagram. With roughly equal numbers of people in all age groups, Sweden can be expected to have very slow growth in poulation, or no growth at all. The United States is a developed country with a decreasing rate of growth (the population is aging).
36. (B) No explanation needed.
37. (D) Car A travels 500 miles at 15 miles to the gallon and therefore uses about 500/15 = 33.3 gallons of gasoline. Car B travels the same 500 miles but uses about 500/30 = 16.7 gallons of gasoline. Car A uses 33.3 – 16.7 = 16.6 more gallons of gasoline than car B. If each gallon of gasoline that is burned contributes 5 pounds of carbon (in the form of CO_2) to the atmosphere, then car A will contribute 5 × 16.6 = 83 *more* pounds of carbon than car B.
38. (E) Earth is actually closest to the sun during the Northern Hemisphere winter.
39. (C) This represents a planetary management philosophy.
40. (A) Besides the effects already listed, additional effects include:
    - Flattens the ocean bottom, filling in holes, leveling humps, and knocking down protruding organisms, all of which reduce biodiversity
    - Initiates erosion, causing long-term sedimentation problems
    - Changes the microhabitat, affecting primary productivity
    - Damages or kills both commercially harvested species and noncommercial species
    - Unlike clear-cutting in forests, the effects are not visible, thereby not getting much attention

41. **(C)** Pressure-treated wood (sometimes called CCA), which usually has a greenish tint to it, contains arsenic, a highly toxic element, as well as chromium and copper. Dog and cat feces may carry diseases that can infect humans. Morning glory, sheep sorrel, ivy, several kinds of grasses, and some other plants can resprout from their roots and/or stems in the compost pile. Bones (and the fat and marrow) are very attractive to pests such as rats. In addition, fatty food wastes can be very slow to break down because the fat can exclude the air that composting microbes need to do their work. Manures typically contain large amounts of nitrogen (the fresher the manure, the more nitrogen it contains) and are considered a green ingredient. Fresh manures can heat a compost pile quickly and will accelerate the decomposition of woody materials, autumn leaves, and other browns.
42. **(A)** Sustainability is a concept long recognized and utilized by many cultures. It results from recognition of the need for a harmonious existence between the environment, society, and economy. Sustainability focuses on improving the quality of life for all people without increasing the use of natural resources beyond the capacity of the environment to supply them indefinitely. In order for hydrogen to be a true, long-term, renewable alternative to fossil fuels (it runs more efficiently than gasoline but it costs twice to three times as much to produce), it will need to be produced with a clean, low-cost electricity source such as hydroelectric power. Most oil-reliant countries simply do not have access to vast amounts of clean electricity, but Iceland and Norway do.
43. **(B)** The organism that can adapt to the environment survives. Organism *Y* has continued to exist since it first appeared. Therefore, organism *Y* has been the most successful in the environment over time.
44. **(E)** Greenbelt advocacy groups would encourage locally owned retail stores located where people live and work. Wal-Marts and Costco only become magnets that draw in more people to a city than can be effectively accommodated. They add to traffic congestion, requiring more parking lots, requiring more roads, and decreasing open space (greenbelts).
45. **(E)** The threshold for pain is at about 120 to 140 dB, but sound begins to damage hearing when it is above 85 dB. Some of the hearing loss after exposure to excessive noise will be permanent. Indication of damage is ringing and noise in the ears (called tinnitus) after sound exposure. Research shows that cumulative exposure to loud sounds, not age, is the major cause of hearing loss.
46. **(D)** Enriched uranium is uranium whose uranium-235 content has been increased through the process of isotope separation. Natural uranium consists mostly of the U-238 isotope, with about 0.72% by weight as U-235, the only isotope existing in nature in any appreciable amount that is fissionable by thermal neutrons. Enriched uranium is a critical component for both nuclear power generation (electricity) and military nuclear weapons. The International Atomic Energy Agency attempts to monitor and control enriched uranium supplies and processes in its efforts to ensure nuclear power generation safety and to curb nuclear weapons proliferation.
47. **(B)** Between 1979 and 1989, the worldwide demand for ivory caused elephant populations to decline to dangerously low levels. During this time period, poaching fueled by ivory sales cut Africa's elephant population in half (from 1.3 million elephants to only 600,000 that still remain). Recently, that number has

stabilized, due in large part to the 1990 Convention on International Trade in Endangered Species (CITES) ban on international ivory sales.

48. (C) Albino voles stand out against a brown background. The increased visibility would have made the voles easy prey for hawks. Predation decreases population.
49. (D) Adaptations encourage the survival of the species in a given environment.
50. (E) The effects of protective coloration, as related to predation, can be evaluated by this experiment because two varieties of voles were subjected to contrasting backgrounds.
51. (B) Both organisms are benefited by the association.
52. (C) The fish is harmed through the loss of its nutritive material, while the clam derives all the benefits from the association.
53. (A) *Natural Capitalism* by Paul Hawken, Amory Lovins, and L. Hunter Lovins introduced four central strategies that are a means to enable countries, companies, and communities to operate by behaving as if all forms of capital were valued: radical resource productivity, biomimicry, service and flow economy, and investing in natural capital.
54. (E) In 1964, Congress established the National Wilderness Preservation System, under the Wilderness Act. The legislation set aside certain federal lands as wilderness areas. These areas, generally 5,000 acres or larger, are wild lands largely in their natural state. The act says that they are areas "where the Earth and its community of life are untrammeled by man, where man himself is a visitor who does not remain." Four federal agencies of the United States government administer the National Wilderness Preservation System, which includes 644 areas, and more than 105 million acres.

    The National Park Service (Department of the Interior) was established to protect the nation's natural, historical, and cultural resources and to provide places for recreation. The Park Service manages 51 national parks. It also oversees more than 300 national monuments, historic sites, memorials, seashores, and battlefields. It manages 13% of federal lands and 42% of the National Wilderness Preservation System.

    The U.S. Forest Service (Department of Agriculture) manages national forests and grasslands. It conducts forestry research and works with forest managers on state and private lands. The Forest Service oversees close to 200 million acres of national forest and other lands. It manages 30% of federal lands and 33% of the National Wilderness Preservation System.

    The Bureau of Land Management (Department of the Interior) manages nearly 270 million acres. Among other activities, the bureau conserves these lands and their historical and cultural resources for the public's use and enjoyment. It manages 42% of federal lands and 5% of the National Wilderness Preservation System.

    The U.S. Fish and Wildlife Service (Department of the Interior) conserves the nation's wild animals and their habitats by managing a system of more than 500 national wildlife refuges and other areas, totaling more than 91 million acres of land and water. It manages 15% of federal lands and 20% of the National Wilderness Preservation System.

55. **(C)** The shaded area, which represents the territory shared by the two pairs, is the area where competition is likely to occur. Competition arises when organisms utilize similar resources that are limited and often involves food or space. The two most common forms of competition are interference and exploitative. Interference includes aggressive interactions in which one individual actively attempts to exclude another. Exploitative competition is indirect competition in which one species uses more of the limited resource or uses the resource more efficiently than another species.
56. **(D)** The most likely reason why the range of pair *B* is larger than that of pair *A* is that pair *B* has less food available near its nest than pair *A*. Pair *B* has extended its range to increase its chance of capturing food.
57. **(B)** The most probable explanation for the increase in the rabbit population after 1950 was that the coyote population declined drastically. The relationship indicated by the graph is one of predator-prey. The coyotes prey upon rabbits, keeping the population in check.
58. **(E)** The most probable explanation for the decline of the coyotes is removal by humans. Since the decline in coyotes occurred with cattle introduction and since cattle and coyotes do not compete for the same food, the individuals that introduced the cattle must have interfered with the coyote population. Because coyotes are predators, they were perceived as dangerous to the cattle and were hunted and poisoned by humans. Attempts to control coyotes by poisoning may deplete the numbers of their natural prey and lead to increasing attacks by coyotes on farm animals.
59. **(A)** Savannas are warm year-round with a prolonged dry season and scattered trees. The environment is intermediate between grassland and forest. An extended dry season followed by a rainy season occurs in Australia; Central, Eastern, and South Africa; India; Madagascar; central South America; Southeast Asia; and Thailand. Savannas consist of grasslands with stands of deciduous shrubs and trees that do not grow more than 30 meters high. Trees and shrubs generally shed leaves during the dry season, which reduces the need for water. Food is limited during the dry season so that many animals migrate during this season. Soils are rich in nutrients. Savannas contain large herds of grazing animals and browsing animals that provide resources for predators.
60. **(D)** The tundra is located at 60° north latitude and farther north. The weather is influenced by the polar cell. Alpine tundra is located in mountainous areas, above the tree line, with well-drained soil and where dominant animals are small rodents and insects. Arctic tundra is frozen treeless plains, low rainfall, low average temperatures (summers average <10°C), and many bogs and ponds. Frozen ground prevents drainage. The growing season lasts 50–60 days. Tundra is found in Alaska, Canada, Europe, Greenland, and Russia. Dominant vegetation includes flowering dwarf shrubs, grasses, lichens, mosses, and sedges. The soil has few nutrients due to low vegetation and little decomposition. There are between 60 to 100 frost-free days per year. Arctic tundra is higher latitude than alpine tundra.
61. **(D)** Biosphere reserves are part of an international network. They are nominated by national governments, must meet a minimal set of criteria, and must adhere to a minimal set of conditions before being admitted into the network.

Each biosphere reserve is intended to fulfill three basic functions, which are complementary and mutually reinforcing:

1. A conservation function—to contribute to the conservation of landscapes, ecosystems, species, and genetic variation
2. A development function—to foster economic and human development that is socioculturally and ecologically sustainable
3. A logistic function—to provide support for research, monitoring, education, and information exchange related to local, national, and global issues of conservation and development

62. (E) Swamps and marshes are the most productive biomes, producing the most biomass per year, while tropical rain forests have the most standing biomass. Extreme deserts are the least-productive biomes. Lakes and streams are equivalent to extreme deserts for having the least standing biomass.
63. (A) The U.S. District Courts are the trial courts of the federal court system. Within limits set by Congress and the Constitution, the district courts have jurisdiction to hear nearly all categories of federal cases, including both civil and criminal matters. There are 94 federal judicial districts.
64. (C) At the point of discharge, there is a sudden increase in the amount of toxins, suspended solids, and dissolved organic compounds. The water becomes unsuitable for any human use. Aerobic bacteria increase rapidly, reducing the organic pollutants but using up dissolved oxygen. As the oxygen level falls, anaerobic bacteria multiply, resulting in an unpleasant smell or stench. Eventually, as their food supply is used up, the decomposers reduce and algae thrive, reoxygenating the water. Fish and other animal life reappear, and the stream returns to its previous condition.

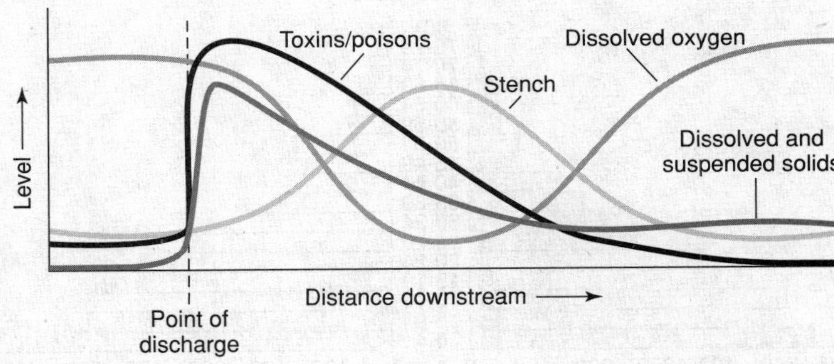

65. (D) Thomas Malthus recognized that once the carrying capacity of an area was exceeded, organisms would die of starvation. Darwin's theory of evolution based on natural selection may have been influenced by Malthus's concept.
66. (E) Note that only 10% of the population of young oysters reaches the 20th unit of life span.
67. (A) Note that almost all of the fruit flies reach the 90th unit of life span before the entire population dies.
68. (C) On a cool, inactive day, the average man loses about 12 eight-ounce cups of water but consumes only about 9 cups of water (about half of that from the water in fruits, vegetables, and other solid foods).

69. (D) Daily indoor per capita water use in the typical single family home with no water-conserving fixtures is 74 gallons. If all U.S. households installed water-saving features, water use would decrease by 30%, saving an estimated 5.4 billion gallons per day.

Use	Gallons per Person	Percentage of Total Daily Use
Showers	12.6	17.3%
Clothes washers	15.1	20.9%
Dishwashers	1.0	1.3%
Toilets	20.1	27.7%
Baths	1.2	2.1%
Leaks	10.0	13.8%
Faucets	11.1	15.3%
Other Domestic Uses	1.5	2.1%
Total	~ 74	~ 100%

70. (A) To calculate doubling time, divide the country's growth rate into the number 70 (actually 69.3 for better accuracy). Thus, a growth rate of 2% will double a population in only 35 years, 1% in 70 years, and so forth.
71. (D) The age structure diagram for Rwanda in 2000 follows:

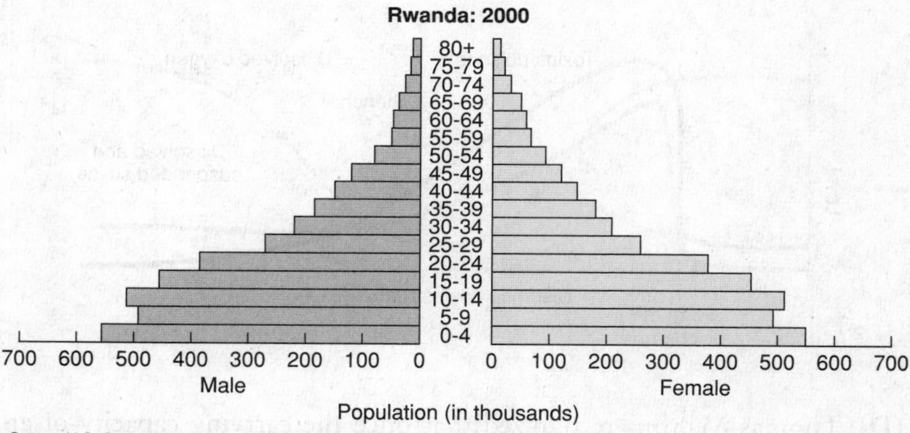

Source: U.S. Census Bureau, International Data Base.

Life expectancy is approximately 41 years. Infant mortality rate (under 5) is 170 per 1,000 live births. The growth rate in Rwanda is 1.16%, lower than in most least-developed countries due to AIDS and emigration due to civil war.

72. (C) When a project is proposed that will affect the environment, laws state that studies be conducted to evaluate the impact that the project will have on land, water, air, plants, animals, humans, and the total environment and balance of nature. These laws are in place for the protection of the environment and species. The reports generated from these studies are called environmental impact reports (EIR).

73. (D) A pure capitalistic economy, in which financial gain is the primary societal motivator, leads to always taking just one more. Be careful of the double negative in the question.
74. (C) Industrial wastes of unknown content are often commingled with domestic wastes in sanitary landfills. Groundwater infiltration and contamination of water supplies with toxic chemicals have recently led to more active control of landfills and industrial waste disposal. Careful management of sanitary landfills, such as providing for leachate and runoff treatment as well as daily coverage with topsoil, has stopped most of the problems of open dumping. In many areas, however, space for landfills is running out, and alternatives must be found.
75. (D) Bringing in sand from other locations does not solve the problem. Although building away from the beach is best, seawalls and breakwaters are built only to protect oceanfront property. When the tide is high, waves hit the seawalls and breakwaters. Often their energy is deflected onto sand in front, contributing to erosion. Over time, this can loosen the wall's footing, causing the seawall to tilt and break up. Restricting access to the beach is draconian and reduces quality of life.
76. (D) About one-third of the United States is forested.
77. (D) The depth at which plankton can exist depends on the penetration of sunlight. Sunlight provides the energy for photosynthesis. Since the phytoplankton is abundant near the surface, the region of sunlight penetration, the consumers are also found in this region.
78. (A) Point-source pollution consists of pollutants discharged from any distinct, identifiable point, or source, including pipes, ditches, channels, sewers, tunnels, or wells. In this case, the electrical generating plant is a distinct, identifiable source.
79. (A) Rill erosion is the collection of sheet erosion water into small channels (rills).
80. (D) John Muir was America's most famous and influential naturalist and preservationist. As a wilderness explorer, he is renowned for his lone excursions in California's Sierra Nevada, among Alaska's glaciers, and in worldwide travels in search of nature's beauty. As a writer, he taught the people of his time (and ours) the importance of experiencing and protecting our natural heritage.
81. (C) President Roosevelt recommended that the CCC (Civilian Conservation Corps) operate in cooperation with and under the technical supervision of the War Department, the Department of the Interior, the Department of Agriculture, and the Department of Labor. Many CCC projects centered on forestry, flood control, prevention of soil erosion, and fighting forest fires.
82. (A) Congress approved several of President Carter's energy proposals, including the deregulation of natural gas prices and incentives for such conservation measures as conversion to coal in industry and fuel-saving improvements in the home. During the last days of his administration, Jimmy Carter signed the Alaska Lands Act, which set aside nearly 103 million acres as national parks, national monuments, wild and scenic rivers, wilderness areas, and wildlife refuges.
83. (E) The Sagebrush Rebellion is known as organized resistance in the West to federal public land policies. One-third of all national land is administered by

the federal government, most of it in the West. When states west of the Mississippi River were admitted to the union, each state had to agree to give up all claims to unsettled lands within its borders. In the late 1970s a movement was started to try and gain the land back from the federal government. The 11 states involved felt the land was rightfully theirs and that they could better utilize the land through extrapolation of resources. The main cause of the defeat of the Sagebrush Rebellion was the states' inability to establish the basic legal claim that the public domain truly belongs to the states. Therefore, the lands remain federal property for use by citizens.

84. **(D)** 700,000,000 × 0.0175 = 12,250,000. Adding this to the original population of 700,000,000 = 712,250,000.
85. **(D)** The Endangered Species Act (ESA), signed into law by President Nixon in 1973, is the nation's principal protection for animal and plant species that are on the brink of extinction. Currently about 950 U.S. species are listed as endangered or threatened under the act.
86. **(C)** From a population of tens of thousands in the 19th century, the bald eagle had declined to only 417 pairs in the lower 48 states by 1963. By 1994, after 20 years of protection through the Endangered Species Act, the population had grown to more than 4,000 breeding pairs.
87. **(D)** Couplings are one-way linkages in which one component of a system affects another component. In positive coupling between components *A* and *B*, a change in *A* causes a change of the same sign in *B*. In negative coupling, a change in *A* causes a change of the opposing sign in *B*. Two or more couplings acting in sequence form a closed loop. In a positive-feedback loop, a positive or negative change in one of the components of the system is amplified by the sequence of couplings. In a negative-feedback loop, the change is reduced in amplitude.
88. **(C)** Globally, approximately 40 million adults and children were living with HIV/AIDS at the end of 2004. Of them, more than 95% were living in low- and middle-income countries. In 2004, 4.9 million people were newly infected with HIV, and there were 3.1 million adult and child deaths due to HIV/AIDS. Since the beginning of the epidemic, there have been more than 20 million AIDS deaths.
89. **(A)** In 2000, beach pollution prompted at least 11,270 closings and swimming advisories in the United States, twice the number that occurred in 1999. The increase was due largely to improved and increased monitoring and reporting. The most frequent pollution sources are polluted runoff and storm water, which led to more than 4,102 closings in 2000.
90. **(B)** Atmospheric carbon dioxide levels are rising rapidly. Currently, they are 25% above where they stood before the Industrial Revolution. Earth's atmosphere now contains some 200 gigatons (GT) more carbon than it did two centuries ago.
91. **(E)**

$$\frac{550 \text{ parts}}{\cancel{\text{million}}} \times \frac{1{,}000 \cancel{\text{ million}}}{\text{billion}} = \frac{550{,}000 \text{ parts}}{\text{billion}}$$

92. (A) Marine sediments contain an estimated 4×10^9 Tg (1 Tg = 10^{12} grams with an estimated turnover time of 2×10^8 years). Turnover time is the time it takes for a complete cycle to occur within a system or the ratio of the mass of a reservoir to the rate of its removal from that reservoir.
93. (D) Shield volcanoes are built almost entirely of fluid lava flows. Flow after flow pours out in all directions from a central summit vent, or group of vents, building a broad, gently sloping cone with a flat, domed shape. They are built up slowly by the accumulation of thousands of flows of highly fluid basaltic lava that spread widely over great distances and then cool as thin, gently dipping sheets.

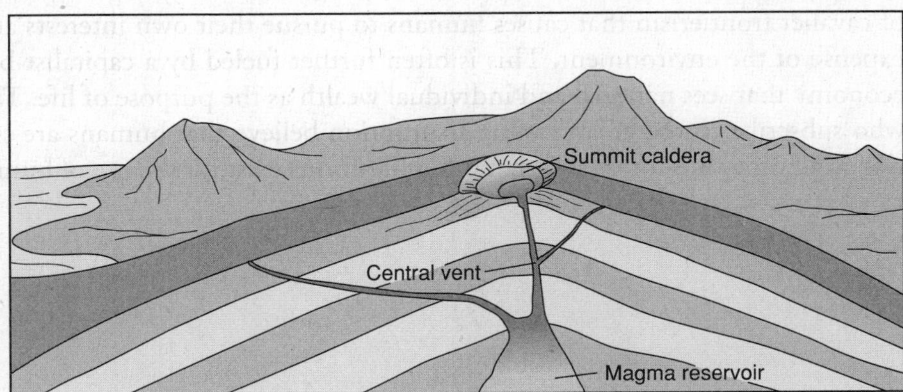

94. (D) In (A) activity is low, and it occurs at very shallow depths. The point is that the lithosphere is very thin and weak at these boundaries, so the strain cannot build up enough to cause large earthquakes. The San Andreas fault is a good example of (B) in which two mature plates are scraping by one another. The friction between the plates can be so great that very large strains can build up before they are periodically relieved by large earthquakes. In (C) one plate is thrust or subducted under the other plate so that a deep ocean trench is produced. In (E) shallow earthquakes are associated with high mountain ranges where intense compression is taking place. In (D) the interiors of the plates themselves are largely free of large earthquakes, that is, they are aseismic.
95. (C) During the Carboniferous, one of the great die-offs or mass extinctions occurred on Earth. The plant and animal life was killed off in such massive quantities that perhaps 90% of all living beings died roughly simultaneously.
96. (B) Differences in air pressure help cause winds and affect air masses. They are also factors in the formation of storms such as thunderstorms, tornadoes, and hurricanes. Differences in air pressure are shown on a weather map with lines called isobars. The weather map illustrates isobars marking areas of high and low pressure. High-pressure areas generally have dry, good weather, and areas of low pressure have precipitation. On the weather map, the only area of low pressure is centered in the Midwest.
97. (E) If the soil was unable to form a ball, it would have been sand or gravel. Had the students been able to form a ribbon, the soil would have contained clay.

98. **(A)** SBS (sick-building syndrome) and BRI (building-related illness) are associated with acute or immediate health problems. Radon and asbestos cause long-term diseases that occur years after exposure and are therefore not considered to be among the causes of sick buildings. Indicators of SBS include headache; eye, nose, or throat irritation; dry cough; dry or itchy skin; dizziness and nausea; difficulty in concentrating; fatigue; and sensitivity to odors.
99. **(D)** Tropospheric ozone would be ozone found in the air that we breathe, generally produced by burning fossil fuels. No relationship has been found that skin cancer is increased by decreases in smog.
100. **(B)** Frontier ethics would be consistent with humans are apart from nature. In many developed nations, human values have traditionally expressed a kind of cavalier frontierism that causes humans to pursue their own interests at the expense of the environment. This is often further fueled by a capitalist-based economy that sees national and individual wealth as the purpose of life. Those who subscribe to the greater value assumption believe that humans are superior to all other life-forms and that other life-forms exist for the use of humans.

FREE-RESPONSE EXPLANATIONS

Question 1

(a) 4 points maximum (5 points possible)

Restatement: Total amount of energy (in Btus) to wash the clothes using the washing machine.

Mass of water:

$$20 \text{ \sout{gallons H}}_2\text{\sout{O}} \times \frac{8 \text{ pounds}}{1 \text{ \sout{gallon H}}_2\text{\sout{O}}} = 160 \text{ pounds } H_2O \quad \text{(1 point)}$$

Energy to heat water:

$$160 \text{ \sout{pounds H}}_2\text{\sout{O}} \times \frac{1 \text{ Btu/\sout{°F}}}{1 \text{ \sout{pound H}}_2\text{\sout{O}}} \times (130°F - 60\text{\sout{°F}}) = 11{,}200 \text{ Btu} \quad \text{(1 point)}$$

Energy to run washing machine (kWh):

$$1500 \text{ \sout{Watts}} \times \frac{1 \text{ kW}}{1000 \text{ \sout{Watts}}} \times 30 \text{ \sout{min}.} \times \frac{1 \text{ hr.}}{60 \text{ \sout{min.}}} = 0.75 \text{ kWh} \quad \text{(1 point)}$$

Energy to run washing machine (Btus):

$$0.75 \text{ \sout{kWh}} \times \frac{3{,}400 \text{ Btu}}{1 \text{ \sout{kWh}}} = 2{,}550 \text{ Btu} \quad \text{(1 point)}$$

Total energy for washing machine:

$$11{,}200 \text{ Btu} + 2{,}550 \text{ Btu} = \mathbf{13{,}750\ Btu} \quad \text{(1 point)}$$

(b) 2 points maximum

Restatement: Total amount of energy (in Btus) to wash clothes by hand.

$$35 \text{ \sout{gallons H}}_2\text{\sout{O}} \times \frac{8 \text{ pounds}}{1 \text{ \sout{gallon H}}_2\text{\sout{O}}} = 280 \text{ pounds } H_2O \quad \text{(1 point)}$$

Energy to heat water:

$$280 \text{ \sout{pounds H}}_2\text{\sout{O}} \times \frac{1 \text{ Btu/\sout{°F}}}{1 \text{ \sout{pound H}}_2\text{\sout{O}}} \times (130\text{\sout{°F}} - 60\text{\sout{°F}}) = \mathbf{19{,}600\ Btu} \quad \text{(1 point)}$$

NOTE:

1. If you do NOT show calculations, no points are awarded.
2. No penalty assessed if you do not show units. However, you risk setting up the problem incorrectly if you do not show units so that they cancel properly.
3. If your setup is correct but you make an arithmetic error, no penalty.

(c) 4 points maximum (13 points possible)

Restatement: Environmental impact of washing clothes.

NOTE: There are no points for just listing ideas. Each idea MUST be explained. The following are ideas that you could use to write your paragraph(s). Take these ideas and create an outline of the order in which you wish to answer them. You do NOT need to use all ideas. Before you begin, decide on the format of how you wish to answer your question (pros versus cons, chart format with explanations within the chart, compare and contrast, and so on). Each numbered point when explained is worth 1 point.

1. Manufacturing washing machines provides jobs. Mining, smelting, designing, engineering, manufacturing, transportation, administrative, advertising, sales.
2. Advantages of repairing washing machines rather than discarding them. Discarding washing machine and landfill issues. Repairing machines provides jobs.
3. Energy efficiency of different models. Labels.
4. Using washing machine during off-peak hours.
5. Use of natural resources in manufacturing washing machine.
6. Type of energy used to produce electricity—renewable energy sources versus nonrenewable. Example, coal, oil, natural gas, solar, hydroelectric.
7. Washing clothes by hand saves (21,400 Btu – 19,600 Btu) = 1,800 Btus
8. Washing clothes by machine saves hot water (35 gallons – 20 gallons) = 15 gallons
9. Washing clothes by machine saves time.
10. Pollution caused by wastewater. Example, phosphates, groundwater contamination, and so on.
11. Air pollution caused by heating water.
12. Water treatment options. Example, using recycled water, using gray water for landscaping use, and others.
13. Choosing a machine that uses the fewest gallons of water per pound of clothes and one that has high, medium, and low water level controls and an automatic cold rinse cycle. When using the washing machine, following these suggestions to save energy:
    - Pre-soak heavily soiled clothes.
    - Fill the washer according to load level.
    - Follow the manufacturer's instructions for adding detergent. Choose a biodegradable detergent.
    - Use as low a water temperature for the washing cycle as will give satisfactory cleaning. Cold and medium water temperatures may vary with the season.
    - Always use a cold rinse cycle.
    - Follow maintenance instructions found in the owner's manual.

Question 2

(a) (i) 2 points maximum

Restatement: Biomass as a function of population density

To determine the optimum population density that produced the greatest biomass would require *dividing the total biomass for each planting by the total amount of seeds planted* per pot. This would provide the average biomass and produce the following results:

Seeds per Pot	Average Biomass (g)
1	5.0
10	7.0
20	3.8

From the data table, it appears that *10 radish seeds per pot produced the largest average biomass.* When graphed, the results look like:

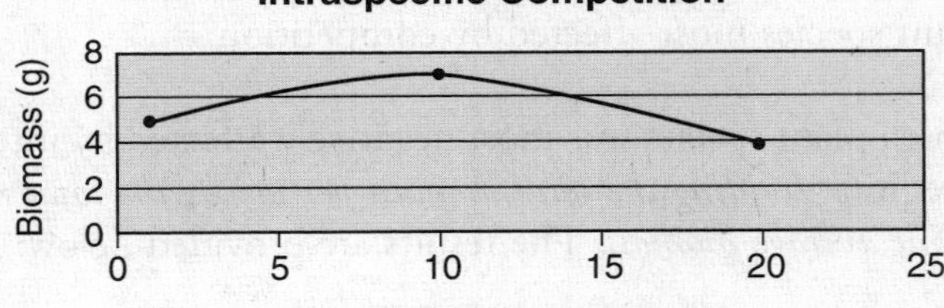

(a) (ii) 2 points maximum

Restatement: Limited resources

The maximum biomass was achieved at 10 seeds per pot. After that, as population density increased, biomass decreased. Factors that may have limited biomass might have included:

1. *Competition for soil nutrients.* As root density increased, less nutrients would have been available for growth.
2. *Competition for light.* As the density increased, less light would have been available for smaller seedlings.
3. *Competition for water.* Given a fixed amount of water provided, as plant density increased, less water would have been available for additional seedlings.

(a) (iii) 2 points maximum

Restatement: Results obtained from the class for Part I in terms of biological laws or principles

The effect of intraspecific competition in plant populations is usually examined by planting the species over a range of densities. The most common result is that a point is reached where the mean weight per plant decreases as density increases, so that the maximum yield approaches some constant value. This result is called the *law of constant yield.* The maximum plant productivity of a particular environment is called the *carrying capacity.* In this case, the carrying capacity was reached when the biomass reached 7.0 grams for the size of the pot. Environmental variables (nutrients, space, water, and light) that restrict the realized niche are called limiting factors. *Liebig's law of the minimum* states that an organism is most limited by the essential factor that is in least supply. In terms of *survivorship,* the radish seeds represent Type III characteristics that include large amounts of seeds being produced with few surviving, small size invaders of disturbed environments, and rapid growth.

(b) (i) 2 points maximum

Restatement: Plant species most affected by competition

To determine which plant species was most negatively affected by interspecific competition would require *dividing the total biomass per pot by the total number of seeds per pot to determine average biomass.* The results are provided below:

Seeds per Pot	Average Biomass (g)
1 radish	5.0
1 wheat	3.0
10 radish	5.0
10 wheat	2.5
20 radish	3.8
20 wheat	2.0

From the table, it appears that the *wheat seedlings were negatively impacted by the presence of the radish seeds.* When graphed, the results are:

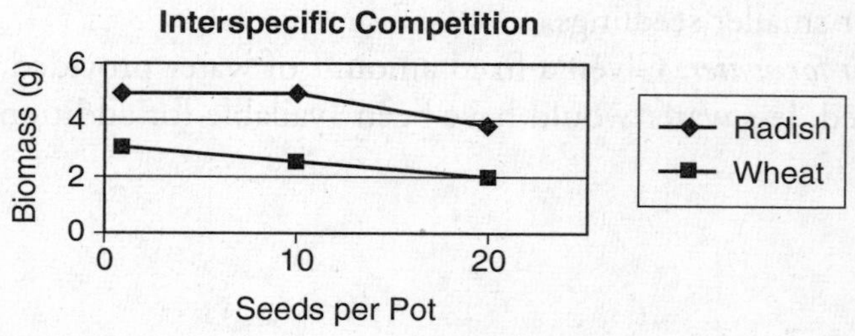

(b) (ii) 2 points maximum

Restatement: Possible reasons why the plant chosen in (b) (i) may have been more successful

Two different species of plants were grown in the *same* medium. *Nutrient requirements may have been different* between the two species. For example, the amount and ratios of minerals contained in the potting soil may have met the metabolic requirements of the radish plants more than the wheat seedlings. *Water and light availability* (either too much or too little) may also have affected outcome. *Differences in growth patterns* (radish may sprout earlier than wheat) may have existed. In this case, the radish seedlings may have *sprouted earlier* and established a population that was taking up a fixed amount of nutrients at a faster rate or had established a canopy that decreased the light available for the wheat seedlings. Simultaneous to the interspecific competition was intraspecific competition. Not only were radish seedlings in competition with wheat seedlings, they were also in competition with other radish seedlings for limited nutrients.

(c) (iii) 2 points maximum

Restatement: Results obtained from the class for Part II in terms of biological laws or principles

In Part II, plant density was held constant in a substitutive or replacement experimental design in which the total density of both species was kept constant while the relative densities were varied (1, 10, and 20). This approach allows investigation of the effects of interspecific competition from the effects of increasing overall plant density on growth or production of a species. However, it requires first quantifying the background effect of intraspecific competition on growth of each species individually. Such an experiment has three possible outcomes. First, one species prospers at the expense of the other (competitive exclusion). Second, one species outperforms the other but only when in higher proportion (coexistence). Third, the two species have no measurable effect on each other (no competition). The latter would be the null hypothesis (H_o) by which the class would judge whether or not one of the other two scenarios happened. Another scenario, which was not tested for, could have been instability—that is, two species are in a constant state of dynamic tension where at one time one species dominates and at other times the other species dominates. Different species of plants are able to coexist within the same biome through resource partitioning, which functions through evolution of different metabolic pathways, having different tolerances to shade, taking up water and nutrients at different depths, and so on.

Question 3

(a) 4 points maximum (13 points possible)

List any two characteristics from the column labeled "Sweden" and any two characteristics from the column labeled "Kenya."

Restatement: Given two age-structure diagrams, Sweden and Kenya for 2000, compare and contrast the diagrams in terms of population dynamics.

Age structure diagrams are basically divided into three major age categories:

- Prereproductive (0–15 years old)
- Reproductive (16–45 years old)
- Postreproductive (46 years old–death)

Population Characteristics

	Affected By	Sweden	Kenya
Birth Rate	• Importance of children as a part of the labor force • Urbanization • Cost of raising and educating children • Educational and employment opportunities for women • Infant mortality rate • Average age at marriage • Pensions • Abortions • Birth control • Religious beliefs	• Population has nearly equal proportions of prereproductive and reproductive individuals. • Little growth over a long period of time will produce a population with about equal numbers of people in all age groups. • Children not required or necessary to support parents. • Availability and acceptance of birth control.	• Population has pyramid–shaped age structures, with large numbers of prereproductive individuals. • Population momentum results from large numbers of prereproductive children becoming reproductive within short period of time. • High population rate due to high birth mortality rates. • Children viewed as status symbol. • Resistance to birth control.
Death Rate	• Increased food supply • Better nutrition • Improved medical and public health technology • Improvements in sanitation and personal hygiene • Safer water supplies	• Elderly survive longer due to advances in medical technology and availability. • Social welfare programs ensure that elderly are taken care of.	• Elderly do not survive due to lack of available medical technology. • Disease (for example, malaria or AIDS) and lack of nutritious food decreases life span.

(b) 4 points maximum (13 points possible)

List any four characteristics from the column labeled "Affected By."

Restatement: Factors that affect both birth and death rates

See preceding chart.

(c) 2 points maximum (6 points possible)

List any two of the six methods mentioned.

Restatement: Methods that have been employed by another country to curb population growth

China: Between 1958 and 1962, an estimated 30 million people died from famine in China. Since then, China has made good progress in trying to feed its people and bring its population growth under control. Much of this reduced population growth was brought about by a drop in the birth rate from 32 to 18 per 1,000 between 1972 and 1985. China instituted one of the most rigorous population control programs in the world at an estimated cost of about $1 per person. Some features of the program included:

1. Strong encouragement for couples to postpone marriage.
2. Providing married couples with free access to sterilization, contraceptives, and abortion.
3. Giving couples who sign pledges to have no more than one child economic rewards such as salary bonuses, extra food, larger pensions, better housing, free medical care and school tuition for their child, and preferential treatment in employment when the child grows up.
4. Requiring those who break the pledge to return all benefits.
5. Exerting pressure on women pregnant with a third child to have abortions.
6. Requiring one of the parents in a two-child family to be sterilized.

Question 4

(a) 3 points

Restatement: Definition and three major categories of pesticides

A pesticide is any substance or mixture of substances intended for preventing, destroying, repelling, or mitigating any pest. Pests can be insects, mice and other animals, unwanted plants (weeds), fungi, or microorganisms like bacteria and viruses. The term pesticide also applies to herbicides, fungicides, and various other substances used to control pests. A pesticide is also any substance or mixture of substances intended for use as a plant regulator, defoliant, or desiccant. Types of pesticides include (*choose any three*):

Algicides	Control algae in lakes, canals, swimming pools, water tanks, and other sites.
Antifouling agents	Kill or repel organisms that attach to underwater surfaces, such as boat bottoms.
Antimicrobials	Kill microorganisms (such as bacteria and viruses).
Attractants	Attract pests (for example, to lure an insect or rodent to a trap). Food is not considered a pesticide when used as an attractant.
Biocides	Kill microorganisms.
Defoliants	Cause leaves or other foliage to drop from a plant, usually to facilitate harvest.
Desiccants	Promote drying of living tissues, such as unwanted plant tops.
Disinfectants and sanitizers	Kill or inactivate disease-producing microorganisms on inanimate objects.
Fungicides	Kill fungi (including blights, mildews, molds, and rusts).
Fumigants	Produce gas or vapor intended to destroy pests in buildings or soil.
Herbicides	Kill weeds and other plants that grow where they are not wanted.
Insect growth regulators	Disrupt the molting, maturing from pupal stage to adult, or other life processes of insects.
Insecticides	Kill insects and other arthropods.

Miticides (acaricides)	Kill mites that feed on plants and animals.
Microbial pesticides	Microorganisms that kill, inhibit, or outcompete pests, including insects or other microorganisms.
Molluscicides	Kill snails and slugs.
Nematicides	Kill nematodes (microscopic, wormlike organisms that feed on plant roots).
Ovicides	Kill eggs of insects and mites.
Pheromones	Biochemicals used to disrupt the mating behavior of insects.
Plant growth regulators	Substances (excluding fertilizers or other plant nutrients) that alter the expected growth, flowering, or reproduction rate of plants.
Repellents	Repel pests, including insects (such as mosquitoes) and birds.
Rodenticides	Control rats, mice, and other rodents.

(b) 2 points maximum

Restatement: Two positive and two negative effects of pesticide use

Positive

(1) Plants are directly and indirectly humankind's main source of food. They are attacked by tens of thousands of diseases caused by viruses, bacteria, fungi, and other organisms. There are over 30,000 kinds of weeds competing with crops worldwide; thousands of nematode species reduce crop vigor; and some 10,000 species of insects devour crops. It is estimated that one-third of the world's food crop is destroyed by these pests annually. *Pesticides increase the world food supply.*

(2) There are an estimated 300–500 million cases of malaria per year. The majority of these occur in Africa. The vast majority of the estimated 1 million annual deaths from the disease occur among children and mainly among poor African children. Malaria is above all a disease of the poor, impacting at least three times more greatly on the poor than any other disease. Africa's GDP would be up to $100 billion greater if malaria had been eliminated years ago. Mosquitoes have been estimated to be responsible for half of all human deaths due to transmission of disease. *Pesticides improve human health by destroying disease-carrying organisms* (West Nile virus, encephalitis, African sleeping sickness, malaria, yellow fever, plague, and so on).

Negative

(1) If a pesticide is continually applied to a population of the pest species, most susceptible individuals will be killed, leaving only resistant individuals. These resistant individuals breed and multiply. Eventually a high proportion of the individuals from that pest species are now resistant to the pesticide. We have simply *caused pest populations with a higher tolerance for poisons to survive and breed*. More toxic pesticides in turn are developed and utilized.

(2) *Pesticides can accumulate in living organisms*. An example of accumulation is the uptake of a water-insoluble pesticide, such as chlordane, by a creature living in water. Since this pesticide is stored in the organism, the pesticide accumulates and increases over time. If this organism is eaten by an organism higher in the food chain that can also store this pesticide, levels can reach higher values in the higher organism than is present in the water in which it lives. Levels in fish, for example, can be tens to hundreds of thousands of times greater than ambient water levels of the same pesticide. This type of accumulation is called bioaccumulation. In this regard, it should be remembered that humans are at the top of the food chain and so may be the most vulnerable to bioaccumulation.

(c) 3 points maximum

Restatement: Three alternatives to the use of pesticides

GROWING PEST-RESISTANT CROPS

When landscaping a yard or planning a garden, choose plant varieties that are native to the region and climate. Hearty, native plants resist disease and infestation, and they often use less water.

CROP ROTATION

Plant rotation and interplanting prevent the buildup and spread of pests in one area or among specific plant types.

BENEFICIAL INSECTS AND ANIMALS

Protect and encourage the presence of insect-feeding birds, bats, spiders, praying mantises, ladybugs, predatory mites, parasitic flies, and wasps. Beneficial insect species can often be purchased in volume.

PHEROMONES

Pheromones are chemical signals produced by animals to communicate with others of the same species. In insects, they consist of highly specific perfumes, generally derivatives of natural fatty acids closely related to the aromas of fruit. They are nontoxic. Pheromones may be used to attract insects to traps or to deter insects from laying eggs. However, the most widespread and effective application of pheromones is for mating disruption.

(d) 2 points

Restatement: Describe one United States federal law or one international treaty that focuses on the use of pesticides.

The U. S. federal government first regulated pesticides when Congress passed the Insecticide Act of 1910. This law was intended to protect farmers from adulterated or misbranded products. Congress broadened the federal government's control of pesticides by passing the original Federal Insecticide, Fungicide and Rodenticide Act (FIFRA) of 1947. FIFRA required the Department of Agriculture to register all pesticides prior to their introduction in interstate commerce. A 1964 amendment authorized the secretary of agriculture to refuse registration to pesticides that were unsafe or ineffective and to remove them from the market. In 1970, Congress transferred the administration of FIFRA to the newly created Environmental Protection Agency (EPA). This was the initiation of a shift in the focus of federal policy from the control of pesticides for reasonably safe use in agricultural production to control of pesticides for reduction of unreasonable risks to humans and the environment. This new policy focus was expanded by the passage of the Federal Environmental Pesticide Control Act of 1972 (FEPCA) that amended FIFRA by specifying methods and standards of control in greater detail. In general, there has been a shift toward greater emphasis on minimizing risks associated with toxicity and environmental degradation and away from pesticide efficacy issues. Under FIFRA, no one may sell, distribute, or use a pesticide unless it is registered by the EPA. Registration includes approval by the EPA of the pesticide's label, which must give detailed instructions for its safe use. The EPA must classify each pesticide as either general use, restricted use, or both. General-use pesticides may be applied by anyone. However, restricted-use pesticides may be applied only by certified applicators or persons working under the direct supervision of a certified applicator. Because there are only limited data for new chemicals, most pesticides are initially classified as restricted use. Applicators are certified by a state if the state operates a certification program approved by the EPA.

Practice Exam 2

With Answers and Analysis

SECTION I (MULTIPLE-CHOICE QUESTIONS)

Time: 90 minutes

100 questions

60% of total grade

No calculators allowed

This section consists of 100 multiple-choice questions. Mark your answers carefully on the answer sheet.

General Instructions

Do not open this booklet until you are told to do so by the proctor.

Be sure to write your answers for Section I on the separate answer sheet. Use the test booklet for your scratch work or notes. Remember, though, that no credit will be given for work, notes, or answers written only in the test booklet. Once you have selected an answer, thoroughly blacken the corresponding circle on the answer sheet. To change an answer, erase your previous mark completely, and then record your new answer. Mark only one answer for each question.

Example **Sample Answer**

The Pacific is Ⓐ Ⓑ ● Ⓓ Ⓔ

(A) a river
(B) a lake
(C) an ocean
(D) a sea
(E) a gulf

To discourage haphazard guessing on this section of the exam, a quarter point is subtracted for every wrong answer, but no points are subtracted if you leave the answer blank. Even so, if you can eliminate one or more of the choices for a question, it may be to your advantage to guess.

Because it is not expected that all test takers will complete this section, do not spend too much time on difficult questions. First answer the questions you can answer readily. Then, if you have time, return to the difficult questions later. Do not get stuck on one question. Work quickly but accurately. Use your time effectively.

Answer Sheet 1

PRACTICE EXAM 2

1 Ⓐ Ⓑ Ⓒ Ⓓ Ⓔ
2 Ⓐ Ⓑ Ⓒ Ⓓ Ⓔ
3 Ⓐ Ⓑ Ⓒ Ⓓ Ⓔ
4 Ⓐ Ⓑ Ⓒ Ⓓ Ⓔ
5 Ⓐ Ⓑ Ⓒ Ⓓ Ⓔ
6 Ⓐ Ⓑ Ⓒ Ⓓ Ⓔ
7 Ⓐ Ⓑ Ⓒ Ⓓ Ⓔ
8 Ⓐ Ⓑ Ⓒ Ⓓ Ⓔ
9 Ⓐ Ⓑ Ⓒ Ⓓ Ⓔ
10 Ⓐ Ⓑ Ⓒ Ⓓ Ⓔ
11 Ⓐ Ⓑ Ⓒ Ⓓ Ⓔ
12 Ⓐ Ⓑ Ⓒ Ⓓ Ⓔ
13 Ⓐ Ⓑ Ⓒ Ⓓ Ⓔ
14 Ⓐ Ⓑ Ⓒ Ⓓ Ⓔ
15 Ⓐ Ⓑ Ⓒ Ⓓ Ⓔ
16 Ⓐ Ⓑ Ⓒ Ⓓ Ⓔ
17 Ⓐ Ⓑ Ⓒ Ⓓ Ⓔ
18 Ⓐ Ⓑ Ⓒ Ⓓ Ⓔ
19 Ⓐ Ⓑ Ⓒ Ⓓ Ⓔ
20 Ⓐ Ⓑ Ⓒ Ⓓ Ⓔ
21 Ⓐ Ⓑ Ⓒ Ⓓ Ⓔ
22 Ⓐ Ⓑ Ⓒ Ⓓ Ⓔ
23 Ⓐ Ⓑ Ⓒ Ⓓ Ⓔ
24 Ⓐ Ⓑ Ⓒ Ⓓ Ⓔ
25 Ⓐ Ⓑ Ⓒ Ⓓ Ⓔ
26 Ⓐ Ⓑ Ⓒ Ⓓ Ⓔ
27 Ⓐ Ⓑ Ⓒ Ⓓ Ⓔ
28 Ⓐ Ⓑ Ⓒ Ⓓ Ⓔ
29 Ⓐ Ⓑ Ⓒ Ⓓ Ⓔ
30 Ⓐ Ⓑ Ⓒ Ⓓ Ⓔ
31 Ⓐ Ⓑ Ⓒ Ⓓ Ⓔ
32 Ⓐ Ⓑ Ⓒ Ⓓ Ⓔ
33 Ⓐ Ⓑ Ⓒ Ⓓ Ⓔ
34 Ⓐ Ⓑ Ⓒ Ⓓ Ⓔ
35 Ⓐ Ⓑ Ⓒ Ⓓ Ⓔ
36 Ⓐ Ⓑ Ⓒ Ⓓ Ⓔ
37 Ⓐ Ⓑ Ⓒ Ⓓ Ⓔ
38 Ⓐ Ⓑ Ⓒ Ⓓ Ⓔ
39 Ⓐ Ⓑ Ⓒ Ⓓ Ⓔ
40 Ⓐ Ⓑ Ⓒ Ⓓ Ⓔ
41 Ⓐ Ⓑ Ⓒ Ⓓ Ⓔ
42 Ⓐ Ⓑ Ⓒ Ⓓ Ⓔ
43 Ⓐ Ⓑ Ⓒ Ⓓ Ⓔ
44 Ⓐ Ⓑ Ⓒ Ⓓ Ⓔ
45 Ⓐ Ⓑ Ⓒ Ⓓ Ⓔ
46 Ⓐ Ⓑ Ⓒ Ⓓ Ⓔ
47 Ⓐ Ⓑ Ⓒ Ⓓ Ⓔ
48 Ⓐ Ⓑ Ⓒ Ⓓ Ⓔ
49 Ⓐ Ⓑ Ⓒ Ⓓ Ⓔ
50 Ⓐ Ⓑ Ⓒ Ⓓ Ⓔ
51 Ⓐ Ⓑ Ⓒ Ⓓ Ⓔ
52 Ⓐ Ⓑ Ⓒ Ⓓ Ⓔ
53 Ⓐ Ⓑ Ⓒ Ⓓ Ⓔ
54 Ⓐ Ⓑ Ⓒ Ⓓ Ⓔ
55 Ⓐ Ⓑ Ⓒ Ⓓ Ⓔ
56 Ⓐ Ⓑ Ⓒ Ⓓ Ⓔ
57 Ⓐ Ⓑ Ⓒ Ⓓ Ⓔ
58 Ⓐ Ⓑ Ⓒ Ⓓ Ⓔ
59 Ⓐ Ⓑ Ⓒ Ⓓ Ⓔ
60 Ⓐ Ⓑ Ⓒ Ⓓ Ⓔ
61 Ⓐ Ⓑ Ⓒ Ⓓ Ⓔ
62 Ⓐ Ⓑ Ⓒ Ⓓ Ⓔ
63 Ⓐ Ⓑ Ⓒ Ⓓ Ⓔ
64 Ⓐ Ⓑ Ⓒ Ⓓ Ⓔ
65 Ⓐ Ⓑ Ⓒ Ⓓ Ⓔ
66 Ⓐ Ⓑ Ⓒ Ⓓ Ⓔ
67 Ⓐ Ⓑ Ⓒ Ⓓ Ⓔ
68 Ⓐ Ⓑ Ⓒ Ⓓ Ⓔ
69 Ⓐ Ⓑ Ⓒ Ⓓ Ⓔ
70 Ⓐ Ⓑ Ⓒ Ⓓ Ⓔ
71 Ⓐ Ⓑ Ⓒ Ⓓ Ⓔ
72 Ⓐ Ⓑ Ⓒ Ⓓ Ⓔ
73 Ⓐ Ⓑ Ⓒ Ⓓ Ⓔ
74 Ⓐ Ⓑ Ⓒ Ⓓ Ⓔ
75 Ⓐ Ⓑ Ⓒ Ⓓ Ⓔ
76 Ⓐ Ⓑ Ⓒ Ⓓ Ⓔ
77 Ⓐ Ⓑ Ⓒ Ⓓ Ⓔ
78 Ⓐ Ⓑ Ⓒ Ⓓ Ⓔ
79 Ⓐ Ⓑ Ⓒ Ⓓ Ⓔ
80 Ⓐ Ⓑ Ⓒ Ⓓ Ⓔ
81 Ⓐ Ⓑ Ⓒ Ⓓ Ⓔ
82 Ⓐ Ⓑ Ⓒ Ⓓ Ⓔ
83 Ⓐ Ⓑ Ⓒ Ⓓ Ⓔ
84 Ⓐ Ⓑ Ⓒ Ⓓ Ⓔ
85 Ⓐ Ⓑ Ⓒ Ⓓ Ⓔ
86 Ⓐ Ⓑ Ⓒ Ⓓ Ⓔ
87 Ⓐ Ⓑ Ⓒ Ⓓ Ⓔ
88 Ⓐ Ⓑ Ⓒ Ⓓ Ⓔ
89 Ⓐ Ⓑ Ⓒ Ⓓ Ⓔ
90 Ⓐ Ⓑ Ⓒ Ⓓ Ⓔ
91 Ⓐ Ⓑ Ⓒ Ⓓ Ⓔ
92 Ⓐ Ⓑ Ⓒ Ⓓ Ⓔ
93 Ⓐ Ⓑ Ⓒ Ⓓ Ⓔ
94 Ⓐ Ⓑ Ⓒ Ⓓ Ⓔ
95 Ⓐ Ⓑ Ⓒ Ⓓ Ⓔ
96 Ⓐ Ⓑ Ⓒ Ⓓ Ⓔ
97 Ⓐ Ⓑ Ⓒ Ⓓ Ⓔ
98 Ⓐ Ⓑ Ⓒ Ⓓ Ⓔ
99 Ⓐ Ⓑ Ⓒ Ⓓ Ⓔ
100 Ⓐ Ⓑ Ⓒ Ⓓ Ⓔ

Directions: For each question or statement, select the one lettered choice that is the best answer and fill in the corresponding circle on the answer sheet.

1. The diagram below represents a phylogenetic tree of the evolution of even-toed ungulates.

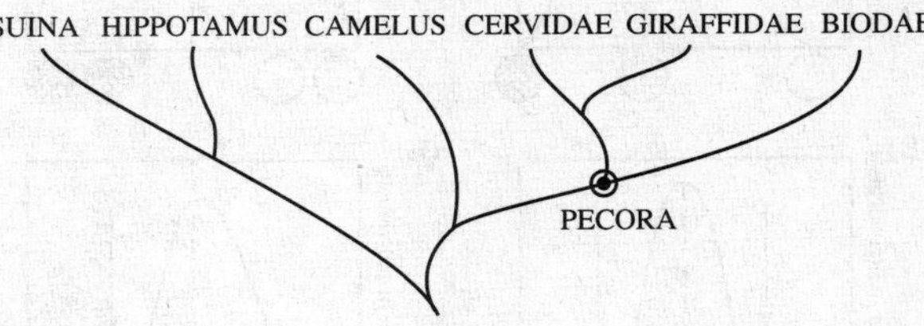

The most likely explanation for the branching pattern seen in the circled region is that

(A) environmental changes caused extinction
(B) inbreeding led to speciation
(C) no speciation occurred
(D) speciation was influenced by environmental change
(E) only the best-adapted organisms survived from generation to generation

2. Two lightbulbs are for sale—a 15-watt fluorescent and a 60-watt incandescent. You know that a 15-watt fluorescent lightbulb will produce the same amount of light as a 60-watt incandescent lightbulb. If electricity costs $0.10 per kWh and you run each lightbulb for 8,000 hours, how much money will you save in the cost of electricity by buying the fluorescent lightbulb?

(A) $4.00
(B) $8.00
(C) $12.00
(D) $36.00
(E) $48.00

3. Winds are primarily caused by

(A) differences in air pressure
(B) the Coriolis effect
(C) ocean currents
(D) seasons
(E) the Earth's rotation on its axis

4. Below are graphs describing the fates of a hypothetical population of organisms in which there is variation in color. The arrows represent selective pressures. Which graph represents a stabilizing mode of selection?

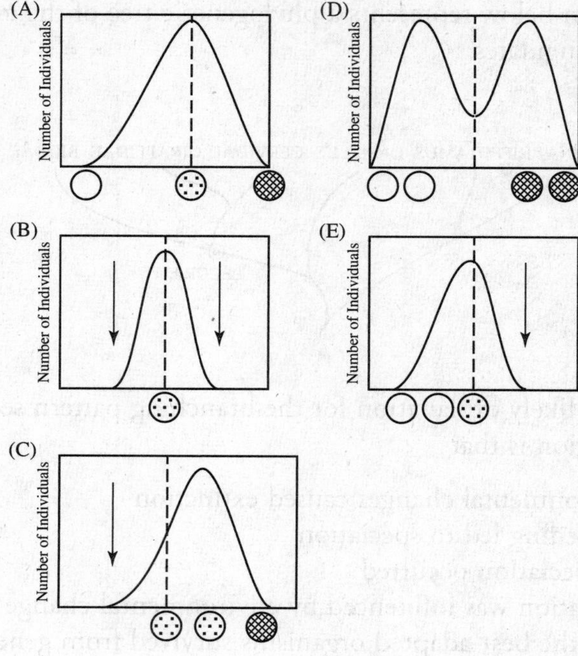

(A) A
(B) B
(C) C
(D) D
(E) E

5. You and a partner run an automobile salvage yard where you offer used automobile parts to customers, offering them substantial savings over buying new parts. Which of the following auto parts in your salvage yard would pose the greatest risk to stratospheric ozone depletion?

(A) Tires, should they catch on fire.
(B) Oil, should it leak from the crank case and enter the groundwater.
(C) Air conditioners, if they should leak Freon™.
(D) Leaking gasoline, should it catch on fire.
(E) Heavy metals from batteries such as lead should it enter the food web.

6. The ecological efficiency at each trophic level of a particular ecosystem is 20%. If the green plants of the ecosystem capture 100 units of energy, about ______ units of energy will be available to support herbivores, and about ______ units of energy will be available to support primary carnivores.

   (A) 120, 140
   (B) 120, 240
   (C) 20, 2
   (D) 20, 4
   (E) 20, 1

7. An experiment with 50 newborn rats was conducted to determine the importance of two nutrients, *A* and *B*, in their diets as possible human supplements. The dashed line shows the normal growth rate of rats based on previous experiments. The solid line shows the growth rate of the 50 newborn rats, which were fed a normal diet containing nutrients *A* and *B* from birth to point *X*. At point *X*, the rats were deprived of both nutrients. At point *Y*, nutrient *A* was again added to the diet. At point *Z*, nutrient *B* was added and nutrient *A* was continued.

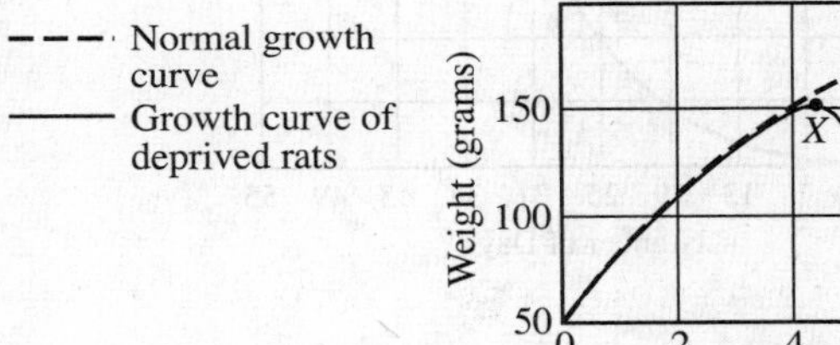

   If the experiment had continued as described except that at point *Z* nutrient *B* had not been returned to the diet of the 50 rats, it is reasonable to conclude that these rats would most likely have

   (A) lived for 4 months and then died
   (B) remained about half the size of normally developed rats
   (C) continued to gain weight, but at a slower rate than the normal rats
   (D) become sexually immature adults
   (E) continuously lost weight

8. A soil test report recommends 8 pounds of 8-0-24 per 1,000 square feet. How much phosphorus does it recommend if the area planted is equal to 10,000 square feet?

   (A) 0 pounds
   (B) 8 pounds
   (C) 24 pounds
   (D) 80 pounds
   (E) 240 pounds

9. Today, most of the world's energy comes from

 (A) natural gas, coal, and oil
 (B) oil, wood, and hydroelectric
 (C) hydroelectric, solar, and biomass
 (D) coal, oil, and nuclear
 (E) natural gas, hydroelectric, and oil

10. An AP Environmental Science class conducted an experiment to illustrate the principles of Thomas Malthus. On day 1, three male and three female fruit flies were placed in a flat-bottom flask that contained a cornmeal/banana medium. No other flies were added or removed during the course of this experiment. The students counted the number of flies in the flask each week. The graph shows the results that the class obtained after 55 days.

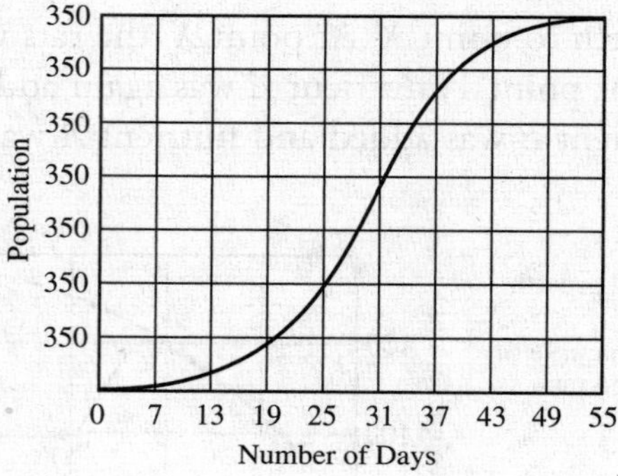

The rate of reproduction is equal to the rate of death on day

 (A) 1
 (B) 7
 (C) 25
 (D) 37
 (E) 49

11. The concept of net primary productivity

 (A) is the rate at which producers manufacture chemical energy through photosynthesis
 (B) is the rate at which producers use chemical energy through respiration
 (C) is the rate of photosynthesis plus the rate of respiration
 (D) can be thought of as the basic food source for decomposers in an ecosystem
 (E) is usually reported as the energy output of an area of producers over a given time period

12. Which of the following is NOT a unit of energy?

(A) Joule
(B) Calorie
(C) Watt
(D) Kilowatt-hour
(E) Btu

13. Which of the following is NOT a phenomenon associated with La Niña?

(A) Unusually cold ocean temperatures in the eastern equatorial Pacific
(B) Wetter-than-normal conditions across the Pacific Northwest
(C) Dryer and warmer-than-normal conditions in the southern states
(D) Warmer-than-normal winter temperatures in the southeastern United States
(E) Substantial decrease in the number of hurricanes

14. Which of the following is NOT a possible cause/effect of increasing ocean temperatures?

(A) A significant increase in the ocean circulation that transports warm water to the North Atlantic
(B) Large reductions in the Greenland and West Antarctic ice sheets
(C) Accelerated global warming
(D) Decreases in upwelling
(E) Releases of terrestrial carbon from permafrost regions and methane from hydrates in coastal sediments

15. In the nitrogen cycle, the bacteria that replenish the atmosphere with nitrogen gas (N_2) are

(A) *Rhizobium*
(B) nitrifying bacteria
(C) denitrifying bacteria
(D) nitrogen-fixing bacteria
(E) *E. coli*

16. The interface where plates move apart in opposite directions is known as a

(A) transform plate boundary
(B) convergent plate boundary
(C) divergent plate boundary
(D) subduction zone
(E) trench

17. Which biome, found primarily in the eastern United States, central Europe, and eastern Asia, is home to some of the world's largest cities and has probably endured the impact of humans more than any other biome?

(A) Desert
(B) Coniferous forest
(C) Temperate deciduous forest
(D) Grassland
(E) Chaparral

18. This type of economy exists when supplies and natural resources seem unlimited.

(A) Frontier economy
(B) Free-market economy
(C) Communal resource management economy
(D) Command economic system
(E) Natural resource economy

19. A country in sub-Saharan Africa decided spray the countryside over several months with massive amounts of DDT to rid the country of mosquitoes that were causing large numbers of citizens to contract malaria. Biologists sampled various quadrats for mosquito numbers after the spraying and the results are presented below:

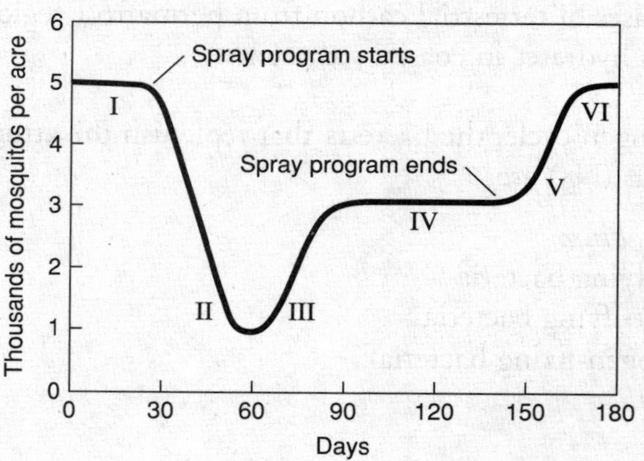

Natural selection is chiefly responsible for the section of the graph labeled

(A) I
(B) II
(C) III
(D) IV
(E) VI

20. Cars, trucks, and buses account for ______ of U.S. greenhouse gas emissions.

(A) less than 5%
(B) between 10% and 20%
(C) between 20% and 33%
(D) between 35% and 75%
(E) more than 75%

Questions 21–22
Choose the appropriate era to answer the questions

(A) Cenozoic
(B) Mesozoic
(C) Paleozoic
(D) Precambrian
(E) Archean

21. Human beings evolved.

22. Land plants appeared.

23. It takes on the order of _____ years for adaptive radiations to rebuild biological diversity after a mass extinction.

(A) 100
(B) 100 thousand
(C) 1 million
(D) 10 million
(E) 1 billion

24. On the leeward side of a coastal mountain range, below 4,000 feet, which of the following types of trees and/or plants would be most likely to occur?

(A) Epiphytes, lianas, bromeliads
(B) Mangrove, mahogany, cedar
(C) Prickly pear, manzanita, scrub oak
(D) Douglas fir, redwood
(E) Ferns, ivy, rhododendron

25. Succession on a sand dune would follow which order?

    (A) Grass, shrubs, beech and maple, cottonwoods, pine and black oak
    (B) Beech and maple, pine and black oak, cottonwoods, shrubs, grass
    (C) Grass, cottonwood, shrubs, beech and maple, pine and black oak
    (D) Grass, shrubs, cottonwoods, pine and black oak, beech and maple
    (E) A mixture of all species listed would occur simultaneously, stronger species replacing weaker species.

26. Certain volcanoes have a bowl-shaped crater at the summit and grow to only about a thousand feet. They are usually made of piles of lava, not ash. During the eruption, blobs of lava are blown into the air and break into small fragments that fall around the opening to the volcano. Parícutin in Mexico and the middle of Crater Lake, Oregon, are examples. These are called

    (A) cinder cones
    (B) shield volcanoes
    (C) composite volcanoes
    (D) mud volcanoes
    (E) spatter cones

27. Which of the following would NOT be considered an ecological service of a marine ecosystem?

    (A) Climate moderation
    (B) Nutrient cycling
    (C) CO_2 absorption
    (D) Genetic resources and biodiversity
    (E) Pharmaceuticals

28. Which continent has the highest deforestation rate?

    (A) Africa
    (B) Asia
    (C) Europe
    (D) South America
    (E) Australia/Oceania

29. Excavating and hauling soil off-site to an approved soil disposal/treatment facility would be an example of

    (A) sustainability
    (B) remediation
    (C) conservation
    (D) preservation
    (E) mitigation

30. A barnacle is a sedentary, highly modified crustacean. Barnacles live by using long, feathering appendages to sweep the surrounding water for small, free-floating organisms. The critical resource for barnacles is a place to stay. Barnacles attach to rocks, ships, shells, and whales. If a barnacle attaches itself to the shell of a scallop, the barnacle gains a place to live, and presumably, the scallop is not harmed by the presence of the barnacle. This relationship would be an example of

    (A) mutualism
    (B) amensalism
    (C) commensalism
    (D) parasitism
    (E) saprotrophism

31. In 1953, Stanley Miller, a graduate student working in the laboratory of Harold Urey, built an apparatus to demonstrate the feasibility of abiotic synthesis. Miller built an apparatus that simulated the presumed conditions of primeval Earth. The conditions included all of the following EXCEPT

    (A) a gaseous phase containing methane, ammonia, water, and hydrogen gas
    (B) electrical energy provided by spark discharge
    (C) ambient temperature between 0°C and 100°C
    (D) sterile conditions (abiotic environment)
    (E) primitive nitrifying bacteria

Questions 32–36
Select from the following locations.

(A) Bhopal, India
(B) Chernobyl, Ukraine
(C) Love Canal, New York
(D) Minamata Bay, Japan
(E) Three Mile Island, Pennsylvania

32. Site of a hazardous chemical dumping ground over which homes and a school were built.

33. Site of mercury poisoning.

34. The most serious commercial nuclear accident in U.S. history.

35. Leakage of poisonous gases from a pesticide manufacturing plant.

36. Nuclear power plant accident that released 30 to 40 times the radiation of the atomic bombs dropped on Hiroshima and Nagasaki.

Questions 37–39
Choose the political party that matches the platform most closely.

(A) Democratic Party
(B) Green Party
(C) Libertarian Party
(D) Natural Law Party
(E) Republican Party

37. "Encourage market-based solutions to environmental problems."

38. "We do not have to choose between economy and environment. Invest in technology and transportation friendly to the Earth. We support grants to Amtrak and the states for improving rail routes. We believe in protecting the coasts and the Arctic National Wildlife Refuge from oil and gas drilling."

39. "We believe in cushioning farmers from the instability unique to agriculture and enabling farmers to better pursue financial profitability. We believe that reducing global warming will help the economy. We believe in creating new jobs in energy conservation."

40. The layer of water in a thermally stratified lake that lies below the thermocline, is noncirculating, and remains perpetually cold is called the

(A) epilimnion
(B) hyperlimnion
(C) hypolimnion
(D) euphotic zone
(E) benthic zone

41. India's family-planning program has yielded disappointing results for all of the following reasons EXCEPT

(A) poor planning and bureaucratic inefficiency
(B) failure to employ sterilization
(C) extreme poverty
(D) a cultural preference for female children
(E) too little administrative and financial support

42. Issues of air and water pollution, noise, pesticides, solid waste management, radiation, and hazardous wastes would be the domain of which executive branch office?

    (A) Department of the Interior
    (B) Department of Health and Human Services
    (C) Council on Environmental Quality
    (D) Environmental Protection Agency
    (E) Office of Management and Budget

43. Which of these threats is NOT one of those that must be decreased to help the survival of the approximately 600 mountain gorillas left in the wild?

    (A) Habitat loss
    (B) Poaching
    (C) War
    (D) Exotic species intrusions
    (E) Disease

44. The largest user of freshwater worldwide is

    (A) mining
    (B) irrigation
    (C) industry
    (D) home use
    (E) production of electrical power

45. Choose the statement that is FALSE.

    (A) Domestic fruits and vegetables are less likely to have pesticide residues than imported ones.
    (B) Cancer may not be the primary risk from chronic, long-term exposure to pesticides.
    (C) When the EPA looks at a pesticide to decide whether to register it for use in the United States, its primary concern is to ensure that there is no significant human health or environmental risk presented by the chemical.
    (D) The federal government does not prohibit the use of pesticides known to cause cancer.
    (E) Washing and peeling fruits and vegetables does not remove all or most pesticide residues.

46. What happens in the market for airline travel when the price of traveling by rail decreases?

    (A) The demand curve shifts left.
    (B) The demand curve shifts right.
    (C) The supply curve shifts left.
    (D) The supply curve shifts right.
    (E) The supply curve intersects with the demand curve at the equilibrium price.

47. When taking into account only price, supply, and demand, if plotted on a graph, the supply and demand curves

    (A) are parallel lines
    (B) are parallel lines running horizontally
    (C) never intersect
    (D) can run in any direction
    (E) intersect at a point called market equilibrium

48. Wind power is a measure of the energy available in the wind, and is a function of the cube (third power) of the wind speed. For example, when wind speed is doubled, wind power increases by a factor of eight. About how much more power is produced by a typical wind turbine at 15 mph than at 12 mph?

    (A) About the same
    (B) About 10% more
    (C) About 25% more
    (D) About twice as much
    (E) About 9 times as much

49. An APES class went on a field trip into a coniferous forest. They discovered a very large section of land that had been completely logged. There were just stumps where large coniferous trees had once stood. There was also very little animal life in the area. Which method of logging had most likely been used in this section of land?

    (A) Strip cutting
    (B) Clear cutting
    (C) Seed-tree cutting
    (D) Shelterwood cutting
    (E) Selective cutting

50. You have been placed in charge of rebuilding a salmon population in a river basin that contains a hydroelectric dam. Which of the following remediation techniques would NOT be effective?

(A) Reduce silt runoff
(B) Build fish ladders
(C) Decrease water flow from the dam
(D) Release juvenile salmon from hatcheries
(E) Use trucks and barges to transport salmon around the dam

51. Which one of the following proposals would NOT increase the sustainability of ocean fisheries management?

(A) Establish fishing quotas based on past harvests.
(B) Set quotas for fisheries well below their estimated maximum sustainable yields.
(C) Sharply reduce fishing subsidies.
(D) Shift the burden of proof to the fishing industry to show that their operations are sustainable.
(E) Strengthen integrated coastal management programs.

52. What amount of cultivated land is used to produce over 95% of the world's food?

(A) About 75%
(B) Between 50% and 75%
(C) About 50%
(D) Between 15% and 50%
(E) Less than 15%

53. The type of succession that begins in an area where the natural community has been disturbed, removed, or destroyed but in which the bottom soil or sediment remains is known as

(A) allogenic
(B) autogenic
(C) primary
(D) secondary
(E) progressive

54. In general, parasites tend to

(A) become more virulent as they live within the host
(B) destroy the host completely
(C) become deactivated as they live within the host
(D) be only mildly pathogenic
(E) require large amounts of oxygen

55. The population of a country in 1994 was 200 million. Its rate of growth was 1.2%. Assuming that the rate of growth remains unchanged and all other factors remain constant, in what year will the population of the country reach 400 million?

(A) 2004
(B) 2024
(C) 2040
(D) 2054
(E) 2104

56. Humans having a finite capacity to manage nature would be consistent with what principle?

(A) Precautionary principle
(B) Integrative principle
(C) Ecological design principle
(D) Humility principle
(E) Environmental justice principle

57. The circulation of air in Hadley cells results in

(A) low pressure and rainfall at the equator
(B) high pressure and rainfall at the equator
(C) low pressure and dry conditions at about 30° north and south of the equator
(D) high pressure and wet conditions at about 30° north and south of the equator
(E) both (A) and (C)

58. All of the following are characteristics of K-strategists EXCEPT

(A) mature slowly
(B) low juvenile mortality rate
(C) niche generalists
(D) Type I or II survivorship curve
(E) intraspecific competition due to density-dependent limiting factors

59. First levels of defensive behaviors, used by both predators and prey, to avoid detection would include all of the following EXCEPT

(A) camouflage
(B) predator swamping
(C) countershading
(D) Batesian mimicry
(E) masquerading

Question 60 refers to the diagram below.

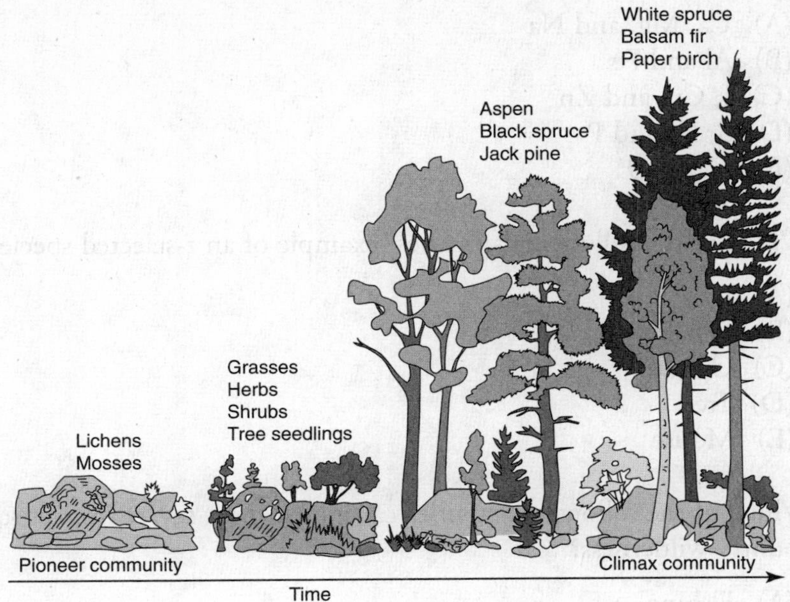

60. The greatest species diversity would be found in

(A) the pioneer community represented by the lichens and mosses.
(B) the community composed primarily of grasses, herbs, shrubs, and tree seedlings.
(C) the community composed primarily of aspen, black spruce, and jack pine.
(D) the community composed primarily of white spruce, balsam fir, and paper birch.
(E) all communities equally.

61. Which of the following is NOT an example of a chronic condition?

(A) Asthma
(B) Measles
(C) Diabetes
(D) Cancer
(E) Malnutrition

62. Which mobile source pollutant cannot be currently controlled by emission control technology?

(A) Ozone-forming hydrocarbons
(B) Carbon monoxide
(C) Carbon dioxide
(D) Air toxics
(E) Particulate matter

63. Which of the following are examples of trace elements necessary in the human diet?

    (A) Ca, Mg, and Na
    (B) Al and Fe
    (C) I, Cu, and Zn
    (D) S, N, and P
    (E) C, H, O

64. Which of the following is the best example of an r-selected species?

    (A) Dog
    (B) Whale
    (C) Condor
    (D) Tree
    (E) Mouse

65. Which of the following activities causes the most severe impact to back-country wilderness?

    (A) Fishing
    (B) Hiking off trails
    (C) Littering
    (D) Building a fire
    (E) Hunting

66. You are going to buy a soda. You see a vending machine that has sodas in aluminum cans, steel cans, plastic bottles, and glass bottles. Which container has the least environmental impact when recycled?

    (A) Aluminum cans
    (B) Steel cans
    (C) Plastic bottles
    (D) Glass bottles
    (E) All have the same negative environmental impact

67. The principle stating that in general, all other things being equal, the higher the price of a good, the greater the quantity of that good sellers will offer for sale over a given period of time is known as (the)

    (A) law of supply
    (B) law of demand
    (C) law of supply and demand
    (D) open access
    (E) neoclassical economics

68. You are a member of a grassroots environmental organization that has successfully lobbied your U.S. congressperson to co-sponsor a bill to create a small wildlife sanctuary for migratory birds on federal land. After your bill was introduced to the Senate, which committee would it likely be referred to for hearings?

(A) Committee on Agriculture
(B) Committee on Energy and Commerce
(C) Committee on Resources
(D) Committee on Wildlife Conservation
(E) Committee on Preservation

69. In 1995, the population of a small island in Malaysia was 40,000. The birthrate was measured at 35 per 1,000 population per year, while the death rate was measured at 10 per 1,000 population per year. Immigration was measured at 100 per year, while emigration was measured at 50 per year. How many people would be on the island after one year?

(A) 39,100
(B) 40,000
(C) 41,050
(D) 42,150
(E) 44,500

Question 70 refers to the diagram below.

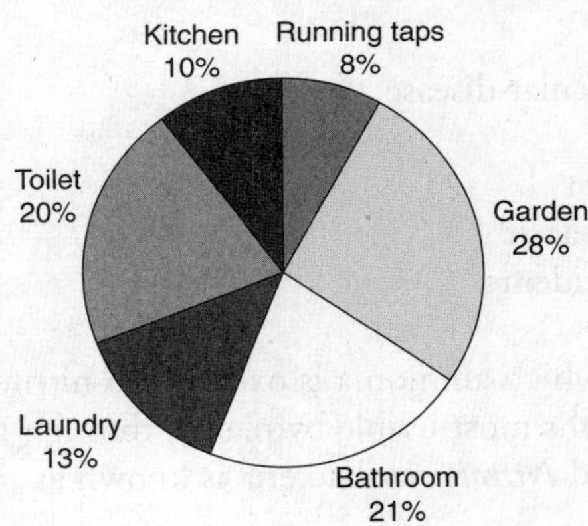

70. According to the information in the graph above, which of the following methods would be the most reasonable and effective means to decrease the amount of domestic water used?

(A) Install flow-restrictors in faucets and shower heads
(B) Wash only full loads of clothes in washing machines
(C) Use landscaping that does not require much water combined with drip irrigation
(D) Check for leaking faucets
(E) Try to use the bathroom less often

71. To evaluate the total impact of a disease by combining premature deaths and disability into data rather than just to quantify the effects or mortality is to use a technique called (a)

(A) risk extrapolation model
(B) comparative risk analysis
(C) epidemiology
(D) risk-based targeting
(E) disability-adjusted life year

72. Clumped spacing patterns of plants are most often associated with

(A) pockets of resources within the population's range
(B) shading and competition for water and minerals
(C) the random distribution of seeds
(D) antagonistic chemicals secreted by plants that inhibit germination and growth of nearby individuals
(E) coincidence

73. Most of the freshwater found on Earth is in the form of

(A) ice (glaciers, polar ice caps, etc.)
(B) rivers and lakes
(C) groundwater
(D) atmosphere (rain, fog, clouds, vapor, etc.)
(E) oceans

74. Among adults, which of the following represents the most preventable cause of death?

(A) Cardiovascular disease
(B) AIDS
(C) Alcoholism
(D) Use of tobacco
(E) Traffic accidents

75. The process in which ammonia is oxidized to nitrite (NO_2^-) and nitrate (NO_3^-), the forms most usable by plants, through the action of *Nitrosomonas* and *Nitrobacter* bacteria is known as

(A) nitrogen fixation
(B) nitrification
(C) assimilation
(D) ammonification
(E) denitrification

76. Which United States president is responsible for creating the National Park System?

(A) Woodrow Wilson
(B) Theodore Roosevelt
(C) Franklin D. Roosevelt
(D) Calvin Coolidge
(E) Herbert Hoover

77. An APES class visited a stream to study biodiversity. The students spent the day prior to visiting the stream learning to identify macroinvertebrates by using a dichotomous key. On the day they visited the stream, one group collected the following macroinvertebrates in the following order:

backswimmer—backswimmer—backswimmer—damselfly—damselfly—midge—mosquito larvae—mosquito larvae—mayfly—mayfly—mayfly—mayfly—damselfly—backswimmer

What is the sequential comparison index?

(A) 0.1
(B) 0.25
(C) 0.35
(D) 0.50
(E) 1.0

78. "We can burn coal to produce electricity to operate a refrigerator" is an example of the ____________ and "If we burn coal to produce electricity to operate a refrigerator, we lose a great deal of energy in the form of heat" is an example of the ____________

(A) first law of thermodynamics, first law of thermodynamics
(B) second law of thermodynamics, first law of thermodynamics
(C) first law of thermodynamics, second law of thermodynamics
(D) first law of thermodynamics, third law of thermodynamics
(E) third law of thermodynamics, first law of thermodynamics

79. Refer to the following age-structure diagram:

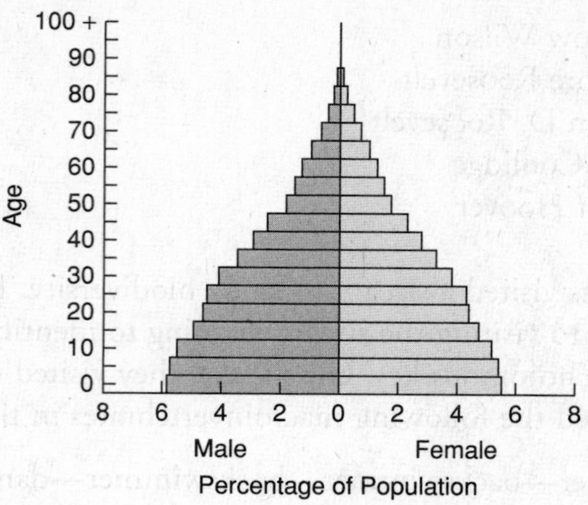

The age-structure diagram above would be typical for what country?

(A) United States
(B) Canada
(C) Mexico
(D) Germany
(E) Japan

80. The largest earthquake in the 20th century occurred

(A) off the coast of Alaska
(B) off the coast of Japan
(C) off the coast of South America
(D) off the coast of California
(E) in Missouri

81. Butterflies and moths both feed on flowers. Butterflies feed during the day, and moths feed at night. This is an example of

(A) r-strategy
(B) K-strategy
(C) resource partitioning
(D) commensalism
(E) mutualism

82. It costs a copper smelter $200 to reduce its emissions by 1 ton and $250 for an additional ton. It costs an electric utility $100 to reduce its emissions by 1 ton and $150 for an additional ton. What is the cheapest way of reducing total emissions by 2 tons?

(A) Legislate that each firm must reduce emissions by two tons.
(B) Charge both firms $251 for every ton they emit.
(C) Allow each firm to buy a permit to pollute that costs $151.
(D) File an injunction to halt production until the firms reduce emissions by 2 tons.
(E) None of the above is correct.

83. An APES class was investigating the estimation of population size. To measure the population density of monarch butterflies occupying a particular park, 100 butterflies were captured, marked with a small dot on a wing, and then released. The next day, another 100 butterflies were captured, including the recapture of 20 marked butterflies. One would correctly estimate the population to be

(A) 100
(B) 200
(C) 500
(D) 2,000
(E) 20,000

84. In the eastern United States, water use law is based on the legal principle of

(A) prior appropriation
(B) private property rights
(C) common property rights
(D) public property rights
(E) riparian rights

85. Demographic transition leads to stabilizing population growth and is generally characterized as having four separate stages. Place the following stages in the proper order as they would most likely occur.

    I. Birthrates equal mortality rates, and zero population growth is achieved. Birthrates and death rates are both relatively low, and the standard of living is much higher than during the earlier periods. In some countries, birthrates may actually fall below mortality rates and result in net losses in population.
    II. Living conditions are severe, medical care is poor or nonexistent, and the food supply is limited due to poor agricultural techniques, poor preservation, and pestilence. Birthrates are high to replace individuals lost through high mortality rates. The net result is little population growth.
    III. Urbanization decreases the economic incentives for large families. The cost of supporting an urban family grows, and parents are more actively discouraged from having large families. Educational and work opportunities for women decrease birth rates. Obtaining food is not a major focus of the day. Leisure time is available, and retirement safety nets are in place, reducing the need for extra children to support parents. In response to these economic pressures, the birthrate starts to drop, ultimately coming close to the death rate.
    IV. Occurs after the start of industrialization. Standards of hygiene and more modern medical techniques begin to drive the death rate down, leading to a significant upward trend in population size. Mortality rates drop as a result of advances in medical care, improved sanitation, cleaner water supplies, vaccination, and higher levels of education. The net result is a rapid increase in population.

    (A) I, II, III, IV
    (B) IV, III, II, I
    (C) I, IV, II, III
    (D) II, IV, III, I
    (E) III, I, II, IV

86. The concentration of which gas can be reduced by preventing forest depletion?

    (A) Carbon dioxide
    (B) Nitrous oxide
    (C) Oxygen
    (D) Methane
    (E) CFCs

87. Environmental lawsuits are limited in results for all of the following reasons EXCEPT

(A) the plaintiff must be directly harmed by an action or lack thereof
(B) corporations may deduct their legal expenses from their federal taxes, whereas public interest lawyers cannot usually recover any fees or tax write-offs
(C) it is usually not difficult to prove that the defendant is liable and responsible for harmful environmental action
(D) the courts may take years to come to a decision and appeals may be submitted to higher courts
(E) bringing a suit is expensive

88. After a recent storm, an APES class took a field trip to a storm drain outlet entering the Pacific Ocean at Ballona Creek, in southern California. They carefully collected water samples and brought the samples back to the lab. Alpha group took 100.00 mL of the collected water and filtered it through a 1.2 μm Millipore glass fiber filter. The filter was carefully transferred to a premassed stainless steel crucible and placed into a 105°C oven for 24 hours. The following is a hypothetical set of data from alpha group:

Weight of crucible + filter + residue after 24 hrs at 105°C = 100.000g
Weight of crucible + filter = 80.000g

What is the total suspended solid amount measured in $mg \cdot L^{-1}$?

(A) $(80.000/100.00) \times 100\%$

(B) $100.00 - 80.00$

(C) $\dfrac{(100.00 - 80.00) \times 1{,}000 \times 1{,}000}{100.00}$

(D) $\dfrac{(100.00 - 80.00) \times 1{,}000}{100}$

(E) $\dfrac{(100.00 + 80.00) \times 100}{1{,}000}$

89. Which of the following is NOT an adaptation of the barn owl that allows it to be a successful predator?

    (A) Eyes that point forward
    (B) Excellent vision and hearing
    (C) Long toes with curved talons
    (D) Excellent sense of smell
    (E) Ability to fly almost silently

Questions 90–93

Diatoms are one-celled, microscopic plants and are distributed throughout the world in aquatic, semiaquatic, and moist habitats. They are found in the sea, estuaries, freshwater lakes, ponds, streams, and ditches. Though individual diatoms are microscopic, masses of diatoms can often be seen on stream bottoms, along the surf zones, and during plankton blooms as brownish-colored waters or films. Base your answers to questions 90–93 on the figure below, dealing with the seasonal fluctuations in the abundance of diatoms in the North Atlantic (solid black line—diatoms; dotted line—light intensity; dashed line—nitrates and phosphates):

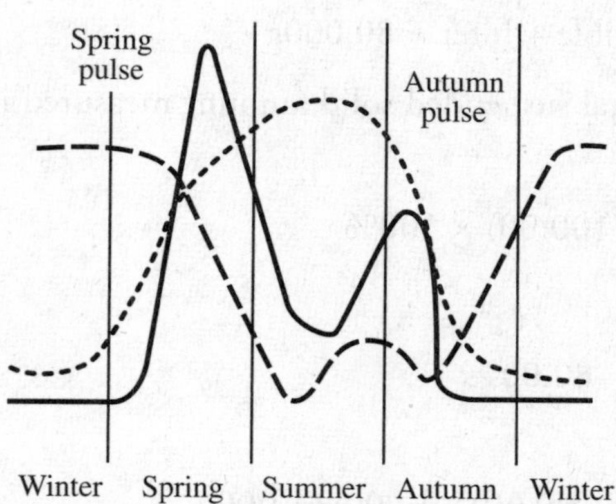

90. Which factor probably contributes to the summer decline of diatoms?

    (A) Decreased concentration of nitrates and phosphates
    (B) Increased concentration of nitrates and phosphates
    (C) Increased intensity of light
    (D) Decreased intensity of light
    (E) Increase of temperature

91. Which is probably the principal source of nitrates and phosphates?

    (A) The water cycle
    (B) Nitrogen fixation of lighting
    (C) Bacterial decay
    (D) Changes in environmental temperature
    (E) Changes in light intensity

92. A probable reason why the autumn pulse is not as great as the spring pulse is that

(A) diatoms undergo metamorphosis in the autumn
(B) temperature and light intensity decrease in the autumn
(C) carnivorous animals increase in the autumn
(D) the diatoms of the spring have used most of the proteins from the environment
(E) bacteria of decay increase in the autumn

93. The low level of the diatom population in winter results partially from the fact that

(A) low temperature slows down metabolism
(B) photosynthesis occurs only in summer
(C) diatoms live in water
(D) diatoms migrate to warm climates during the winter
(E) in winter, there is a decrease in available nitrates and phosphates

94. What pulls warm Atlantic water north in the summer?

(A) La Niña
(B) Convection
(C) El Niño
(D) Coriolis effect
(E) Gulf Stream

95. Harmful effects of increased UV radiation on Earth's surface include all of the following EXCEPT

(A) reduction in crop production
(B) reduction in the growth of phytoplankton and its cumulative effect on food webs
(C) cooling of the stratosphere
(D) warming of the stratosphere
(E) increases in sunburns and damage to the skin

96. Which greenhouse gas has the most negative chemical properties with respect to global warming?

(A) CO_2
(B) CFCs
(C) SF_6
(D) N_2O
(E) H_2O

97. Which of the following indoor pollutants would NOT be contributed by carpeting?

(A) Formaldehyde
(B) Styrene
(C) Mold
(D) Methylene chloride
(E) Mites

98. A woman reading a book of poems turned a page, causing a tiny air current to slip out the open window, nudging a passing breeze. That breeze in turn nudged another breeze and eventually was the cause of a tornado in the next state. This series of conceptual events is referred to as

(A) the Gaia hypothesis
(B) the butterfly effect
(C) chaos
(D) law of common cause
(E) natural sequencing

99. All of the following are correct statements about the regulation of populations EXCEPT

(A) a logistic equation reflects the effect of density-dependent factors, which can ultimately stabilize a population around the carrying capacity
(B) density-independent factors have a greater effect as a population's density increases
(C) high densities in a population may cause physiological changes that inhibit reproduction
(D) because of the overlapping nature of population-regulating factors, it is often difficult to determine their cause-and-effect relationships precisely
(E) the occurrence of population cycles in some populations may be the result of crowding or lag times in the response to density-dependent factors

100. Which of the following energy sources has the lowest quality?

(A) High-velocity water flow
(B) Fuelwood
(C) Food
(D) Dispersed geothermal energy
(E) Saudi Arabian oil deposits

SECTION II (FREE-RESPONSE QUESTIONS)

Time: 90 minutes

No calculators allowed

4 questions

Directions: Answer all four questions, which are weighted equally. The suggested time is about 22 minutes for answering each question. Write all your answers on the pages following the questions in the pink booklet. Where calculations are required, clearly show how you arrived at your answer. Where explanation or discussion is required, support your answers with relevant information and/or specific examples.

1. A family is building a new home in Buffalo, New York. Buffalo experiences severe winters. Assume the following.

   - The house has 4,000 square feet.
   - 100,000 Btus of heat per square foot are required to heat the house for the winter.
   - Natural gas sells for $5.00 per thousand cubic feet.
   - 1 cubic foot of natural gas supplies 1,000 Btus of heat energy.
   - 1 kilowatt-hour of electricity supplies 10,000 Btus of heat energy.
   - Electricity costs $50 per 500 kWh.

   (a) Calculate the following, showing all the steps of your calculations, including units.

   (i) The number of cubic feet of natural gas required to heat the house for the winter.
   (ii) The cost of heating the house using natural gas.
   (iii) The cost of heating the house using electricity.

   (b) The homeowners are discussing with their architect the possibility of using either active or passive solar design to reduce their heating and/or cooling costs. Compare these two techniques.

   (c) The homeowners wish to incorporate green design. Discuss five techniques that the homeowners could adopt to make their home green. (Green does not refer to the color of the house!)

2.

Legal Notice

Initial Public Offering of Common Stock

The Louisiana Shrimp Company, headquartered in New Orleans, Louisiana, is offering 5,000,000 shares of common stock at $3.00 per share to investors for the purpose of raising capital to construct shrimp farms in Southeast Asia. Raising shrimp in controlled environments, known as shrimp aquaculture or shrimp farms, provides needed food and employment for the local population. Excess shrimp that is raised will be sold in open trading in world markets, adding to the profitability of the operation. Shrimp farming does not negatively impact the environment since shrimp are native to the area. Shrimp also supply a valuable source of income and food supply (protein) to the local population. No guarantee of return on investment is implied. For further information on this investment opportunity, contact L. D. Breckenridge, Ltd. of Shreveport, LA.

(a) Describe any negative and/or positive environmental impacts of shrimp aquaculture.

(b) Describe any parallels that have occurred in history in the development of aquaculture.

(c) Comment on the social implications of large-scale aquaculture, and suggest possible modifications that might make it more sustainable.

3. Life on Earth has been punctuated by several mass extinctions. Humans are playing a role in another mass extinction, potentially the largest ever. As we attempt to create a sustainable future, efforts are being taken to slow the loss of endangered species.

   (a) How can scientists assess the current population size of a species? Explain how tag and recapture methods could be used to estimate the number of monarch butterflies in an area too large to sample exhaustively.

   (b) Give one example each of a direct and an indirect threat to biodiversity.

   (c) Explain a piece of legislation designed to preserve biodiversity.

   (d) After a population of organisms has been reduced to a small size, it has greater risk of going extinct, even if the population returns to its original size (a phenomenon sometimes referred to as the bottleneck effect). Why would the same-size population be more likely to go extinct after a population bottleneck?

4. In 1900, the average American could expect to live just 47 years. Today, the average American lives 75 years—an increase of nearly three decades of life in just a century's time.

   (a) Explain two possible things that have occurred within the last 100 years that can account for the increase in human life span.

   (b) What are the greatest current and projected risks to human health?

   (c) Explain the concept and components of risk analysis. How would information gained from studies of risk analysis increase the human life span?

ANSWER KEY

Practice Test 2

Section I (Multiple-Choice Questions)

1. D	26. A	51. A	76. A
2. D	27. E	52. E	77. D
3. A	28. A	53. D	78. C
4. B	29. B	54. D	79. C
5. C	30. C	55. D	80. C
6. D	31. E	56. D	81. C
7. B	32. C	57. A	82. C
8. A	33. D	58. C	83. C
9. A	34. E	59. B	84. E
10. E	35. A	60. D	85. D
11. E	36. B	61. B	86. A
12. C	37. E	62. C	87. C
13. E	38. A	63. C	88. C
14. A	39. D	64. E	89. D
15. C	40. C	65. D	90. A
16. C	41. D	66. A	91. C
17. C	42. D	67. A	92. B
18. A	43. D	68. C	93. A
19. C	44. B	69. C	94. E
20. C	45. A	70. C	95. D
21. A	46. A	71. E	96. C
22. C	47. E	72. A	97. D
23. D	48. D	73. C	98. B
24. C	49. B	74. D	99. B
25. D	50. C	75. B	100. D

MULTIPLE-CHOICE EXPLANATIONS

1. **(D)** Speciation occurs when environmental factors change the composition of the gene pool.

2. **(D)** Incandescent: $\frac{60\ \cancel{\text{Watt}}}{1} \times \frac{\$0.10}{1{,}000\ \cancel{\text{Watt}} \bullet \cancel{\text{hr}}} \times \frac{8{,}000\ \cancel{\text{hr}}}{1} = \$48.00$

   Flourescent: $\frac{15\ \cancel{\text{Watt}}}{1} \times \frac{\$0.10}{1{,}000\ \cancel{\text{Watt}} \bullet \cancel{\text{hr}}} \times \frac{8{,}000\ \cancel{\text{hr}}}{1} = \$12.00$

   Therefore, the saving is $48.00 – $12.00 = $36.00

3. **(A)** Wind is primarily caused by the uneven heating of Earth's surface; i.e., the equator receives more solar radiation than the poles. This uneven heating results in differences in air pressure and sets convection currents (winds) in motion.

4. **(B)** Stabilizing selection operates on the extremes. Stabilizing selection reduces phenotypic variations but maintains the status quo in the gene pool. An example of stabilizing selection found in the human population is phenylketonuria, a genetic disorder that inhibits the proper processing of dietary protein and, without proper dietary control, causes brain damage. In directional selection, a population may find itself in circumstances where individuals occupying one extreme in the range of phenotypes are favored over the others. In other circumstances, individuals at both extremes of a range of phenotypes are favored over those in the middle. This is called disruptive selection.

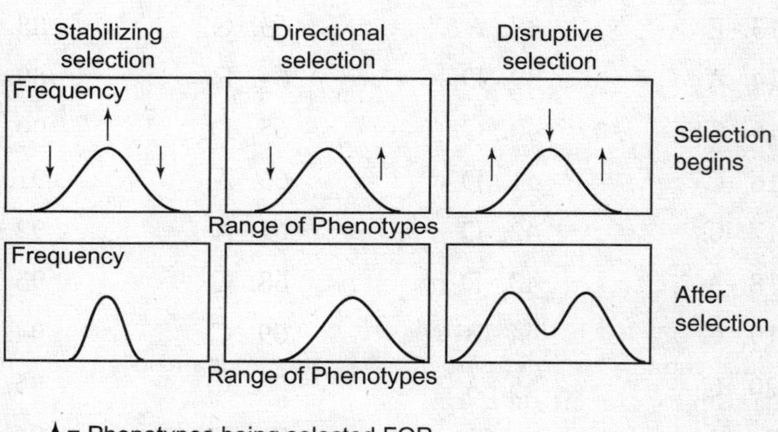

5. **(C)** Air conditioners that contain chlorine-fluorine based refrigerants are known as CFCs (chloro-fluorocarbons). One chlorine atom released from a CFC can ultimately destroy over 100,000 stratospheric ozone molecules.

6. **(D)** 100 units of energy are found in the green plants. $100 \times 0.20 = 20$ units available for herbivores. Of the 20 units now available, $20 \times 0.20 = 4$ units are left for the primary carnivores.

7. (B) The rats would most likely have remained about half the size of normally developed rats if nutrient *B* had not been returned at point *Z*. The rats would continue to weigh less than 100 grams.
8. (A) The three numbers on the side of the fertilizer bag refer to the amount of nitrogen, phosphorus, and potassium that the fertilizer contains. The rest of the bag contains minor nutrients and filler material. Since the number representing phosphorus is 0, no phosphorus is required.
9. (A) Notice that all three forms of energy are nonrenewable sources derived from fossil fuels. Oil is the primary energy source for the world today with the United States at approximately 40% dependency on this single source. Coal and natural gas are each around 22% dependency in the United States and the world.
10. (E) After the 49th day, there is no increase or decrease in the population. The graph levels off.
11. (E) Primary productivity is the amount of biomass produced through photosynthesis per unit area per time by plants. Gross primary productivity is the total energy fixed by plants through photosynthesis. A portion of the energy of gross primary productivity is used by plants for respiration. Respiration provides a plant with the energy needed for various activities. Subtracting respiration from gross primary productivity gives net primary productivity, which represents the rate of production of biomass that is available for consumption by heterotrophic organisms.
12. (C) A watt is a unit of power.
13. (E) The effects of La Niña are opposite to those of El Niño. A La Niña effect may be defined as a drop in average sea-surface temperatures to more than 0.7°F (0.4°C) below normal, lasting at least six months, across parts of the eastern tropical Pacific. When La Niña forms, the hurricane season is affected as the cooling water creates dramatic changes in the upper-level air currents that play a major role in storm development. During La Niña, high-level westerly winds either weaken or shift to come from the east, allowing *more* storms to develop.
14. (A) In order to balance the excess heating near the equator and cooling at the poles of Earth, both atmosphere and ocean transport heat from low to high latitudes. Warmer surface water is cooled at high latitudes, releasing heat to the atmosphere, which is then radiated away to space. This heat engine operates to reduce equator-to-pole temperature differences and is a prime moderating mechanism for climate on Earth.
15. (C) Denitrifying bacteria are any strain of bacteria able to utilize nitrate or nitrite in an energy-yielding metabolic sequence that eventually produces nitrogen gas. When colonies of these bacteria occur on croplands, they may deplete the soil nutrients and make it difficult for crops to grow.
16. (C) A divergent plate boundary is the interface where plates move apart in opposite directions; i.e., mid-ocean ridges. A new ocean basin is created when a tectonic plate carrying a continent literally splits apart. In this process the heat from underlying magma wells up from deep within the Earth, weakening and stretching the overlying continental crust. The brittle crust then fractures on each side of the stressed area, allowing sections to drop. As a young ocean widens and matures, the undersea rift develops a ridge of lava mountains on the trailing edge of each plate. The Mid-Atlantic Ridge, for example, rises

where the American continents are separating from Europe and Africa. Other Mid-Ocean Ridges include the East Pacific Rise, several hundred miles off the western coast of South America, and the Indian Ridge, off the Eastern coast of Africa, south of India.

17. (C) Temperate deciduous forests have warm summers and cold winters. Deciduous trees escape these winters by losing their leaves. Typical mammals are bears, badgers, squirrels, woodchucks, insectivores, rodents, wolves, wildcats, and deer. These forests are rich in birds. The climate that is suitable for temperate deciduous forests is most suited for humans—hence forest destruction.
18. (A) The term "frontier economy" refers to a time in U.S. history when resources were almost limitless and when the challenge was to harvest them. Today, what used to be limitless has become finite. Our challenge is to conserve what remains and to embrace opportunities that the new economy has to offer.
19. (C) Prior to the development of DDT, it is hypothesized that some mosquitoes had an allele that could break down DDT and render it harmless. The allele was just a random mutation of a gene for an enzyme. Mosquitoes with that allele may also have had somewhat lower reproductive rates. Because there was no DDT in the environment, these mosquitoes did not receive an advantage from the mutation. Consequently, they were selected against and the allele was maintained at a very low frequency in the environment. However, when DDT was introduced into the environment, the mosquitoes with that allele were selectively favored because they survived and reproduced at a much higher rate than the mosquitoes without the allele. The result was the increase in the frequency of the DDT-resistant allele. Remember, the allele in question was already present in the population—the addition of DDT did *not* cause the allele to arise. Selection acts on only preexisting variations, it does not create adaptive variations. Furthermore, the allele in question may have been actually selected against in one set of environmental circumstances and favored in another.

    Over the last 50 years, 400 species of insects, 50 species of fungi, and several species of weeds have become resistant to pesticides that had previously killed them. Unfortunately, in areas that have been sprayed to kill malaria mosquitoes, up to 43 species are now immune to pesticides. In section III of the graph, the population of mosquitoes that were naturally immune to the effects of DDT are beginning to reproduce at a disproportionately higher rate than those that were susceptible or weakened by DDT.
20. (C) Automobiles and other forms of transportation are responsible for approximately one-third of human-made nitrogen oxide and volatile organic compound emissions, one-fifth of particulate emissions, two-thirds of carbon monoxide emissions, and less than 5% of sulfur dioxide emissions.
21. (A) Around 1 million years ago, the ancestors of *Homo sapiens* became dominant.
22. (C) The Paleozoic era lasted from about 570 to 250 million years ago. The 320 million years of the Paleozoic era saw many important events, including the development of most invertebrate groups; life's conquest of land; the evolution of fish, reptiles, insects, and vascular plants; and the formation of the supercontinent of Pangaea.

23. **(D)** A mass extinction is an opportunity for adaptive radiation. Perhaps the most dramatic example is the rise of the mammals. Ancestral mammals were small, undifferentiated scavengers. Within 10 million years, after the demise of the dinosaurs, all of the major orders of mammals (and of birds as well) had differentiated.
24. **(C)** The leeward side of a coastal mountain range is characteristically dry due to the rain shadow effect. Water is the limiting factor. You would expect to find plants that require little water—those that would be classified as xeric and typically found in chaparral or desert biomes.
25. **(D)** After grasses become established, which hold down loose sand, winds bring seeds of more complex plants (shrubs, alders, and willows) to the area. In this open canopy, sun-requiring trees such as cottonwoods and some spruce become established. Next, larger trees like pines, black oak, aspen, and birch begin to grow and dominate. After shade develops from the first large trees, other types of trees, which are more shade resistant, begin to grow in the area. What has now formed is called a broad-leaf forest. The pines, aspens, and birches have grown rapidly, and sun required for their saplings to grow has been shut out. New trees cannot grow in this shade. As a result, the forest understory is conducive to shade-tolerant oak, tulip trees, beech, maple, ash, hemlock, and others. Slowly, the pines, aspens, and birches die out, leaving the climax forest.
26. **(A)** Cinder cones are one of the most common types of volcanoes. A steep, conical hill of volcanic fragments called cinders accumulates around a vent. The rock fragments, often called cinders, are glassy and contain numerous gas bubbles "frozen" into place as magma explodes into the air and then cools quickly. Cinder cones range in size from tens to hundreds of meters tall and usually occur in groups.
27. **(E)** Ecological services are benefits directly provided by ecosystems. Pharmaceuticals are not directly provided by marine ecosystems. Pharmaceuticals would be an economic service.
28. **(A)** Forests are disappearing most rapidly in Africa and Latin America. In Asia, though, the reduction of natural forests is largely compensated for by new plantation forests (which reduce biodiversity). In Europe and North America, the forest area is increasing. Overall, the world contains around 6,000 square meters of forest for each person, which is reducing by 12 square meters every year. In Africa, annual deforestation is approximately 0.8%, making it the continent with the highest deforestation rate. The country suffering the highest deforestation rate is Burundi (Africa), which has an incredible annual deforestation rate of 9.0%.
29. **(B)** Remediation technologies are those that render harmful or hazardous substances harmless after they enter the environment.
30. **(C)** Commensalism is a relationship between two species where one species derives a benefit from the relationship and the second species is unaffected by it.
31. **(E)** The observed products were common amino acids, fatty acids, and other organic molecules. The contemporary conclusion from this effort is that life originated through spontaneous, inanimate processes and that they took place under the conditions that existed on a primitive Earth.

32. (C) During the 1940s and 1950s, the Hooker Chemical Company dumped approximately 21,000 tons of organic solvents, acids, and pesticides as well as their by-products, many of them carcinogenic (cancer causing) or teratogenic (creating birth defects) into an abandoned canal in New York State (near Niagara Falls). A school and homes were built over the site. Chemicals began to leak from the ground and cause illness. Since the disaster, various levels of government have spent around $250 million (Superfund) and 20 years cleaning up the site, but all the waste is still buried there. New York State has since rebuilt homes in the area at reduced prices to attract new residents.
33. (D) From 1932 to 1968, Chisso Corporation (a petrochemical and plastics manufacturer) dumped an estimated 27 tons of mercury compounds into Minamata Bay, Japan. Thousands of people, whose normal diet included fish from the bay, unexpectedly developed symptoms of methyl mercury poisoning. The illness became known as Minamata Disease. Victims were diagnosed as having a degeneration of their nervous systems, numbness occurring in their limbs and lips, their speech becoming slurred, and their vision being constricted; 12,615 people have been officially recognized as patients affected by mercury.
34. (E) On March 28, 1979, a minor malfunction occurred in the system that fed water to the steam generators at the Three Mile Island Nuclear Generating Station near Harrisburg, Pennsylvania. This event eventually led to the most serious commercial nuclear accident in U.S. history and caused fundamental changes in the way nuclear power plants were operated and regulated. Despite the severity of the damage, no injuries due to radiation occurred. Eleven days after the events of Three Mile Island, the movie *The China Syndrome* was released—a film about a nuclear accident.
35. (A) On the night of December 2–3, 1984, 40 tons of cyanide and other lethal gases began spewing from Union Carbide Corporation's pesticide factory in Bhopal, India. Over half a million people were exposed to the deadly gases. The gases burned the tissues of the eyes and lungs, crossed into the bloodstream, and damaged almost every system in the body. With an estimated 10–15 people continuing to die each month, the number of deaths to date is put at close to 20,000. Today more than 120,000 people are still in need of urgent medical attention.
36. (B) On April 26, 1986, a reactor exploded in the town of Chernobyl in the Ukraine and released 30 to 40 times the radioactivity of the atomic bombs dropped on Hiroshima and Nagasaki. Hundreds of thousands of Ukrainians, Russians, and Belorussians had to abandon entire cities and settlements within a 20-mile zone of extreme contamination. Some 3 million people are still living in contaminated areas.
37. (E) No explanation needed.
38. (A) No explanation needed.
39. (D) No explanation needed.
40. (C) The epilimnion is the upper layer of a lake. The term "hyperlimnion" does not exist. The euphotic zone is the upper layer of water that is penetrated by sunlight and contains waters rich in minerals and organic nutrients that often promote a proliferation of plant life, especially algae. The algae may reduce the dissolved oxygen content and often causes the death of other organisms. The benthic zone would be the deepest layer of ocean water.

41. **(D)** The risk of dying between ages 1 and 5 is 43% higher for girls than boys. India has less than 93 women for every 100 men against the world average of 105. The main reason for the widespread female infanticide in parts of India is the dowry system, which although long prohibited by law, continues to play a significant role in Indian society. Dowries and wedding expenses regularly run to more than a million rupees ($35,000) in a country where the average civil servant earns about 100,000 rupees ($3,500) a year. Added to this is the low status of women in rural India, where they perform the menial tasks of the family such as carrying water and firewood and seeing to the feeding of the animals. India is estimated to have some 432 million illiterate people. Sixty-four percent of Indian men are literate, but fewer than 40% of women can read and write. About 41% of Indian girls under the age of 14 do not attend school.
42. **(D)** No explanation needed.
43. **(D)** Mountain gorillas are found in Rwanda. The population of Rwanda has more than doubled since the early 1970s. With a continued growth rate of about 3% per year, the population is projected to double approximately every 25 years. The rarest and largest of the great apes, mountain gorillas, are among our closest relatives yet one of the most endangered mammals on Earth. They are threatened by poaching, loss of habitat, disease, and war. The gorilla's only known enemies are leopards and humans. Gorillas are commonly hunted for meat or in retaliation for crop raiding. They have been the victims of snares and traps set for antelope and other animals and are also subject to human disease. Poachers have also destroyed entire family groups in their attempts to capture infant gorillas for zoos. Others are killed to sell their heads and hands as trophies. Consider helping these beautiful animals by having your APES class adopt a gorilla—visit the mountain gorilla conservation fund at *www.mgcf.net*.
44. **(B)** Almost 60% of all the world's freshwater withdrawals go toward irrigation purposes. Rice production uses the most water; soybeans and oats use the least. Producing electrical power is also a major use of water in the United States. In 1995, 189,700 million gallons of water each day were used to produce electricity—to cool the power-producing equipment.
45. **(A)** In (A), a 1999 study by *Consumer Reports* found that two-thirds of domestic produce had more toxic pesticide residues than those products that were imported. In (B), risks to the human immune, reproductive, and endocrine systems, as well as neurotoxicity, may be equally or even more significant than cancer. (C) is true as written. The legal standard for registration set down by the Federal Insecticide, Fungicide, and Rodenticide Act (FIFRA) is a risk-benefit standard. The EPA must register pesticides if they do not pose "unreasonable risk to man or the environment, taking into account the economic, social and environmental costs and benefits of the use of any pesticide." In (D), 12 of the 26 most widely used pesticides in the United States are classified as possible or probable carcinogens by the EPA based on studies of laboratory animals. In (E), a study found carcinogens, neurotoxins, and pesticides that disrupt the endocrine or reproductive system in fruits and vegetables that had been washed, peeled, and prepared for consumption. The foods most likely to be contaminated were (in declining order) peaches, apples, celery, potatoes, grapes, and oranges.

46. (A) The law of demand holds that all other things being equal, as the price of a good or service rises, the quantity demanded falls. The reverse is also true. As the price of a good or service falls, the quantity demanded increases. Think of your trips to the grocery store. When the price of beef rises, you buy less of it.

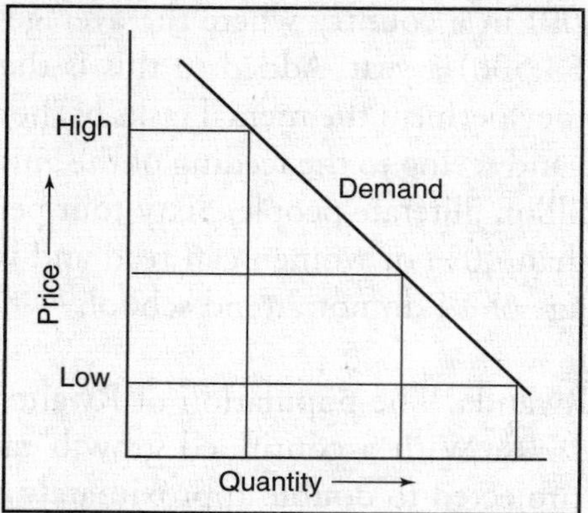

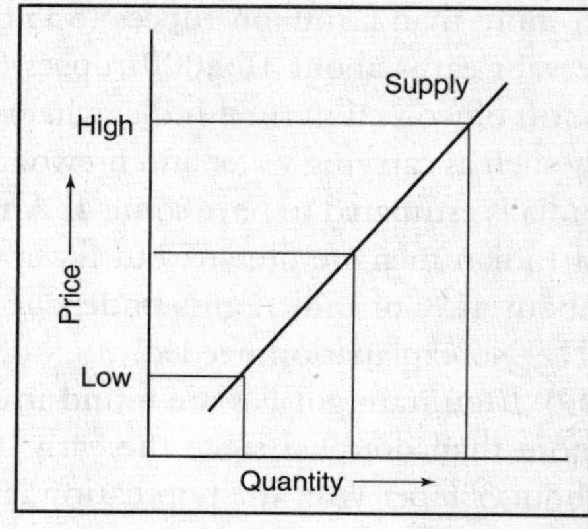

47. (E) At the market equilibrium (where supply intersects with demand), every consumer who wishes to purchase a product at the market price is able to do so, and the supplier is not left with any unwanted inventory.

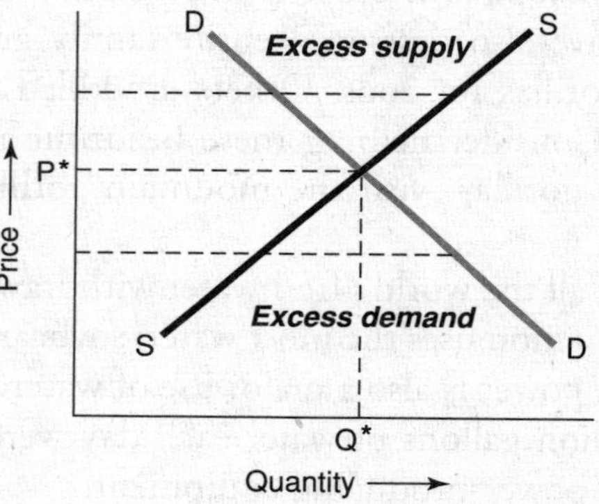

48. (D) If the wind speed is 15 mph, the relative power produced is 3,375 (15^3). For a wind turbine in a 12 mph wind, the relative power is 1,728 (12^3). 3,375 is about twice 1,728. Wind speeds in this range are classified as Class 3 wind sites, with an energy density of 300–400 watts per square meter. Characteristic of much of the Midwestern United States, Class 3 sites cover 13% of the total land area of the United States.

49. (B) Clear-cutting removes *all* trees from an area in a single cutting. Seed-tree cutting occurs when loggers harvest nearly all of a stand's trees in one cutting but leave a few uniformly distributed seed-producing trees to regenerate the stand. Shelterwood cutting removes all mature trees in an area in two or three cuttings over a period of time. Selective cutting occurs when intermediate-aged or mature trees in an uneven-aged forest are cut singly or in small groups.

50. (C) Dams create a series of lakes, slowing the current and delaying downstream migration. The delay interferes with internal biological changes that enable the young salmon to survive in saltwater. In addition, the slow water

exposes them to predators and disease. To increase salmon populations, one should release extra water from dams to wash juvenile salmon downstream.

51. (A) Today, 3 million fishing boats operate and are greatly depleting the supply of fish and other aquatic life-forms. Over the past 40 years, fishing quotas have more than tripled. In 1950, 20 million tons of fish and marine products were harvested. By 1990, this amount had increased to 100 million tons per year. The depletion of fish stocks has led to overfishing in all oceans and other bodies of water. Past catches were generally higher than those of today due to overfishing. Therefore, to base today's harvest limits on harvest numbers that were probably higher in the past would only accelerate depletion of current fishing stocks.
52. (E) Cultivated land occupies only 11% of the world's land resources but produces about 95% of the world's food.
53. (D) Secondary succession begins in habitats where communities are entirely or partially destroyed by some kind of damaging event. For example, secondary succession begins in habitats damaged by fire, floods, insect devastations, overgrazing, and forest clear-cutting and in disturbed areas such as abandoned agricultural fields, vacant lots, roadsides, and construction sites. Because these habitats previously supported life, secondary succession, unlike primary succession, begins on substrates that already contain soil. In addition, the soil contains a native seed bank.
54. (D) Virulence is the harm that parasites and diseases cause to their host (e.g., parasite-induced host mortality or reduced fecundity). Parasite virulence is, in general, proportional to the degree that the parasite exploits the host. Parasite offspring are produced by exploiting the host; therefore, some virulence is inevitable. However, too-strong host exploitation leads to high virulence that jeopardizes survival of the host and the parasite itself. Thus, there should be an optimal level of host exploitation and virulence by the parasite.
55. (D) To find the doubling time of a population at any given annual rate of growth, divide 72 by the annual growth rate (in this case 1.2): $72 \div 1.2 = 60$ years. $1994 + 60 = 2054$.
56. (D) The precautionary principle says to be cautious when making decisions about something that we do not understand and that could have potentially serious side effects. The integrative policy says to make decisions that involve integrated solutions to environmental and other problems. The ecological design principle says to incorporate concepts of good ecological design into decisions and laws. The environmental justice principle says to develop policies so that no one group bears a disproportionate share of harmful environmental risks. The answer, (D), the humility principle, reminds us of the limits of human knowledge and, by extension, the limits of our capacity to manage and control Earth.
57. (A) Large-scale circulations develop in Earth's atmosphere due to uneven heating of its surface by the sun's rays. Daytime solar heating is greatest near Earth's equator, where incoming sunlight is nearly perpendicular to the ground. Near the poles, heat lost to space by radiation exceeds the heat gained from sunlight, so air near the poles is losing heat. Conversely, heat gained from sunlight near the equator exceeds heat losses, so air near the equator is gaining heat. The heated air near the equator expands and rises, while the cooled air

near the poles contracts and sinks. Rising air creates low pressure at the equator. Air cools as it rises, causing water vapor to condense (rain) as the air cools with increasing altitude. As an air mass cools, it increases in density and descends back to the surface in the subtropics, creating high pressure.

58. **(C)** See the table in Chapter 6, "Populations" that shows differences between r-strategists and K-strategists.
59. **(B)** Predator swamping occurs among some organisms that produce huge numbers of offspring—predators are simply swamped and cannot eat them all. It is not a defensive behavior that relies on avoiding detection. Some squids, wildebeests, springboks, and 17-year cicadas use predator swamping. Camouflage or crypsis occurs when animals match backgrounds for color, shape, and size. Masquerading is a type of crypsis in which the organism resembles something inedible. Countershading occurs when animals are darker on the top and lighter on the bottom, which counteracts the effect of sunlight striking the top surface. Some squids have luminescent organs on the ventral (bottom) side to prevent fish from seeing them from below. Batesian mimicry (common in snakes and butterflies) involves three species—predator, model, and mimic. The model species is noxious or dangerous, so predators avoid it. The mimic species has evolved a resemblance to the model, but is not itself noxious. Thus, it engages in false advertising but often gains because the predator is fooled and avoids the mimic. In Müellerian mimicry, both the model and the mimic are distasteful.
60. **(D)** Climax communities are characterized by: stability, high species diversity, low competitive interaction, limited niche overlap, large body size, few offspring per year, one reproductive cycle per year, and K-selective species.
61. **(B)** Acute—having a rapid onset, severe symptoms, and a short course (less than six weeks) duration. Chronic—an illness that has been or is expected to be a condition that affects the individual for an extended period of time and that historically is an illness that is not expected to be resolved through regular medical treatment or the passage of time. About 1 million children die worldwide from measles each year.
62. **(C)** Carbon dioxide is the ultimate result of perfect combustion of any carbon-based fuel. The only ways to reduce carbon dioxide emissions are to make vehicles more fuel efficient and/or drive less, to use a noncarbon fuel such as hydrogen, or to use a green fuel such as ethanol that is produced from crops that absorb carbon dioxide as they grow.
63. **(C)** Trace elements are minerals that the body requires in amounts of 100 mg or less per day. For some, including iodine, proper dosage may be as small as one-tenth of 1 mg. Minuscule as these amounts are, insufficient intake of trace elements can seriously impair health. Iodine is used by the thyroid gland to produce hormones essential for growth, reproduction, nerve and bone formation, and mental health. Copper is necessary for the formation of blood cells and connective tissue. It is also involved in producing the skin pigment melanin. Zinc is involved in the structure and function of all cell membranes as well as the production of more than 200 enzymes. It also is essential for proper wound healing.
64. **(E)** r-selected species have high biotic potential, have high mortality, are short-lived, usually are pioneer species, and have a small build. A weed would be an r-selected species. As conditions such as soil, water, and light change,

r-selected plant species are gradually replaced by K-selected plant species. These more stable species include perennial grasses, herbs, shrubs, and trees. The K-selected species live longer, and therefore their environmental effects slow down the rate of succession.

65. (D) According to the U.S. Department of Agriculture, fire building has the biggest impact, especially if not done with care. Blackened roots, charred wood, limbs broken from trees, and garbage left in fire pits are impacts commonly found. Hunting and fishing are regulated and monitored by both state and federal agencies.

66. (A) Initially, aluminum cans take a lot of energy to produce. By recycling soda cans, energy requirements and air pollution is cut by 95%. Each recycled aluminum soda can saves energy equal to half a can of gasoline! Aluminum containers can contain 100% recycled content, and their light weight conserves energy during transportation. Steel cans, just in U.S. landfills, weighed about 2.5 million tons in 1986. These cans have an outer coating of tin, which is very expensive and imported into the United States. This tin can be recovered and resold or used to make new cans. Glass bottles comprise 8% of our garbage. Two types of glass bottles are manufactured—refillable and nonrefillable. The refillable type is heavier and sturdier, and it just needs a thorough cleaning for it to be recycled up to 30 times. However, this option requires significant amounts of energy. The downside to glass containers is that they weigh a lot, so shipping expenses per unit of volume may be more than that of products packaged in lighter materials. Plastics make up 7% of our waste by weight but 32% by volume. Plastics take hundreds or thousands of years to decompose, so they are essentially permanent in the landfills. Plastic is currently made from petroleum and natural gas, but future plastics are most likely to come from plant and animal matter. In all but a few cases, plastic has not been approved to be recycled. This barrier significantly reduces the recyclability of these products.

67. (A) Price is an important determinant of the quantity of a good supplied. The law of supply states that the amount offered for sale rises as the price is higher.

68. (C) Any member in the House of Representatives may introduce a bill at any time while the House is in session by simply placing it into the hopper. A public bill may have an unlimited number of cosponsoring members. The bill is referred to the appropriate committee by the Speaker. An important phase of the legislative process is the action taken by committees. It is during committee action that the most intense consideration and fact-finding are given to the proposed measures; this is also the time when the people are given their opportunity to be heard.

69. (C) $N_1 = N_o + B - D + I - E$

$$N_1 = 40{,}000 + 35(40) - 10(40) + 100 - 50$$

$$N_1 = 41{,}050$$

70. (C) A well-designed drip irrigation system loses practically no water to runoff, deep percolation, or evaporation. Drip irrigation reduces water contact with crop leaves, stems, and fruit; therefore, conditions may be less favorable for the onset of plant diseases.

71. **(E)** The disability-adjusted life year or DALY has emerged as a measure of the burden of disease. It reflects the total amount of healthy life lost to all causes, whether from premature mortality or from some degree of disability during a period of time. The intended use of the DALY is to assist in setting health service priorities, identifying disadvantaged groups and targeting of health interventions, and providing a comparable measure of output for intervention, program and sector evaluation, and planning. The number of DALYs estimated at any moment reflects the amount of health care already being provided to the population as well as the effects of all other actions that protect or damage health.

72. **(A)** There are basically three patterns of plant distribution—random, regular, and clumped. Random distribution implies neutral interrelations among individuals and no limiting resources. Regular distribution implies negative interactions among individuals and competition for some limiting resource. Clumped distribution implies attraction between individuals or to a common resource.

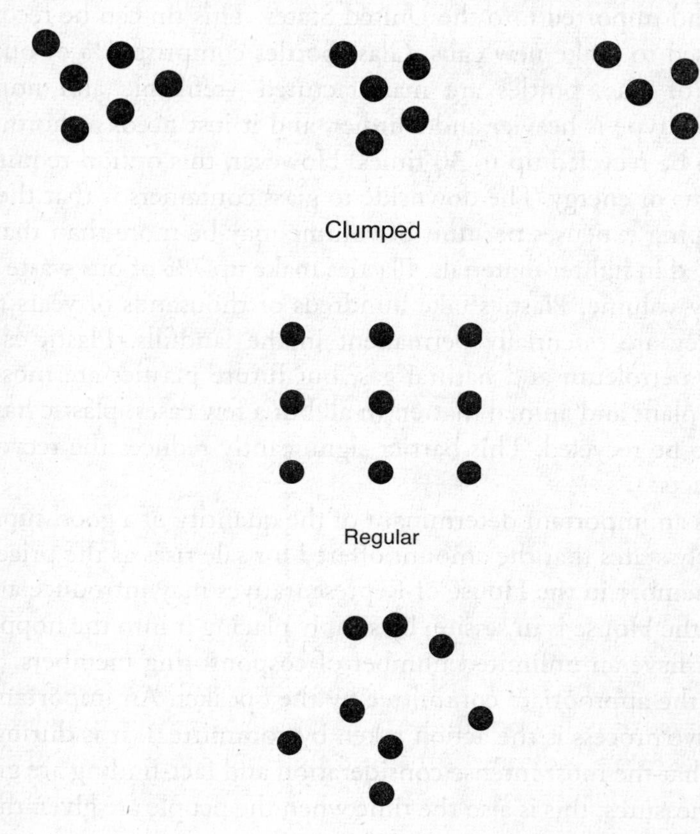

73. **(C)** Most of the freshwater found on Earth is in the form of groundwater. The estimated and actual relative amounts of freshwater are groundwater (66.6%), ice (33.3%), rivers/lakes (less than 0.01%), atmosphere (less than 0.001%).

74. **(D)** Everyday, 3,000 Americans under the age of 18 become regular smokers, and a third of them will eventually die of nicotine-related causes. Two out of three 12 to 17 year olds who smoked nicotine in the last year show signs of addiction. Everyday, over 1,000 Americans die as a result of nicotine addition.

In 1900, 4 billion cigarettes were produced in the United States. By the year 2000, the number had increased to 720 billion.

75. **(B)** Nitrification is a microbial process by which reduced nitrogen compounds (primarily ammonia) are sequentially oxidized to nitrite and nitrate. The nitrification process is primarily accomplished by two groups of autotrophic nitrifying bacteria that can build organic molecules using energy obtained from inorganic sources, in this case ammonia or nitrite. In the first step of nitrification, ammonia-oxidizing bacteria (*Nitrosomonas*) oxidize ammonia to nitrite $NH_3 + O_2 \rightarrow NO_2^- + 3H^+ + 2e^-$. In the second step of the process, nitrite-oxidizing bacteria (*Nitrobacter*) oxidize nitrite to nitrate $NO_2^- + H_2O \rightarrow NO_3^- + 2H^+ + 2e^-$.

76. **(A)** President Woodrow Wilson created the National Park System with the Organic Act of 1916. It was designed to "conserve the scenery and the natural and historic objects and leave them unimpaired for the enjoyment of future generations." There are now 77.5 million acres of land preserved in the park system. Teddy Roosevelt is often wrongly cited as the father of this system. Teddy Roosevelt created the National Wildlife Refuge System in 1903. During his tenure as president, he placed over 230 million acres under federal protection.

77. **(D)** A sequential comparison index (SCI) is a relative number (no units) that can be used to assess species diversity and water quality. A new run is started when a different species appears in the identification sequence. The SCI is computed as follows:

$$SCI = \frac{\text{number of runs}}{\text{number of individuals}}$$

For this sample, let backswimmers = *A*; damselflies = *B*; midge = *C*; mosquito larvae = *D*; and mayfly = *E*.

$$SCI = \frac{AAA}{\text{run 1}}\ \frac{BB}{\text{run 2}}\ \frac{C}{\text{run 3}}\ \frac{DD}{\text{run 4}}\ \frac{EEEE}{\text{run 5}}\ \frac{B}{\text{run 6}}\ \frac{A}{\text{run 7}}$$

$$SCI = \frac{\text{number of runs}}{\text{number of individuals}} = \frac{7}{14} = 0.5$$

SCI values range from 0.1 to 1.0. An SCI of 1.0 indicates highest biodiversity, dominance of pollution-intolerant species, and outstanding water quality. An SCI value of 0.5 indicates average or moderate biodiversity and average or moderate water quality. An SCI value of 0.1 indicates lowest biodiversity, dominance of pollution-tolerant species, and degraded water quality. Even though this method is fairly simple, care must be taken so that the samples are representative of the area of the stream in which you are interested. For example, vegetated areas at stream margins tend to have higher indexes than the middle stream. Another factor influencing the SCI index is season. For example, the diversity of a stream may be lower in the early summer than in the winter owing to the emergence of many aquatic insects.

78. **(C)** The first law of thermodynamics states that energy can be changed from one form to another but it cannot be created or destroyed. The total amount of energy and matter in the universe remains constant, merely changing from one form to another. The second law of thermodynamics states that in all

energy exchanges, if no energy enters or leaves the system, the potential energy of the state will always be less than that of the initial state. This is also commonly referred to as entropy.

79. (C) Less-developed countries typically have a large proportion of their population in the pre-reproductive age category.
80. (C) The 1960 Chilean earthquake, which occurred off the coast of South America, was estimated to be 9.5. The earthquake created a deadly tsunami more than 30 feet in height along the Chilean coast, eliminating entire villages. The tsunami continued across the Pacific, striking Hawaii, where it killed 61 people despite a warning that had been issued five hours earlier. Some hours later, the tsunami killed hundreds more in Japan, more than 8,000 miles from the epicenter. The earthquake that occurred at sea, which triggered the tsunami that occurred in Indonesia in December 2004, was estimated to be 9.0. The New Madrid, Missouri, earthquake in 1812 was the largest earthquake in the continental United States, estimated to be higher than 8.0, and it changed the course of the Mississippi River. The 1964 Prince William Sound, Alaska, earthquake, at 9.2 on the Richter scale, is the largest earthquake to occur in the United States so far. *(Author's note: I live 2 miles from the epicenter of the Northridge earthquake that occurred in 1994. I'll never forget it; neither will my house!)*
81. (C) Although communities are variable and dynamic in both space and time, the interactions between species lead to patterns of community structure that are characteristic of particular community types. For example, North American forests have vertical layers—canopy, midstory, and understory—that are composed of different life forms: trees, shrubs, herbs, and so on. Midwestern prairies have species that are active in the spring and fall (cool-season species) and others that are active in midsummer (warm-season species). In the northeast, many communities have some species that bloom in spring, some in summer, and some in fall. These other structures are thought to be a result of resource partitioning. This evolutionary strategy allows species that are potential competitors for a resource (space, light, pollinators) to coexist by specializing in different aspects of the resource.
82. (C) Selling pollution permits is essentially the same as charging each firm for every ton that they emit. Essentially, it is charging them in advance. If the pollution permit costs more than abatement, then they will abate. If it costs less, they will buy the permit and continue to pollute. In (A), the copper smelter would have to spend $450 and the electric utility would have to spend $250. In (B), the minimum each company would pay would be $251. In (C), each firm would be charged $151. In (D), halting production of an entire plant would be prohibitive.
83. (C) $20/100 = 100/x \quad x = 500$
84. (E) The doctrine of riparian rights defines the rights relating to the bank of a watercourse. It says that a landowner adjacent to a stream has the right to the water in that stream. This places the responsibility on the upstream users and protects private rights in streams and lakes. The concept of appropriation as a right to use water came out of the western United States where public land was parceled out to individuals but gave them no control over streams. This left water to be treated as though it belonged to no one and could be appropriated

in a manner similar to that of a gold claim. In the absence of public control, people took water from streams and used it, that is, they appropriated it. When water laws were enacted, this appropriation practice was legalized and the basis of such laws became known as the doctrine of prior appropriation. The doctrine was simply that the first user on a stream has a better right to the supply in times of shortage. Population growth and increased demand for water have produced many limitations and modifications to the doctrine.

85. **(D)** Refer to "Demographic Transition" in Chapter 6.
86. **(A)** Carbon dioxide and nitrous oxide levels are also linked to industry, automobile usage, and the use of fossil fuels to generate power at electrical plants. Nitrous oxide emissions can also be reduced by decreasing the amount of nitrogen-based fertilizers.
87. **(C)** The burden of proof is with the defendant to prove that the plaintiff caused harmful action.
88. **(C)** Solids refer to matter suspended or dissolved in water or wastewater. Solids may affect water or effluent quality in a number of ways. Water with high dissolved solids is generally of inferior palatability and may cause health problems. Highly mineralized waters are unsuitable for many industrial applications. High suspended solids content can also be detrimental to aquatic plants and animals by limiting light and deteriorating habitat. Total dissolved solids are the amount of filterable solids in a water sample. To calculate total suspended solids:

    Let A = weight of crucible + filter + residue after 24 hrs at 105°C (mg)
    Let B = weight of crucible + filter (mg)

$$\text{Total suspended solids} = \frac{(A - B) \times 1{,}000 \text{ mL/L}}{\text{sample volume (mL)}} \times \frac{1{,}000 \text{ mg}}{\text{g}}$$

$$= \frac{(100.00 \text{ g} - 80.00 \text{ g}) \times 1{,}000 \cancel{\text{mL}}/\text{L}}{100.00 \cancel{\text{mL}}} \times \frac{1{,}000 \text{ mg}}{1 \text{ g}}$$

89. **(D)** Owls are raptors, or birds of prey, which means they hunt other living things for their food. The forward-facing aspect of the eyes give it a wide range of binocular vision (seeing an object with both eyes at the same time). This means the owl can see objects in three dimensions (height, width, and depth) and can judge distances in a similar way to humans. Owls are able to tell direction through hearing because of the minute time difference in which the sound is perceived in the left and right ear. This is important at night when light is low or nonexistent. Owls can also tell if the sound is higher or lower than the position of the owl by using the asymmetrical or uneven ear openings—the left ear opening is higher than the right so a sound coming from below the owl's line of site will reach the right ear first. This left-right and up-down positioning system gives the owl a perfect fix on the location of prey. The comblike leading edge of their primary wing feathers effectively muffles the sound of the air rushing over the wing surface, allowing the owl to fly silently.
90. **(A)** According to the graph, nitrates and phosphates are depleted during the summer following a surge in the numbers of diatoms during the spring pulse.

A decrease in nitrate and phosphate nutrient levels would limit the population growth capacity of diatoms during the summer. Even though light intensity increases during the summer months and would be expected to cause an increase in the diatom population (since they are plants), the limiting factor of low nutrient levels reduces their numbers.

91. **(C)** Phosphates and nitrates are required for photosynthesis and vary in concentration due to biological activity. In surface waters, where plants are actively involved in the process of photosynthesis, nitrates and phosphates can be in short supply, limiting the amount of biological activity that can take place. The cycling of materials (in this case, nitrates and phosphates) is the function of organisms of decay. Microscopic bacteria in seawater can contain up to 1,000,000 bacterial cells per cubic centimeter.
92. **(B)** According to the graph, light intensity is a factor in autumn growth. It is to be assumed that temperature is also an important factor since the data indicate the seasonal aspect of the cycle.
93. **(A)** There is an optimum temperature for metabolic activities. Low temperatures slow down the speed of chemical reactions.
94. **(E)** In summer months, the Gulf Stream pulls warm water up from the tropics. Hurricanes often follow this same path because they require warm water in order to form. Off the West Coast, the current is pulling cooler water down from the north, suppressing most hurricanes and pushing them farther out to sea.
95. **(D)** The more ozone in a given parcel of the stratosphere, the more heat it retains. Ozone generates heat in the stratosphere, both by absorbing the sun's ultraviolet radiation and by absorbing upwelling infrared radiation from the lower atmosphere (troposphere). Consequently, decreased ozone in the stratosphere results in lower stratospheric temperatures and increased UV radiation reaching the surface of Earth. Observations show that over recent decades, the mid to upper stratosphere (from 30 to 50 km above Earth's surface) has cooled by 2°F to 11°F (1°C to 6°C). This stratospheric cooling has taken place at the same time that greenhouse gas amounts in the lower atmosphere (troposphere) have risen.
96. **(C)** SF_6, sulfur hexafluoride, has a warming potential 24,000 greater than carbon dioxide, and its average lifetime in the troposphere is extremely long, averaging over 3,000 years. However, when considering the *impact* of various greenhouse gases, the total amount of gas released into the environment must be taken into account.
97. **(D)** Most indoor air pollution experts agree that carpet should be avoided whenever feasible. The reason for this is that carpet is made up of some 120 different chemicals, many of which can cause health problems. Once installed, carpet can collect dust and even lead (tracked in from shoes) and can grow mold and dust mites. Methylene chloride is given off by paint thinners and paint stippers. Methylene chloride has a fairly long atmospheric half-life, 3 to 4 months, indicating the fairly long persistence typical of transported pollutants.
98. **(B)** The butterfly effect, based on theories of chaos, is the propensity of a system to be sensitive to initial conditions. Such systems over time become unpredictable. This idea came from a 1979 paper entitled "Does the Flap of a Butterfly's Wings in Brazil Set Off a Tornado in Texas?" which proposed that small differences and changes in the environment can make a massive differ-

ence further down the line. Its technical name is sensitive dependence on initial conditions. It affects any complex, dynamic system.

99. **(B)** A density-independent factor is one where the effect of the factor on the size of the population is independent of and does not depend upon the original density or size of the population. The effect of weather is an example of a density-independent factor. A severe storm and flood coming through an area can just as easily wipe out a large population as a small one.
100. **(D)** High-quality energy is capable of performing a large amount of work, while low-quality energy is capable of performing less work. Energy always changes from high to low quality when work is performed. During the change, some energy is lost in the form of heat, which cannot do work (second law of thermodynamics). The amount of energy lost as heat is often as high as 90% of the total energy involved. The reason that (D) is the answer is because of the word "dispersed." Dispersed energy is not concentrated, therefore it is of low quality. An analogy would be ore deposits. A rich gold mine has a lot of gold in concentrated form. A stream might also have a lot of gold (maybe even more than the mine). However, if there are only a few small nuggets every few hundred feet, it is not as rich.

FREE-RESPONSE EXPLANATIONS

Question 1

(a) 3 points maximum

(i) 1 point

Restatement: Number of cubic feet of natural gas required to heat the house for the winter.

$$4{,}000 \cancel{\text{ square feet}} \times \frac{100{,}000 \text{ Btus}}{1 \cancel{\text{ square foot}}} = \mathbf{400{,}000{,}000 \text{ Btus}}$$

(ii) 1 point

Restatement: Cost of heating the home for one winter using natural gas.

$$400{,}000{,}000 \cancel{\text{ Btus}} \times \frac{1 \cancel{\text{ cubic foot natural gas}}}{1{,}000 \cancel{\text{ Btus}}} \times \frac{\$5.00}{1{,}000 \cancel{\text{ cubic feet}}} = \mathbf{\$2{,}000}$$

(iii) 1 point

Restatement: Cost of heating the home for one winter using electricity.

$$400{,}000{,}000 \cancel{\text{ Btus}} \times \frac{1 \cancel{\text{ kWh}}}{10{,}000 \cancel{\text{ Btus}}} \times \frac{\$50}{500 \cancel{\text{ kWh}}} = \mathbf{\$4{,}000}$$

NOTE:

1. If you do NOT show calculations, no points are awarded.
2. No penalty assessed if you do not show units. However, you risk setting up the problem incorrectly if you do not show units so that they cancel properly.
3. If your setup is correct but you make an arithmetic error, no penalty is assessed.

NOTE: For Parts (b) and (c), there are no points for just listing ideas. Each idea MUST be explained. The following are ideas that you could use to write your paragraph(s). Take these ideas, and create an outline of the order in which you wish to answer them. You do NOT need to use all ideas. Before you begin, decide on the format of how you wish to answer your question (pros versus cons, chart format with explanations within the chart, compare and contrast, and so on). Each numbered point when explained is worth 1 point.

(b) 4 points maximum

Restatement: Active versus passive solar design.

Active Solar System (2 points)

- Collectors that collect and absorb solar radiation. Optimum collector orientation is true south (the highest apparent point in the sky that the sun reaches during the day—not necessarily magnetic south). Collector orientation may deviate up to 20° from true south without significantly reducing the performance of the system. Collectors should be tilted at an angle equal to latitude plus 15° for optimum performance. A collector receives the most solar radiation between 9:00 A.M. and 3:00 P.M. Trees, buildings, hills, or other obstructions that shade collectors reduce their ability to collect solar radiation. Even partial shading will reduce heat output.
- Electric fans or pumps to transfer and distribute the solar heat in a fluid (liquid or air) from the collectors.
- Storage system to provide heat when the sun is not shining.
- May be more flexible than a passive system in terms of location and installation.
- Usually most economical to design an active system to provide 40% to 80% of the home's heating needs.
- Liquid systems heat water or an antifreeze solution in a hydronic collector, whereas air systems heat air in an air collector.
- Liquid solar collectors are most appropriate for central heating. They are the same as those used in solar domestic water heating systems. Flat-plate collectors are the most common, but evacuated tube and concentrating collectors are also available. In the collector, a heat transfer or working fluid such as water, antifreeze (usually nontoxic propylene glycol), or other type of liquid absorbs the solar heat. At the appropriate time, a controller operates a circulating pump to move the fluid through the collector. The liquid flows rapidly through the collectors, so its temperature increases only 10–20°F (5.6–11°C) as it moves through the collector. The liquid flows to either a storage tank or a heat exchanger for immediate use. Other system components include piping, pumps, valves, an expansion tank, a heat exchanger, a storage tank, and controls.
- Air collectors produce heat earlier and later in the day than liquid systems. Therefore, air systems may produce more usable energy over a heating season than a liquid system of the same size. Also, unlike liquid systems, air systems do not freeze, and minor leaks in the collector or distribution ducts will not cause problems. Air collectors can be installed on a roof or an exterior (south-facing) wall for heating one or more rooms. These systems are easier and less expensive to install than a central heating system. They do not have a dedicated storage system or extensive ductwork. The floors, walls, and furniture will absorb some of the solar heat, which will help keep the room warm for a few hours after sunset. Masonry walls and tile floors will provide more thermal mass and thus provide heat for longer periods. A well-insulated house will make a solar room air heater more effective. Factory-built collectors and do-it-yourself for on-site installation are available. The collector has an airtight and insulated metal or wood frame and a black metal plate for absorbing heat with glazing in front of it. Solar radiation heats the plate that, in turn, heats the air in the collector. An

electrically powered fan or blower pulls air from the room through the collector and blows it into the room(s). Roof-mounted collectors require ducts for supplying air from the room(s) to the collector and for distribution of the warm air into the room(s). Wall-mounted collectors are placed directly on a south-facing wall. Holes are cut through the wall for the collector air inlet and outlets. Simple window box collectors fit in an existing window opening. They can be active (using a fan) or passive. A baffle or damper keeps the room air from flowing back into the panel (reverse thermosiphoning) when the sun is not shining. These systems provide only a small amount of heat, since the collector area is relatively small.

- Local covenants may restrict options. For example, homeowner associations may not allow installation of solar collectors on certain parts of the house.

Passive Solar System (2 points)

- Basic idea of passive solar design is to allow daylight, heat, and airflow into a building only when beneficial. The objectives are to control the entrance of sunlight and air flows into the building at appropriate times and to store and distribute the heat and cool air so it is available when needed.
- Four basic approaches to passive systems: First, direct gain—solar energy is transmitted through south-facing glazing. Works best when the south window area is double-glazed and the building has considerable thermal mass in the form of concrete floors and masonry walls insulated on the outside. Second, indirect gain—a storage mass collects and stores heat directly from the sun and then transfers heat to the living space. The sun's rays do not travel through the occupied space to reach the storage mass. Third, isolated gain—passive solar concept, solar collection and storage are thermally isolated from the occupied areas of the building. This allows the collector and storage to function independently of the building. Fourth, remote collection—includes a collector space, which intercedes between the direct sun and the living space and is distinct from the building structure.
- Collectors usually receive the most sunlight when placed onto the roof. South-facing walls may also work.
- Passive system does not use a mechanical device to distribute solar heat from a collector.
- Example of a passive system for space heating is a sunspace or solar greenhouse on the south side of the house.
- Passive system is simpler in design and less expensive to build.
- May not be possible depending upon location of site. For example, building an effective sunspace may not be possible due to trees, other buildings in the way, etc.
- Local climate, the type and efficiency of the collector(s), and the collector area determine how much heat a solar heating system can provide.
- Most building codes and mortgage lenders require a backup heating system. Supplementary or backup systems supply heat when the solar system cannot meet heating requirements. They range from a wood stove to a conventional central heating system.
- Passive solar buildings use 47% less energy than conventional new buildings and 60% less than comparable older buildings.

(c) 3 points maximum

Restatement: Five techniques that the homeowners could adopt to make their home a green building.

Definition (1 point): A green building focuses on a whole system perspective, including energy conservation, resource-efficient building techniques and materials, indoor air quality, water conservation, and designs that minimize waste while utilizing recycled materials. Green buildings are a product of a good design that minimizes a building's energy needs while reducing construction and maintenance costs over the life cycle of a building.

Techniques (2 points based on choosing any five and discussing each one).

- Solar collectors for space heating.
- Solar collectors for water heating.
- Photovoltaics to supply electrical energy.
- Hybrid systems that incorporate more than one power source such as wind or microhydro.
- Energy-efficient appliances (energy star: *http://www.energystar.gov*).
- Products made from environmentally attractive materials, including products that reduce materials use. For example, drywall clips that eliminate the need for corner studs, salvaged materials, recycled materials (sprayed cellulose, a recycled material, used for insulation), products made from agricultural waste, and certified wood products that carry an FSC stamp, indicating they meet the high standards set by the Forestry Stewardship Council.
- Products that do not contain toxins. For example, products that substitute for polyvinyl chloride (PVC), ozone-depleting chemicals, and conventional pressure-treated lumber.
- Products that reduce the environmental impacts of construction, renovation, and demolition. For example, erosion control products and exterior stains with low emissions of volatile organic compounds (VOCs).
- Products that reduce the environmental impacts of building operation, including products that reduce energy and water use, reduce the need for pesticide treatments (for example, physical termite barriers), or have unusual durability or require little maintenance.
- Products that contribute to a safe, healthy indoor environment, including products that do not release significant pollutants (for example, low-VOC paints, caulks, and adhesives), products that remove indoor pollutants (for example, certain ventilation products), and products that warn inhabitants of health hazards (for example, carbon monoxide detectors and lead-paint testing kits).
- Superinsulated houses are constructed to be airtight, have a higher level of insulation compared with conventional houses, and have a ventilation system to control air quality.
- Green landscaping—landscaping that provides shade during summer and allows sunlight to warm the house during winter (deciduous trees). Large deciduous trees planted at least 15 to 25 feet away from a house on the south and west sides provide afternoon summer shade while still allowing cooling winds to pass through the tree canopy. Yet they also permit sun to pass through bare branches for solar gain in winter. In addition, they can reduce cooling costs by shading air-conditioning units. Finally, plants that do not require a lot of water can be used.

Question 2

Question (a): 4 points maximum
Question (b): 3 points maximum
Question (c): 3 points maximum

Total: 10 points possible

Aquaculture, or the raising of marine organisms in confined areas and once promoted as a panacea for world hunger, has resulted in unexpected social and environmental consequences. Aquaculture is not something new. In fact, it has existed for hundreds of years on a small-scale, local level. What once were small, family operations that supplied food for small communities of people or was sold in small, local markets and relied on the natural tidal action of water to flush out the ponds has changed significantly due to the current large-scale operation. Most aquaculture enterprises in less-developed countries today are funded by large, multinational corporations with extensive distribution facilities and financial resources. Shrimp farming, as an example of aquaculture and monoculture, utilizes vast coastal areas and is often promoted and financially supported by less-developed third-world governments in lucrative contracts with these multinational corporations.

The shrimp that is raised, including the species known in the United States as tiger prawns, is generally not raised to feed the people in the country in which the shrimp are raised. Instead, the shrimp are exported to countries in western Europe, the United States, and Japan, where market demand and available capital make the shrimp a readily available luxury commodity.

In 1990, Asia produced close to half a million metric tons of commercially raised shrimp, which made up about 80% of the world output. The environmental cost to produce such large numbers of shrimp was almost 1 million hectares of land that is forever lost. Most of the wetlands that are used for shrimp-farming purposes were once prime wetlands consisting of mangroves and other highly productive biomes. Besides the serious environmental consequences are the disruptions to the communities of people that make their living in these highly productive biomes by fishing or operating small farms, particularly rice farming—which incidentally, increases the level of atmospheric methane, a greenhouse gas. Many of these small farmers and fishermen were forced from the land that they had traditionally used and lived on for many years. This parallels in many ways the exploitation of Native Americans who were forced against their will to move to the least desirable of all lands and placed on reservations.

Shrimp farms consist of very large ponds that are built near the ocean. Into these confined ponds (or tanks), which are filled with both seawater and groundwater to create an artificial brackish environment, the farmers add pesticides, antibiotics, food, and other chemicals to increase production. So much groundwater is extracted that land subsidence has been noted. Other methods used to increase shrimp production include increasing the number of shrimp in the bond (density load). Yet, as with humans, increasing density has serious drawbacks. As a response to the stress of increased density, antibiotics and pesticides are added to the water. High densities of shrimp also result in higher waste buildup in the water with consequently lower oxygen levels. As a result of these factors, the water must be changed periodically. The farmers either release this large amount of contaminated water back into the sea (near the coastline where it can do the most environmental damage) or allow it to escape onto prime agricultural lands to infiltrate back into the ground and eventually the groundwater. In areas near the coastline, the shrimp farms have interfered with the daily routines of the fishing community. Furthermore, active capturing of small shrimp from native waters to "seed" the ponds decreases the amount of shrimp in the natural environment. This has serious food web consequences along with decreasing genetic variation in both natural and farm-raised shrimp. The ground that the water is discharged into increases in salinity to such an extent that the productivity of rice drastically decreases. The salt that is in the water also increases salt concentrations in groundwater. Aquatic life living near where the shrimp water effluent is discharged also suffers from added pollutants, salinity tolerance, and oxygen level issues.

Social disharmony is rampant where the shrimp farms have been located. One sees the parallel of this in examples that occurred in the United States Midwest region during the late 1800s and early 1900s between the farmers who wanted barbed wire and the ranchers who wanted open ranges.

Recently, the productivity of shrimp farms has decreased significantly. In Taiwan, almost 100,000 metric tons of shrimp were produced in 1987. One year later, less than 50,000 metric tons were produced—primarily the result of epidemic increases in pathogenic bacteria, viruses, and protozoans. During the mid-1990s, a virus invaded the shrimp farms in India and destroyed the majority of the stock. Many less-developed countries have reexamined the issue of shrimp aquaculture.

An alternative to raising shrimp or fish near the coasts may be to raise it inland, far from the coastal waters where the wild species feed and breed. Tilapia, a type of plant-eating fish, are easy to raise. They produce protein for people without using wild fish as feed. Catfish and trout are raised inland in the United States. Carp have been pond raised for centuries in China and Europe.

Will the little shrimp feed the world? The answer is clear—not until the entire aquaculture industry is re-examined in terms of its effect on the environment and the lives of indigenous people it disrupts and displaces.

Question 3

(a) One way to estimate the size of a large population is to tag and recapture individuals. This process involves catching several individuals from the population—the more the better—and marking them in some way that you can identify later. It is important for this mark not to interfere with the functioning or likelihood of survival. In the case of monarch butterflies, a small dot of nail polish applied to the top of the thorax would suffice. After marking the butterflies, you would then release them into the population and wait long enough for them to become randomly dispersed. Next, you would capture several butterflies and count how many of them have your mark. You would repeat this procedure—releasing, waiting, and capturing the same number of butterflies until you had several pieces of data to average. Based on the average percentage of marked butterflies you captured, you can then estimate the total population size. If for example, 5% of the butterflies you capture are marked, then the number you originally marked represents about 5% of the total population size.

(b) A direct threat to biodiversity is something that affects organisms by interfering with them. For example, cars might run over them. An indirect threat to biodiversity impacts them through a chain of events. If urban development fragments the habitat of a species of large mammal, then the mobility of the animals will be decreased, limiting their access to resources and mating partners. If the habitat fragmentation or loss is severe enough, extinction of the population may result.

(c) The Endangered Species Act is a piece of legislation that makes it illegal to injure, kill, or collect any species listed as threatened or endangered within the United States. This legislation also prohibits the import (for any purpose other than preservation or research) of any endangered species or any product made of endangered species. This clause is intended to keep organisms in their native habitats.

(d) When the size of a population is severely reduced, many of the genetic variations present in the original population will be lost. For example, a world population of 20 humans could not represent all of present human diversity. Even if the bottlenecked population reproduces enough to return to the original population size, the genetic diversity will still be limited to that of the parents since new genetic variations will arise very slowly. Thus, the population will be more vulnerable to selection pressures such as disease because the likelihood of the population having resistant individuals is much lower. This would make the population more likely to become extinct.

Question 4

(a) 3 points maximum

Restatement: Two possible reasons that human life span has increased dramatically within the last 100 years.

During the 20th century, the health and life expectancy of persons residing in the United States has improved dramatically. Since 1900, the average life span of persons in the United States has lengthened by greater than 30 years; 25 years of this gain are attributable to advances in public health. The two public health interventions that have had the greatest impact within the last 100 years on world health and human life span have been the availability of clean water and sanitation, and the development of vaccines. Nearly a third to half of all babies born in the United States in 1900 did not survive past five years old. Whooping cough, diphtheria, rheumatic fever, and scarlet fever were major killers at that time. Major diseases that have been controlled to a greater or lesser extent through the use of vaccines within the last 100 years include plague, smallpox, diphtheria, tuberculosis, polio, measles, and hepatitis B. In terms of the single most infectious carrier, it is estimated that the *Anopheles* mosquito has been responsible for half of all human deaths since the Stone Age.

In 1900, it was common practice to dispose of garbage, industrial wastes, and raw sewage by dumping them into waterways. Few municipalities treated wastewater because it was widely believed that running water purified itself. Typhoid alone killed more than 150 per 100,000 people annually, including Wilbur Wright. Dysentery and diarrhea—the most common waterborne diseases—were the third largest cause of death in the nation 100 years ago. Infections such as typhoid and cholera transmitted by contaminated water, a major cause of illness and death in America early in the 20th century, have been reduced dramatically by improved sanitation. As a result of disinfection of public drinking water and improved sanitation methods, the major waterborne diseases all but ceased to exist in the United States by World War II. Worldwide, the median reduction in deaths from water-related diseases is approximately 70% among people with access to potable water and proper sanitation. Yet, waterborne diseases continue to be major killers in less-developed countries with half of all deaths of children in poor countries caused by waterborne diseases—one child every eight seconds.

(b) 3 points maximum

Restatement: Greatest current and projected risks to human health.

Answers may include:

- Within 25 years, smoking will become the single largest cause of death and disability in the world. The World Health Organization estimates that by 2020, tobacco-related illness will be responsible for more deaths than tuberculosis, AIDS, car accidents, and homicides together. Smoking, because of its impact on heart disease, lung cancer, and other disorders, caused 3 million deaths in 1990. That total will nearly triple to 8.4 million deaths in 2020.

- Noncommunicable diseases such as cancer, heart disease, and diabetes already cause more deaths in the developing world than infectious diseases. About 56% of all deaths are from noncommunicable diseases, a proportion that is expected to jump to 73% by 2020. More people die of heart disease than any other cause. Of the 6.3 million who died of heart disease in 1990, about 1/3 were from developed countries. In the same year, strokes killed 1.5 million people in developed countries and a total of 4.5 million worldwide. Worldwide, pneumonia killed 4.5 million people and diarrheal disease 3 million, nearly all of them in the developing countries.
- Depression, thought to be largely associated with affluence, accounts for a full 10% of productive years lost throughout the world. By 2020, depression will account for 15% of the total disease burden.
- Largely as a result of smoking, alcoholism, and accidents, men in the former Soviet Union face a 28% risk of death between the ages of 15 and 60, the highest risk anywhere outside sub-Saharan Africa.
- Worldwide, one out of every three people dies from communicable diseases (50,000 each day), childbirth, or malnutrition. Virtually all of those deaths are in developing regions. One of 10 deaths result from injuries caused by accidents, wars, suicides, and homicides.
- By 2020, car accidents will be the world's fifth-leading cause of death and disability as developing nations build more roads and the number of young adults—those most often killed in traffic mishaps—increases.
- Tuberculosis, pneumonia, and diarrheal diseases currently account for nearly 20% of the global disease burden. Annually, diarrheal diseases—including cholera, typhoid, and dysentery, which spread chiefly by contaminated water or food—kill 3 million, most of them children. Tuberculosis kills almost 3 million, mostly adults.
- About 1.5 billion people are infected with intestinal worms.
- In 1990, the major diseases and causes of disability were ranked as:
    1. Lower-respiratory infections
    2. Diarrheal diseases
    3. Conditions arising during prenatal period
    4. Unipolar major depression
    5. Ischemic heart disease
    6. Cerebrovascular disease
    7. Tuberculosis
    8. Measles
    9. Road traffic accidents
    10. Congenital anomalies
- In 2020, the ranking worldwide is projected to change to:
    1. Ischemic heart disease
    2. Unipolar major depression
    3. Road traffic accidents
    4. Cerebrovascular disease
    5. Chronic obstructive pulmonary disease
    6. Lower-respiratory infections
    7. Tuberculosis

8. War
9. Diarrheal diseases
10. HIV

(c) 4 points maximum (6 points possible)

Restatement: Concept of risk analysis. How would information gained from studies of risk analysis increase the human life span?

Definition (1 point): Risk, in the context of human health, is the probability of injury, disease, or death from exposure to environmental hazards. In quantitative terms, risk is expressed in values ranging from 0 (representing the certainty that harm will not occur) to 1 (representing the certainty that harm will occur). The process of risk assessment and analysis involves four major steps: hazard identification, dose-response assessment, exposure assessment, and risk characterization.

Describe each component (1 point each):

Hazard identification is the collection and evaluation of data on the types of health injury or disease that may be produced by a chemical and on the conditions of exposure under which injury or disease is produced. It may also involve characterization of the behavior of a chemical within the body and the interactions it undergoes with organs, cells, or even parts of cells. Data of the latter types may be of value in answering the ultimate question of whether the forms of toxicity known to be produced by a substance in one population group or in experimental settings are also likely to be produced in humans.

Dose-response assessment involves describing the quantitative relationship between the amount of exposure to a substance and the extent of toxic injury or disease. Data are derived from animal studies or, less frequently, from studies in exposed human populations. There may be many different dose-response relationships for a substance if it produces different toxic effects under different conditions of exposure.

Exposure assessment involves describing the nature and size of the population exposed to a substance and the magnitude and duration of their exposure. The evaluation could concern past or current exposures, or exposures anticipated in the future.

Risk characterization generally involves the integration of the data and analysis of the first three components of the risk assessment process (hazard identification, dose-response assessment, and exposure assessment) to determine the likelihood that humans will experience any of the various forms of toxicity associated with a substance.

Apply concept of risk analysis to real-world examples (1 point): Risk analysis information is used in the risk management process in deciding how to protect public health and extend human life. Examples of risk management actions include deciding how much of a chemical a company may discharge into a river; deciding which substances may be stored at a hazardous waste disposal facility; deciding to what extent a hazardous waste site must be cleaned up; setting permit levels for discharge, storage, or transport; establishing levels for air emissions; and determining allowable levels of contamination in drinking water. Essentially, risk assessment provides information on the health risk, and risk management is the action taken based on that information.

Index

BARRON'S COLLEGE GUIDES

AMERICA'S #1 RESOURCE FOR EDUCATION PLANNING

PROFILES OF AMERICAN COLLEGES, 27th Edition, w/CD-ROM
Compiled and Edited by the College Division of Barron's Educational Series, Inc. Today's number one college guide comes with computer software to help with forms and applications. Book includes profiles plus Barron's INDEX OF COLLEGE MAJORS! Vital information on majors, admissions, tuition, and more.
$28.99, Can. $41.99, (0-7641-7903-9)

COMPACT GUIDE TO COLLEGES, 15th Edition
A concise, fact-filled volume that presents all the essential facts about 400 of America's best-known, most popular schools. Admissions requirements, student body, faculty, campus environment, academic programs, and so much more are highlighted.
$9.99, Can. $13.99, (0-7641-3370-5)

BEST BUYS IN COLLEGE EDUCATION, 9th Edition
Lucia Solorzano
Here are detailed descriptions—with tuitions and fees listed—of 300 of the finest colleges and universities in America judged on a value-for-your-dollar basis.
$18.99, Can. $25.99, (0-7641-3369-1)

GUIDE TO THE MOST COMPETITIVE COLLEGES, 4th Edition
Barron's latest and most innovative college guide describes and examines America's top 50 schools. What makes this guide unique is its special "insider" information about each school including commentaries on faculty, and more.
$21.99, Can. $31.00, (0-7641-3197-4)

PROFILES OF AMERICAN COLLEGES: THE NORTHEAST, 17th Edition
Comprehensive data specifically for students interested in schools in Connecticut, Delaware, D.C., Maine, Maryland, Massachusetts, New Hampshire, New Jersey, New York, Pennsylvania, Rhode Island, or Vermont.
$16.99, Can. $23.99, (0-7641-3368-3)

THE ALL-IN-ONE COLLEGE GUIDE
Marty Nemko, Ph.D.
This book tells even the most serious college-bound student and family everything they need to know to choose, get into, find the money for, and make the most of college.
$12.99, Can. $17.99, (0-7641-2298-3)

HEAD START TO COLLEGE PLANNING
Susan C. Chiarolanzio, M.A.
Here's the book to help parents plan for their child's academic success. It explains the many kinds of college options available for students having different interests and aptitudes.
$11.95, Can. $17.50, (0-7641-2697-0)

BARRON'S EDUCATIONAL SERIES, INC.
250 Wireless Blvd., Hauppauge, NY 11788
In Canada: Georgetown Book Warehouse
34 Armstrong Ave., Georgetown, Ont. L7G 4R9
$ = U.S. Dollars Can.$ = Canadian Dollars

Prices subject to change without notice. Books may be purchased at your bookstore, or by mail from Barron's. Enclose check or money order for total amount plus 18% for postage and handling (minimum charge of $5.95). New York, New Jersey, Michigan, Tennessee, and California residents add sales tax to total. All books are paperback editions.

Visit us at www.barronseduc.com

(#8) R10/06

Success on Advanced Placement Tests Starts with Help from Barron's